Biomedical Visualization

Volume 8

Series Editor

Paul M. Rea, Anatomy Facility, School of Medicine, Dentistry and Nursing, College of Medical, Veterinary and Life Sciences, University of Glasgow, Glasgow, UK

The Biomedical Visualization book series invites contributions related to visualization and imaging in the biomedical sciences and related fields such as medicine, dentistry, veterinary surgery, informatics, and the allied health professions.

This can encompass work in image analysis, workflow methodologies, photogrammetry in science, animations, digital reconstructions and applications, big data and visualizations, educational methodologies, usability and evaluations, augmented and virtual reality, and 3D and 4D technologies, including 3D printing, as well as informatics, e-tutorials, MOOCs, HCI and public engagement. It can be from macroscopic to microscopic but show how we can view data and information related to the biomedical field in a much more accessible, innovative and engaging way using technology.

This series is intended for researchers, clinicians, students, and teaching staff in biomedical and medical schools who use and develop visualization and imaging techniques in medical education, patient care, and biomedical and medical research.

Rachael Door
Editor

Accessibility, Inclusivity and Diversity in Education and Beyond

Springer

Editor
Rachael Door
Keele Anatomy and Surgical Training Centre
Keele University
Staffordshire, UK

ISSN 2731-6130 ISSN 2731-6149 (electronic)
Biomedical Visualization
ISBN 978-3-031-81794-6 ISBN 978-3-031-81795-3 (eBook)
https://doi.org/10.1007/978-3-031-81795-3

Cover Illustration: Catherine MacRobbie.

This Springer imprint is published by the registered company Springer Nature Switzerland AG
The registered company address is: Gewerbestrasse 11, 6330 Cham, Switzerland

Preface

Higher Education (HE) and the sector of Biomedicine more broadly has a responsibility to create accessible, inclusive, and diverse learning and working environments. This volume explores the current and future challenges for those striving toward creating such spaces and offers practical advice and case studies of how to overcome these hurdles.

Initially, we explore how practitioners can create accessible environments for their current and future learners. In this section, we discuss the effect of mandatory and assessed attendance on the learner, and how educators can build a sense of belonging within their student community. We also explore play-based learning in HE with practical advice and case studies, and how psychology can be made accessible to students in medical education.

In the second section, we focus on inclusive practice in HE and beyond. Chapters include an exploration of best practice in establishing representative research environments, a toolkit for early career anatomists aiming to create inclusive learning environments, and a discussion on the pedagogy, psychology, and culture of question-asking in class.

The final section centers on improving diversity in the learning environment that prepares students for professional practice. Chapters include an evaluation on the use of human remains as pedagogic aids in the modern context, and the growing world of cine-meducation; the use of television and film to support effective science communication.

Staffordshire, UK Rachael Door

Contents

Contributors

Scott Abbott-Paterson School of Anatomy, University of Bristol, Bristol, UK

Karen M. Adams School of Medicine, Keele University, Keele, UK

Christopher Aris School of Chemical and Physical Sciences, Keele University, Keele, UK

Carissa Birznieks Integrative Physiology Program, University of Colorado Anschutz Medical Campus, Aurora, CO, USA

Andrew S. Cale Department of Anatomy, Cell Biology, & Physiology, Indiana University School of Medicine, Indianapolis, IN, USA

Shannon N. Curran Kansas College of Osteopathic Medicine, Wichita, KS, USA

Noelia Dominguez Falcon School of Biological Sciences, University of East Anglia, Norwich, UK

Kelly Edmunds School of Biological Sciences, University of East Anglia, Norwich, UK

Lily A. Evans Peninsula Medical School, University of Plymouth, Plymouth, UK

Rebekah Freed University of Washington, Seattle, Seattle, WA, USA

Georgina A. Goodison School of Chemical and Physical Sciences, Keele University, Keele, UK

Munesh Pal Khamuani School of Medicine, Faculty of Health Sciences and Wellbeing, University of Sunderland, Sunderland, UK

Siobhan M. Moyes Peninsula Medical School, University of Plymouth, Plymouth, UK

Dominic O'Brien Brighton and Sussex Medical School, Brighton, UK

Elena Patera Human Anatomy Resource Center, University of Liverpool, Liverpool, UK

Section of Anatomy, St George's University of London, London, UK

Amanda M. Root Kansas College of Osteopathic Medicine, Wichita, KS, USA

Katie M. Snow Peninsula Medical School, University of Plymouth, Plymouth, UK

School of English, Drama and Film, University College Dublin, Dublin, Ireland

Sara Sulaiman Royal College of Surgeons in Ireland Medical University of Bahrain, School of Medicine, Busaiteen, Kingdom of Bahrain

School of Anatomy, University of Bristol, Bristol, UK

Hailey Wilkinson Modern Human Anatomy Program, University of Colorado Anschutz Medical Campus, Aurora, CO, USA

Rebecca L. Wisner Department of Anatomy, Cell Biology, & Physiology, Indiana University School of Medicine, Indianapolis, IN, USA

About the Editor

About the Editor

Rachael Door Rachael is a Senior Lecturer in Anatomy at Keele University. She is Deputy Director of Scholarship for the School of Medicine, a member of the Equality, Diversity and Inclusion Team in the School, and a Person Designated under the Human Tissue Authority.

Rachael has an Undergraduate Degree in Anatomy and Human Biology, a PhD in Cellular and Molecular Physiology, and an MA in Learning and Teaching in Higher Education. She is a Senior Fellow of the Higher Education Academy and contributes to the mentoring and assessing of others on this route.

Rachael has presented her research and scholarship at national and international conferences and delivered numerous invited workshops based on her scholarship around the use of artificial intelligence (AI) in HE.

Rachael is committed to creating inclusive educational environments and passionate about outreach. This is reflected in her role as a STEM ambassador and World Anatomy Day events for the public.

About the Series Editor

Paul M. Rea Paul is a Professor of Digital and Anatomical Education at the University of Glasgow. He is Director of Innovation, Engagement and Enterprise within the School of Medicine, Dentistry and Nursing. He is also a Senate Assessor for Student Conduct, coordinates the day-to-day running of the Body Donor Program, and is a Licensed Teacher of Anatomy, licensed by the Scottish Parliament.

He is qualified with a medical degree (MBChB), a MSc (by research) in craniofacial anatomy/surgery, a PhD in neuroscience, the Diploma in Forensic Medical Science (DipFMS), and an MEd with Merit (Learning and Teaching in Higher Education). He is a Senior Fellow of the Higher Education Academy, a Fellow of the Institute of Medical Illustrators (FIMI), and a registered medical illustrator with the Academy for Healthcare Science.

Paul has published widely and presented at many national and international meetings, including invited talks. He has been the lead Editor for the first 17 volumes as Biomedical Visualisation, part of Advances in Experimental

Medicine and Biology. In 2022, this then became its own standalone volume becoming Biomedical Visualization, for which Paul is the Series Editor. This has now published 5 additional volumes, with several currently in press. These volumes have resulted in over 180,000 online accesses, more than 940 citations, and over 470 social media mentions. It has published 212 chapters, from 434 different authors across 133 different institutes in 23 countries across the globe, and with readership in 79 countries. He is also Associate Editor for the European Journal of Anatomy and has reviewed for 25 different journals/publishers.

He is the Public Engagement and Outreach lead for anatomy, coordinating collaborative projects with the Glasgow Science Centre, NHS, and the Royal College of Physicians and Surgeons of Glasgow. Paul is also a STEM ambassador and has visited numerous schools to undertake outreach work.

His research involves a long-standing strategic partnership with the School of Innovation and Technology, The Glasgow School of Art. This has led to multi-million pound investment in creating world leading 3D digital datasets to be used in undergraduate and postgraduate teaching to enhance learning and assessment. This successful collaboration resulted in the creation of the world's first taught MSc Medical Visualisation and Human Anatomy combining anatomy and digital technologies of which he was the Founding Director. The Institute of Medical Illustrators also accredits it. It has created college-wide, industry, multi-institutional, and NHS research linked projects for students.

Part I

Creating Accessible Environments

1 Turning Up: The Effect of Mandatory and Assessed Attendance on Educational Experiences and Student Wellbeing

Dominic O'Brien

Abstract

Mandatory attendance in education is commonplace, with certain courses within higher education extending this to being an assessed component. Medicine is one such example of this, where guidance from the General Medical Council states that students must achieve an 80% attendance rate to pass relevant modules.

The thinking behind this enforced attendance is that it promotes engagement from the students and thus better chances at achieving the learning outcomes. Attendance at the various session types within the medical curriculum (i.e. lectures, clinical skills, and symposia) also allows students to learn aspects of the hidden curriculum; developing skills such as professionalism and improving their interpersonal skills through interactions with colleagues, clinicians, and academics.

Whilst there is literature and my own personal experience to support these arguments, there are a myriad of other factors to be considered. Adult learners such as medical students can often have complicated personal situations outside of their academic careers which make attendance challenging. Factors such as physical and mental health, and family responsibilities such as children or carer roles are seldom considered when the attendance policies are enforced. This threat of failure to students, who in many cases are unable to meet the attendance requirements can cause or exacerbate issues of stress and pose risks to their wellbeing. The removal of student autonomy to learn how they feel best suits them is also a problem which warrants consideration, as this should be seen as a vital skill learned while attending university.

The subject of attendance is complex, with no 'one size fits all' solution. However, evidence shows the enforcement of mandatory attendance policies can be actively detrimental to students, and an individual's presence in a room does not equate to academic ability—the evaluation of which should be reserved for assessments.

Keywords

Attendance · Assessment · Attainment · Professionalism · Mental health · Medicine

1.1 Introduction

Compulsory attendance is an accepted facet of education beginning at the primary level and extending across secondary education, before

D. O'Brien (✉)
Brighton and Sussex Medical School, Brighton, UK
e-mail: D.O'Brien@bsms.ac.uk

R. Door (ed.), *Accessibility, Inclusivity and Diversity in Education and Beyond*, Biomedical Visualization 8, https://doi.org/10.1007/978-3-031-81795-3_1

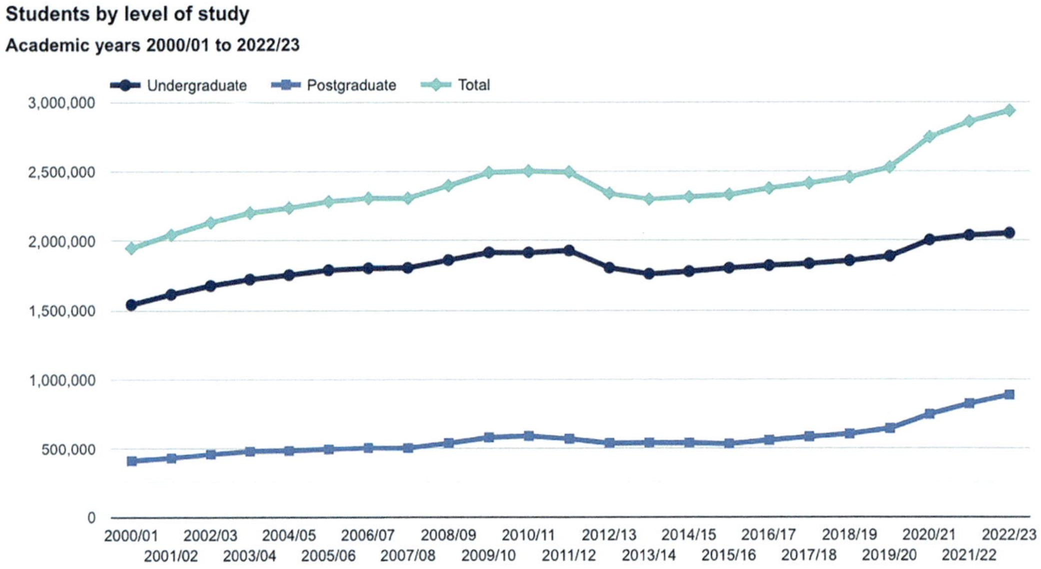

Fig. 1.1 University student numbers between 2000/2001 and 2022/2023 academic years

leniency is seen at further and higher education institutions (Jackson 1990, pp. 1–11). In the later years of education, the level of learner autonomy increases, as students near and enter adulthood and take more responsibility for their education.

Not all programmes and institutions abide by this thinking however, such as in medical education where attendance remains a compulsory and assessed portion of the programme (General Medical Council (GMC) 2011, 2016; BSMS.ac.uk 2022). Both the literature and my own opinions on this are conflicted, as there are clear benefits to encouraging student attendance; aiming to ensure they get the most from their education and develop as individuals in their relevant learning environment. Unfortunately, this can often be seen by students as restrictive and imposes certain rigidity around their learning experience, which cannot benefit all students.

Numbers of university attendees have risen by almost 1 million between the 2000/2001 and 2022/2023 academic years (Fig. 1.1), meaning that significantly more students are being subject to the attendance policies of university institutions (HESA 2024).

Figure 1.1 shows the number of students attending university at undergraduate, postgraduate and a combined total from the 2000/2001 and 2022/2023 academic years, showing a steady overall increase in the period (HESA 2024).

This trend is also particularly worth noting in combination with current plans for the increases in medical student numbers proposed by the NHS Long Term Workforce Plan, which aims to double medical student places to 15,000 by 2031/2032 (NHS 2023a). This will subject exponentially more students to the most stringent of attendance policies over this time, in what is already considered to be a challenging degree programme.

It is with a vested interest that this topic felt worth exploring, which prompted not only consulting the current literature on the effects of mandatory attendance, but also conducting my own research. This was undertaken with a view to answering the following question: Can we force students to learn, or should we encourage a more 'free range' approach?

1.2 Professionalism

Professionalism is one of the nebulous non-academic skills expected to be developed at medical school (GMC 2016) and has been the subject of scrutiny in medicine (Thistlethwaite and

Spencer 2008). Definitions vary, but medical professionalism specifically is said to be based upon three fundamental principles and ten professional responsibilities (Table 1.1) and must be taught to prospective clinicians, not being intrinsically known to students (Hilton and Slotnick 2005, pp. 58–65).

The way this is taught is highly debatable, as the issues surrounding the 'hidden curriculum' within medicine are rife and difficult to navigate (Lempp and Seale 2004, pp. 770–773).

Hochberg et al. (2010, pp. 86–93) evaluated if it would be possible to directly teach and assess professionalism to a group of surgical residents by distilling professionalism into the following categories: information gathering, rapport building, patient education, delivering bad news, responding to emotion, and interdisciplinary respect. This study demonstrated that the residents performed better in post-curricular Objective Structured Clinical Examination (OSCE) evaluations, suggesting that once professionalism can be defined, it can be taught with appropriate curriculum design. This type of study creates a strong argument for ensuring that students attend all aspects of their medical school teaching, as it would seem there is tangible evidence that professionalism can be actively imparted to students from educators.

Table 1.1 The three overarching fundamental principles and ten professional responsibilities of medical professional (ABIM Foundation et al. 2002, pp. 243–246)

Three fundamental principles	Ten professional responsibilities
– Primacy of patient welfare – Primacy of patient autonomy – Primacy of social justice	– Professional competence – Honesty with patients – Patient confidentiality – Maintain appropriate relationships with patients – Improving quality of care – Improving access to care – Just distribution of resources – Scientific knowledge – Maintaining trust by managing conflicts of interest – Professional responsibilities

Professionalism can also be exercised through collaborative working, which has shown to be beneficial for simulated clinical outcomes among students (Shrader et al. 2013), as well as improving healthcare. This spans not only collegial relationships and the use of healthcare services and resources, but also improves communication with patients throughout their care (Task Force Two 2005; Kates and Ackerman 2002; Nolte and Tremblay 2005); this in turn results in higher satisfaction on the delivery of healthcare (Kates and Ackerman 2002; Marriott and Mable 2002).

The NHS themselves continue to impress the importance of professionalism in the workplace with a dedicated toolkit provided to institutions with materials such as Fig. 1.2 to be displayed within the workplace in common areas and presented to staff during team meetings (NHS 2020a). By utilising these visual reminders, it is hoped that the importance of staff acting with the titular civility and respect towards one another will be stressed and that staff will be more conscious of their interactions with others to reduce instances of workplace bullying, providing a better experience for staff and ultimately result in better patient care.

Figure 1.2 is an example of a poster resource provided by the NHS to promote professionalism and better treatment of staff in the workplace (NHS 2020b).

Provision of this type of resource, if adequately deployed, will create a useful point of reference and visual reminder for staff on the expectations of their conduct, and will also benefit medical students when undertaking clinical placements where they will encounter these resources and hopefully appreciate and adhere to the messaging. This does require attendance to these areas however, otherwise these resources may not be seen by individuals and thus their benefits not realised. This does however beg the question that if these resources are needed in the workplace, why has professionalism not been achieved to the requisite standards during staff training at the university level—even with a compulsory attendance policy? According to the NHS (2023a, b) Staff Survey, 9.94% and 17.66% of staff had experienced bullying, harassment, or

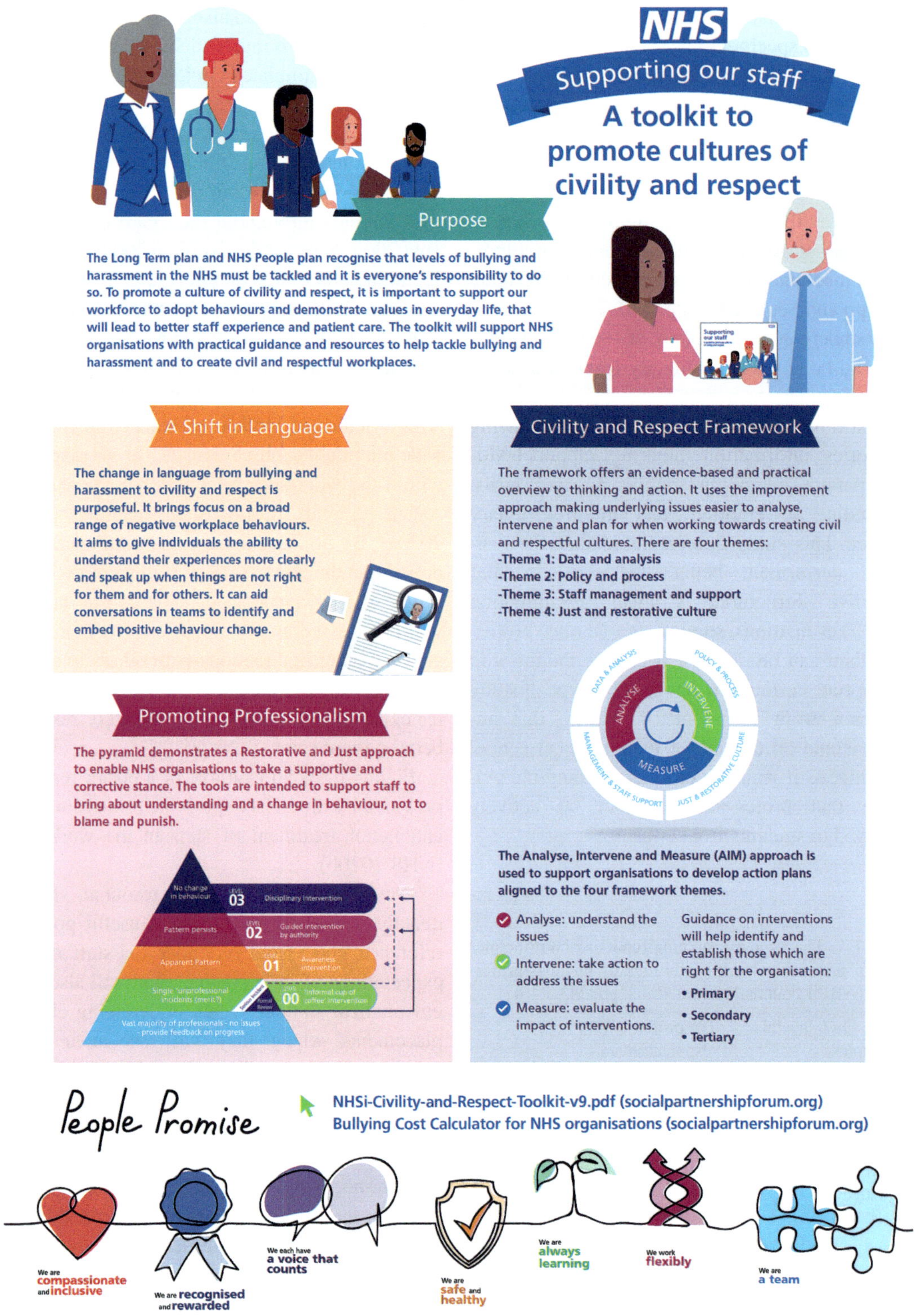

Fig. 1.2 NHS: supporting our staff poster

abuse from managers and other members of staff, respectively, in the last 12 months alone (NHS 2023a, b). Could this be due to lacking attendance at the university level, or does the attendance and teaching of professionalism not have the desired effect?

Considering these points, it seems difficult to rationalise how many of these skills could be obtained without being physically present at all aspects of medical school teaching, and thus compulsory attendance should be supported. However, there are counterpoints which need to be deliberated.

Presenteeism is generally defined as the act of being physically present whilst not fully able to function—be that through physical or mental issues, such as illness (Hemp 2004, pp. 49–58). Awareness has risen of presenteeism in recent years with regard to the private sector; the American Productivity Audit (Stewart et al. 2003, pp. 1234–1246) undertook a mass survey which suggested that presenteeism costs the US economy in excess of $150 billion per annum. The impact on education is not as quantifiable as this but presents in different ways such as levels of student engagement and disruption.

Classroom disruption appears commonplace in primary and secondary education, with 99% of UK teachers having dealt with disruptive pupils in their careers (BBC News, 2006) and this is spreading into higher education worldwide (Seeman 2010; James Madison University Counselling and Student Development Center 2020). Disruptive behaviour robs other pupils of the opportunity to learn effectively and creates an unpleasant environment for both students and teachers. The causes behind disruptive behaviour vary widely, including problems with the teaching environment, problems relating to the teaching method, or teacher and health problems (Latif et al. 2016, pp. 44–52).

Even noting the benefits of enforcing attendance, the fact that disruptive students are incentivised to attend misuses their time as well as detracts from the learning opportunities of others. This creates a more challenging teaching environment for staff, which is a considerable enough detriment to so many that it warrants weighing against the aforementioned advantages.

Obligate attendance also deprives students of an opportunity to take responsibility for their learning. The process of students feeling accountable for their learning is a necessary component of metacognition—a vital skill in higher education (Devlin 2002, pp. 125–138). This claim is supported by a study which showed improvements in several areas of student performance in cooperative tasks, such as critical thinking, problem-solving, and self-evaluation after a 3-month programme to aid students in taking responsibility for their education (Bujan 1999). This is evidence that giving students control of their education could be beneficial long term, rather than incentivising attendance with 'fear of failure'.

1.3 Attainment

Student attainment is often seen as the overarching measure of educational success, being more quantifiable than skills such as communication and professionalism and being the metric for many progression avenues (such as further study or job applications).

In favour of compulsory attendance is the fact that by having contact with students, educators can assess their capabilities formatively. Students attending interactive teaching sessions allows for exploration of dialogic teaching (Alexander 2005). This can be beneficial for both educator and student, allowing a more engaged learning experience for the student and enabling active information processing with the opportunity to seek guidance on areas of uncertainty (Caine and Caine 1994). For the educator, conversing with students can be enlightening as to how they learn and how teaching is perceived. If the opportunity is taken, educators can use this to learn how to enable and enrich students' learning experiences most effectively. With a lack of contact, this is almost impossible to do, and so compulsory attendance ensures these benefits can be gained for all.

With regard to medical school, many practical skills cannot be learnt without attending the requisite teaching sessions. Clinical skills such as venepuncture and auscultation are necessary practical skills according to the GMC guidance for medical professionals (GMC 2019). It is unlikely that students would be able to obtain these skills without the instruction given at specifically designed clinical sessions, and so it feels there is little room to argue that attendance in this case is rightly compulsory. Nevertheless, it seems unfair to assess attendance based on a single facet.

Research is unclear as to how attendance affects student grades, largely due to attendance being used in relation to the grades themselves. Moore and Jensen (2008, pp. 64–71) demonstrated this by stating higher attendance equated to higher grades in lab-based teaching for students, and that attendance levels could be seen as a predictor attainment. However, there was a reduction of 7% in the student grade for each absence; this is a confounding factor of the findings, exemplifying the problem in claims that attendance improves grades when attendance itself is graded.

Research by Brocato (1989, pp. 2–6) and Jones (1984, pp. 133–136) agreed that attendance was a useful predictor of academic achievement, showing a correlation with grades earned. Moore (2005, pp. 26–28, 30, 32) asserts that correlation does not show causation in this area, and the results are potentially more indicative that students who are more engaged and attend regularly are likely to achieve higher grades than those who are absent for any number of reasons. Moore (2005, pp. 26–28, 30, 32) also demonstrates the interesting point that they found both performance and attendance improved in groups who had the importance of attendance explained to them. The interplay here is much more nuanced, and the more impactful factor from this study is the student's taking responsibility for their learning (Moore 2005, pp. 26–28, 30, 32).

Snyder et al.'s (2014, pp. 433–440) findings were contrary to earlier research and found that the indicator for academic attainment was not based on attendance, but on previous attainment. Those with better grades achieve better grades, regardless of attendance. This lends further credence to the idea that more committed and academically inclined students can achieve highly (in Snyder et al.'s study, this was defined as students with a GPA of 3.2 and higher, equivalent to a score of 60% and higher (Hastings-Woodhouse 2022)) without compulsory attendance.

The form teaching takes also has a sway over how useful a session is to individual students. Learning styles vary widely, with literature citing over one hundred styles and theories (Bates 2015); it is therefore highly unlikely that every teaching session can address the needs of each individual learner. Rudolphi-Solero et al. (2021) exhibited this quantitatively, where a previously voluntary session—based around an online formative assessment in a video game—was made compulsory. Data showed that not only did the average score in the assessment decrease, but feedback and general opinions of the resource also became less favourable. This demonstrated that enforcing the attendance and engagement of students does nothing for either students or educators. These results will have left many students feeling disheartened at their performance, and their time was lost using an unenjoyable resource; as for educators, the feedback has the potential to be unnecessarily harsh, when this particular type of session was never going to appeal to certain students. Whilst I do believe teaching sessions should be accessible and useful to all, it is naïve to think that one session will fit all needs, which is where the variety of teaching offered becomes of paramount importance.

The COVID-19 pandemic created many challenges for the education sector, but also provides interesting insights into student attendance and their attainment. Due to the closure of educational institutions throughout the lockdown periods of 2020, all teaching was transitioned to online methodologies, be they synchronous (Teams or Zoom calls) or asynchronous (prerecorded content). The perceived effects that this period had on university attendance and engagement was demonstrated in data collected by Times Higher Education (THE) in 2022 via a survey of academics (Fig. 1.3).

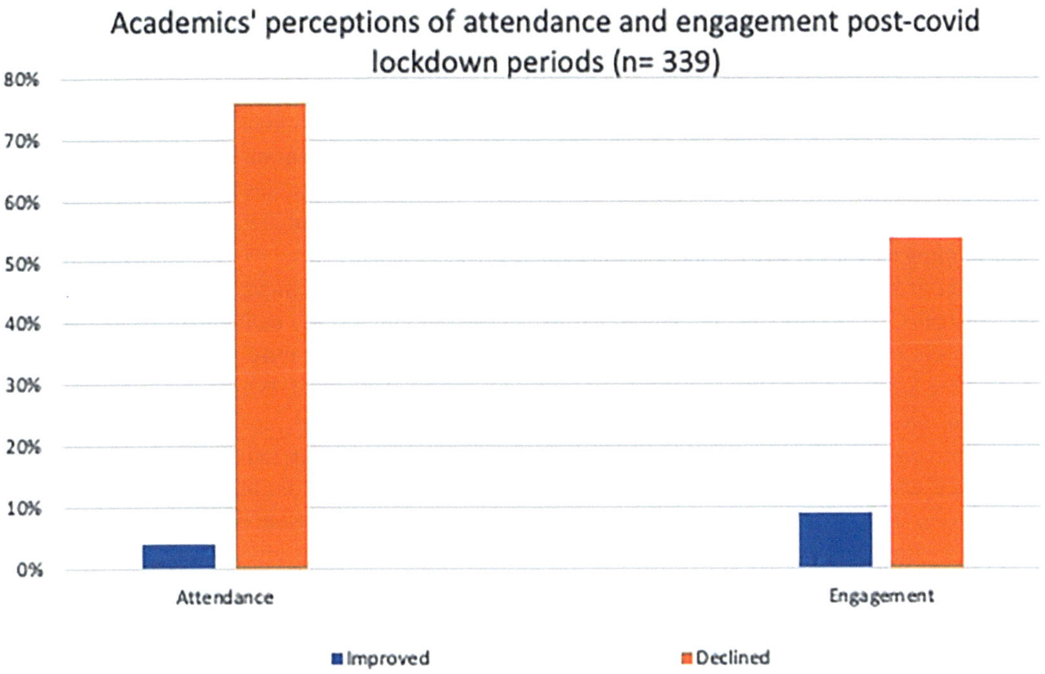

Fig. 1.3 Academics' perceptions of university attendance and engagement post COVID-19 lockdowns

Figure 1.3 displays data from THE on academics' perceptions of changes in the attendance and engagement levels at in-person lectures and in-person seminars after the return to in-person teaching in 2022 (Times Higher Education 2022).

THE's data shows a significant perceived reduction to in-person attendance at the university level, which stands to reason when considering how long, and how well, the transition to online and asynchronous teaching was handled. An obvious challenge to this would be to explore student attainment comparing these periods, and a number of institutions did so. Multiple studies compared their respective student cohorts assessment results both during and after the lockdown measures and showed that students either performed similarly or even better, which included medical students (Engelhardt et al. 2021; Ryan et al. 2022). These results support that the transition to learning without being physically present at teaching sessions did not harm the level of knowledge or capability of students on their courses. This could be due to many reasons, it could be that the new resources offered and methods of teaching applied were inherently better than older legacy teaching materials, or perhaps students being given more freedom to learn as they chose and when they chose proved beneficial across the board. Irrespective of the reasoning, when it came to the tangible measure of assessments, the reduction in in-person attendance was proven to have no detriment to student attainment.

A quote from Davis' (1993) paper (p. 138) sums up my feelings exactly here: 'attendance should not be mandatory or a factor in your grading policy. Rather, grades should be based on students' mastery of the course content and not on such non-academic factors as attendance'. Academia is here to assess student's abilities, so how they achieve the knowledge and skills at a higher education level should be their decision. The concern for many seems to be that it is known that students are less likely to attend when they are not rewarded for attendance, but when attendance seems arbitrary, who does this benefit? (Launius 1997, pp. 86–92; Moore 2004, pp. 24–39).

1.4 Personal Growth

An important part of anyone's university career is personal growth. For many undergraduates, university will be their first time experiencing true independence and autonomy as they begin their adult lives.

According to Kolb's (2014) experiential learning model, it is good sense to make as much of the attendance compulsory as possible to ensure these learning opportunities are taken advantage of; be they academic content, interpersonal and communication skills, or otherwise. A particular experience offered at university is the ability to experience wider diversity which may not have been present at students' previous places of study. To quantify this, Whitla et al. (2003) conducted a survey in which they asked Harvard Medical School and University of California, San Francisco, School of Medicine students about their experiences of encountering individuals of differing races and ethnicities during their schooling; and showed that interactions rose from 50% during high school to 85% during medical school. This increase in diversity of their peers is an experience of paramount importance for medics, one which has been proven to not only enhance students' educational experiences, but also provides better representations of society and the people who they will inevitably treat (Whitla et al. 2003, pp. 460–466; Kai et al. 1999, pp. 616–623).

Interacting with others also allows an individual to better assess their strengths and weaknesses. The Johari Window model (Table 1.2) succinctly explains the benefit of this in relation to the idea of a 'blindspot'.

Table 1.2 The Johari Window model is a matrix which displays the impact an individual may have as they begin to initiate a relationship with another, based on the information and impressions shared between the parties involved (Newstrom and Rubenfeld 1983)

	Known to self	Unknown to self
Known to others	Arena	Blindspot
Unknown to others	Façade	Unknown

This model highlights the importance of self-awareness in our interactions with others and shows how we can use these opportunities to grow by identifying areas of ourselves we are otherwise oblivious to (Newstrom and Rubenfeld 1983). Without the opportunities to meet and collaborate with others—as is offered at many teaching sessions, especially with the previously mentioned higher diversity at universities—there is a real possibility that some individuals may never have any of their views, habits, or preconceptions identified and challenged. Whilst this may not necessarily be disastrous, it is hugely valuable to educate more tolerant individuals and help to rectify any misconceptions held; one could say this is vital when considering the medical profession. These instances also offer an opportunity for the practice of peer-peer teaching, a methodology which is praised in literature for improving student performance (Bennett et al. 2018; Schuetz et al. 2017). From personal experience in my own teaching, asking students to explain concepts to each other is a fantastic way of highlighting how well they know a topic, as any gaps in their understanding become apparent when asked to share their knowledge with their colleagues. This practice encourages students to think about how to best convey information so that it is easily digestible by others, who may have different learning preferences or backgrounds from their own. By prompting this, students are forced to consider the manner they share information on potentially complicated matters, even creating a bespoke style for the individual, not unlike what would be necessary in certain patient-facing clinical situations (Schuetz et al. 2017).

Addressing one's personal blind spots of course requires that the individual is happy to receive constructive criticism and acknowledges their responsibility for their thoughts and actions—a theme within education already discussed. Personally, I feel that enforcing attendance robs students of the opportunity to feel they are fully in control of their education. Compulsory attendance by its very nature removes a level of student autonomy, making uncomfortably close links to Jackson's (1990, pp. 1–11) comparisons of classrooms and prisons. Whilst I would take Jackson's comments as more tongue-in-cheek, it still bears considering that many students are forced to endure numerous sessions that they obtain nothing from, and given more freedom, could utilise their time more effectively. Providing students with more independence allows them time and space to explore how they are best able to learn, which is an invaluable skill on its own. As it is said: 'The most important learning is learning to learn' (Nisbet and Shucksmith 2017, p. 1).

Students often get the impression that the way they are taught is the only way to learn. This may not be an issue if it happens to align with their preferred learning style, but they should also be given the freedom to explore their individual learning process. Thinking creatively in relation to learning has been proven effective in bettering academic performance, therefore stifling student's opportunities to do so seems counterintuitive to creating the lifelong learners that universities promise and is essential for a professional degree such as Medicine (Fatmawati et al. 2019, p. 012070).

There must be an appropriate middle ground between the warden-like policing of students' attendance and complete apathy; an environment where students use their university careers to develop on a personal as well as academic level.

1.5 Mental Health

Current estimates state that one in six individuals will experience some form of mental health problem each week (McManus et al. 2016). This varies significantly depending on the year of university study, but at times exceeds the national average (Macaskill 2012, pp. 426–441). Medicine is often seen as a more stressful and intensive programme, and so mental health difficulties are thought to be more prevalent amongst these cohorts. However, literature in this space is conflicted, and mental wellbeing may be more dependent on the institution's treatment and attitude towards students (Carson et al. 2000, pp. 115–116; Shariati et al. 2002, pp. 29–37). Nonetheless, the prevalence of mental health issues in modern society warrants that universities need to take the subject seriously within their student cohorts.

In relation to attendance, there are multiple reasons that enforcing a compulsory attendance model is beneficial for the health of students. By keeping records of attendance habits of students, this can be used to easily track their engagement with their studies, which as was discussed earlier, can be negatively impacted by poor health (Latif et al. 2016, pp. 44–52). This offers an opportunity for potential wellbeing issues to be spotted by staff, who can then reach out to and offer their support if needed. For struggling students, who do not feel they can ask for help when needed, or know who to turn to, this can be a vital step in receiving aid. Intervening as early as possible could also prevent the worsening of the student's situation. In addition, by attending and interacting with colleagues, students can form working relationships and friendships with others, which allows them to develop a support network. Having relationships with others who are more likely to understand your experiences can help with managing stress and other difficulties (Shirey 2004, pp. 291, 313–319).

On the other hand, attendance monitoring is also possible without the compulsory and assessed aspects. By making attendance part of the assessed criteria, assumptions are being made that all students have the same capacity to be present at teaching sessions. Many seem to think that university courses and employment should be the highest priorities of individuals, and so be fulfilled ahead of other commitments; this however is called into question when considering Maslow's Hierarchy of Needs (Fig. 1.4).

Maslow's hierarchy of needs/motivation model depicts motivational theory. Needs at the lower portions of the pyramid need to be addressed before those higher up the pyramid can be achieved (Adapted from Wikimedia Commons 2009; McLeod 2024).

Translating this into an educational setting, it would be unlikely that a student can effectively learn should they not have their 'Basic' and 'Psychological' needs met (i.e. needs for food, water, and emotional wellbeing) (Maslow 1943,

Fig. 1.4 Maslow's hierarchy of needs/ Maslow's motivational model

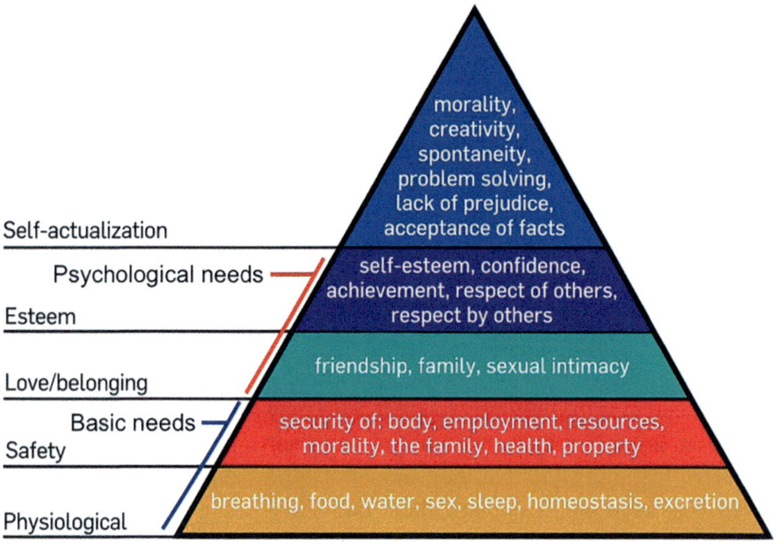

1954, 1987). Reasons for this can be wide ranging and not easily addressed. From my own experience, students have reported difficulties dealing with grief, housing issues, physical injury, and mental health problems, all preventing them from taking full advantage of the session they are attending. In each of these cases, the student felt obligated to attend, as they knew their attendance was being assessed, which inherently added to their already heightened levels of stress. Had it not been assessed, these students could have used their time more effectively, actively working towards solving their problem or simply caring for themselves, rather than foregoing their physiological and psychological needs (McLeod 2024).

This is an issue that is currently being exacerbated by government messaging within the Department for Education's (DfE) attendance campaign. This campaign provides materials and messaging to educational institutions with students at stages as early as primary school, aiming to encourage students to attend in spite of any negative feelings towards their school, or even illness at the start of their day (Figs. 1.5a, b).

Figures 1.5a, b are poster examples of materials that were distributed to schools by the DfE with guidance on how they can be used to encourage school attendance in spite of how the students' may feel at the start of their day. The faces of the children have been blurred to obscure their identities, as the individuals may not have consented to their likeness being utilised outside of an NHS environment (Department for Education 2024).

The effects of marketing and advertising have been explored extensively in the literature and have spawned regulations on products that have restricted advertising capabilities due to worries about their harmful effects. The most common examples of this include substances such as tobacco and alcohol (Lapierre et al. 2017). This was particularly pertinent when concerning children and adolescents, it was shown that exposure to this type of advertising risked the increase of adolescents beginning drinking at an earlier age, or exacerbating current drinking habits (Anderson et al. 2009). The susceptibility of children and adolescents to visual advertising is well documented, largely in the food and drink industry. A literature review by Kraak and Story (2015) showed that across 11 studies, brands utilising mascots or recognisable characters in their marketing were able to effectively influence the choices of children 12 and under with regard to their diet. Results showed increases in favouring choices of fruit and vegetables, but were also seen to be more effective in 'energy-dense' or

Fig. 1.5 (**a**, **b**) Examples of campaign materials from the Department for Education attendance campaign guidelines 2024

'nutrient-poor' options. However, this review still shows that the way the marketing is constructed is most important with regard to having an effect on younger individuals, regardless of the product or message. The ease with which children and adolescents are swayed by marketing can be attributed to their inability to think critically of the advertising being shown to them while so early in their cognitive development, or that they fail to even recognise when they are being advertised to (John 1999; Moses and Baldwin 2005). The utilisation of poster-based media has also been proven to be effective with relation to healthcare activities, with Noviani et al.'s (2023) study showing that this type of advertising aimed at primary school children was effective in promoting changes in their dental care habits. Their lack of awareness in this area can lead to a lack of questioning the messaging and regarding it as fact, and depending on the message, this can ultimately have negative impacts—particularly psychologically (Han 2022). With these potential impacts affecting children so young and impressionable, it stands to reason that these issues could persist throughout adolescence and into adulthood.

This becomes all the more concerning with regard to the messaging being disseminated by the DfE such as in Fig. 1.5, as they are specifically targeting student perceptions of their physical and mental health. DfE's materials utilising imagery of individuals of similar ages to students they are targeting with this campaign can be seen to be utilising the premise of peer influence to add weight to their advertising. By including images of those students can easily identify as peers, this can add to the influential potential of the adverts, as it can easily prey on the desire for conformity and social inclusion which is often heightened in younger individuals (Laursen and Veenstra 2021). The DfE explicitly includes anxiety as a target for their campaign (2024) (Fig. 1.5b), and so recognise this is a reason for many students not attending their institutions. The available statistics agree that anxiety and other mental health issues are prevalent in school age individuals, with the CDC (2022) showing that 9.4% of US children aged 3–17 years old experienced mental health issues between 2016 and 2019, with some states reporting rates as high as 14.7%. Attendance whilst ill, mentally or physically, also raises the question of how well the student will be able to engage with their studies. Here the DfE are asking almost one in ten students to ignore their wellbeing and attend school, actively encouraging presenteeism, and preventing students from addressing the health issues they are facing. Which as Maslow (Fig. 1.4) demonstrates, prevents students from achieving their basic and psychological needs, and these are necessary to gain the intended benefits from their education.

This theory is also supported by the premise of 'Cognitive Load', which refers to the capacity of working memory taken up during task completion. Studies show that anxiety levels can significantly impact cognitive load, hampering performance in test environments (Chen and Chang 2009, pp. 729–746). Students have attested, albeit anecdotally, that attendance requirements of courses add to their overall stress and anxiety, and whilst mitigation measures are in place to prevent disadvantaging students with good reason for absences, this is often a lengthy process which involves giving students the burden of evidencing the validity of their absence. This inherently adds to their cognitive load and takes up their time outside of mandated teaching sessions.

There is always an element of pastoral care in education, and so personally, I feel that students' mental health falls within the remit of the institution to actively care for, be it by directly offering support services or indirectly by not supporting potentially harmful practices where the benefit is not directly tangible.

1.6 Conclusion

All points considered, notwithstanding the benefits of enforcing attendance seen in the literature, the inherent risk of increasing levels of stress, enabling presenteeism, and removing students' autonomy outweighs the need to impose assessed attendance. Not only due to these detrimental possibilities, but also the fact that improvements in attendance were seen after simply explaining the benefits to students, which in turn preserved their autonomy as adult learners to decide what is best for them in their education.

On reflection, Davis' quote: 'grades should be based on students' mastery of the course content…' (1993, p. 138) rings most true. Their presence in a particular room on a particular day will be inconsequential years into their career, treating patient after patient with the skills and knowledge obtained and validated through assessments.

To rework an old adage—you can lead a medical student to teaching, but you can't make them learn.

References

ABIM Foundation, ACP-ASIM Foundation & European Federation of Internal Medicine (2002) Medical professionalism in the new millennium: a physician charter. Ann Intern Med 136:243–246

Alexander R (2005) Towards dialogic teaching. Dialogos, York

Anderson P, de Bruijn A, Angus K, Gordon R, Hastings G (2009) Impact of alcohol advertising and media exposure on adolescent alcohol use: a systematic review of longitudinal studies. Alcohol Alcohol 44(3):229–243

Bates B (2015) Learning theories simplified: …and how to apply them to teaching. Sage, London

BBC News (2006) BBC News Education Report – Association of Teachers and Lecturers, UK survey

Bennett SR, Morris SR, Mirza S (2018) Medical students teaching medical students surgical skills: the benefits of peer-assisted learning. J Surg Educ 75(6):1471–1474

Brocato J (1989) How much does coming to class matter? Some evidence of class attendance and grade performance. Educ Res Q 13(3):2–6

BSMS.ac.uk (2022) Assessments - BSMS. [online]. https://www.bsms.ac.uk/undergraduate/our-course/assessments.aspx. Accessed 20 Jan 2022

Bujan J (1999) Increasing students' responsibility for their own learning. [online]. Illinois. https://eric.ed.gov/?id=ED400072. Accessed 24 Jan 2022

Caine R, Caine G (1994) Making connections: teaching and the human brain. Addison Wesley, Somerset, NJ

Carson AJ, Dias S, Johnston A, McLoughlin MA, O'Connor M, Robinson BL, Sellar RS, Trewavas JJC, Wojcik W (2000) Mental health in medical students. A case control study using the 60 item general health questionnaire. Scott Med J 45(4):115–116

Centers for Disease Control and Prevention (U.S.) (2022) Mental health surveillance among children — U.S. 2013–2019. Published: 24th February 2022. https://

stacks.cdc.gov/view/cdc/113924. Accessed 17 Oct 2024

Chen I, Chang CC (2009) Cognitive load theory: an empirical study of anxiety and task performance in language learning. Electron J Res Educ Psychol 7(2):729–746

Davis JR (1993) Better teaching, more learning: strategies for success in postsecondary settings. Oryx, Phoenix, AZ, p 138

Department for Education (2024) DfE attendance campaign guidelines 2024. https://dfegovukassets.blob.core.windows.net/assets/Attendance%20campaign/DfE%20Attendance%20campaign%20guidelines%202024.pdf. Accessed 16 Oct 2024

Devlin M (2002) Taking responsibility for learning isn't everything: a case for developing tertiary students' conceptions of learning. Teach High Educ 7(2):125–138

Engelhardt B, Johnson M, Meder ME (2021) Learning in the time of Covid-19: some preliminary findings. Int Rev Econ Educ 37:100215

Fatmawati A, Zubaidah S, Mahanal S, Sutopo S (2019) Critical thinking, creative thinking, and learning achievement: how they are related. J Phys Conf Ser 1417:012070

General Medical Council (2011) Assessment in undergraduate medical education. General Medical Council/Medical Schools Council, London

General Medical Council (2016) Achieving good medical practice. General Medical Council/Medical Schools Council, London

General Medical Council (2019) Practical skills and procedures. General Medical Council/Medical Schools Council, London

Han J-I (2022) The role of visual communication for emotional marketing strategy. J Ind Distrib Bus 13(11):39–46. https://doi.org/10.13106/JIDB.2022.VOL13.NO11.39

Hastings-Woodhouse S (2022) GPA and other grading systems—a guide. https://www.findamasters.com/guides/gpa-grade-point-average. Accessed 21 Oct 2024

Hemp P (2004) Presenteeism: at work-but out of it. Harv Bus Rev 82(10):49–58

HESA (2024) Who's studying in HE? https://www.hesa.ac.uk/data-and-analysis/students/whos-in-he. Accessed 18 Oct 2024

Hilton SR, Slotnick HB (2005) Proto-professionalism: how professionalisation occurs across the continuum of medical education. Med Educ 39:58–65

Hochberg MS, Kalet A, Zabar S, Kachur E, Gillespie C, Berman RS (2010) Can professionalism be taught? Encouraging evidence. Am J Surg 199(1):86–93

Jackson PW (1990) Life in classrooms. Teachers College Press, New York, pp 1–11

James Madison University Counselling and Student Development Center (2020) Coping with Disruptive Students. jmu.edu/counselingctr/audience-resources/faculty-staff/disruptive.shtml. Accessed: 27th Jan 2025

John DR (1999) Consumer socialization of children: a retrospective look at twenty-five years of research. J Consum Res 26(3):183–213

Jones CH (1984) Interaction of absences and grades in a college course. J Psychol 116:133–136

Kai J, Spencer J, Wilkes M, Gill P (1999) Learning to value ethnic diversity–what, why and how? Med Educ 33(8):616–623

Kates N, Ackerman S (2002) Shared mental healthcare in Canada. A compendium of current projects. Canadian Psychiatric Association; and Mississauga, College of Family Physicians of Canada Collaborative Working Group on Shared Mental Health, Ottawa

Kolb DA (2014) Experiential learning: experience as the source of learning and development. Pearson Education, London

Kraak VI, Story M (2015) Influence of food companies' brand mascots and entertainment companies' cartoon media characters on children's diet and health: a systematic review and research needs. Obes Rev 16(2):107–126

Lapierre MA, Fleming-Milici F, Rozendaal E, McAlister AR, Castonguay J (2017) The effect of advertising on children and adolescents. Pediatrics 140(Suppl 2):S152–S156

Latif M, Khan UA, Khan AN (2016) Causes of students' disruptive classroom behaviour: a comparative study. Gomal Univ J Res 32(1):44–52

Launius MH (1997) College student attendance: attitudes and academic performance. Coll Stud J 31:86–92

Laursen B, Veenstra R (2021) Toward understanding the functions of peer influence: a summary and synthesis of recent empirical research. J Res Adoles 31(4):889–907

Lempp H, Seale C (2004) The hidden curriculum in undergraduate medical education: qualitative study of medical students' perceptions of teaching. Br Med J 329(7469):770–773

Macaskill A (2012) The mental health of university students in the United Kingdom. Br J Guid Couns 41(4):426–441

Marriott A, Mable J (2002) Sharing the learning: the health transition fund synthesis series: primary healthcare. Health Canada, Ottawa

Maslow AH (1943) A theory of human motivation. Psychol Rev 50(4):370–396

Maslow AH (1954) Motivation and personality. Harper & Row, New York

Maslow AH (1987) Motivation and personality, 3rd edn. Pearson Education, Delhi, India

McLeod SA (2024) Maslow's hierarchy of needs. https://www.simplypsychology.org/maslow.html. Accessed 19 Aug 2024

McManus S, Bebbington P, Jenkins R, Brugha T (2016) Mental health and wellbeing in England: adult psychiatric morbidity survey 2014. NHS Digital, Leeds. http://content.digital.nhs.uk/catalogue/PUB21748/apms-2014-full-rpt.pdf. Accessed 27 Jan 2022

Moore R (2004) Does improving developmental education students' understanding of the importance of class attendance improve students' attendance and academic performance? Res Teach Dev Educ 20(2):24–39
Moore R (2005) Attendance: are penalties more effective than rewards? J Dev Educ 29(2):26–28, 30, 32
Moore R, Jensen PA (2008) Do policies that encourage better attendance in lab change students' academic behaviors and performances in introductory science courses? Sci Educ 17:64–71
Moses LJ, Baldwin DA (2005) What can the study of cognitive development reveal about children's ability to appreciate and cope with advertising? J Public Policy Mark 24(2):186–201. https://doi.org/10.1509/jppm.2005.24.2.186
Newstrom JW, Rubenfeld SA (1983) The Johari window: a reconceptualization. In: Developments in business simulation and experiential learning: proceedings of the annual ABSEL conference, vol 10
NHS (2020a) NHS Civility and respect toolkit. https://www.socialpartnershipforum.org/system/files/2021-10/NHSi-Civility-and-Respect-Toolkit-v9.pdf. Accessed 18 Oct 2024
NHS (2020b) NHS - supporting our staff - a toolkit to promote civility and respect poster. https://www.england.nhs.uk/wp-content/uploads/2022/07/NHS-Supporting-our-staff-poster-A-toolkit-to-promote-culutures-of-civility-and-respect.jpg. Accessed 18 Oct 2024
NHS (2023a) NHS long term workforce plan. https://www.england.nhs.uk/long-read/nhs-long-term-workforce-plan-2/. Accessed 18 Oct 2024
NHS (2023b) NHS staff survey national results briefing 2023. https://www.nhsstaffsurveys.com/results/national-results/. Accessed 18 Oct 2024
Nisbet J, Shucksmith J (2017) Learning strategies. Taylor & Francis, London
Nolte J, Tremblay M (2005) Enhancing interdisciplinary collaboration in primary healthcare in Canada. Enhancing interdisciplinary collaboration in primary healthcare initiative, Ottawa
Noviani N, Ngatemi N, Fadjeri I, Purnama T, Restuning S (2023) Stair poster media as an effort to improve the dental health behavior of elementary school children. Asian J Dent Health Sci 3(3):14–18. https://doi.org/10.22270/ajdhs.v3i3.48
Rudolphi-Solero T, Lorenzo-Alvarez R, Ruiz-Gomez MJ, Sendra-Portero F (2021) Impact of compulsory participation of medical students in a multiuser online game to learn radiological anatomy and radiological signs within the virtual world second life. Anat Sci Educ
Ryan MS, Holmboe ES, Chandra S (2022) Competency-based medical education: considering its past, present, and a post-COVID-19 era. Acad Med 97(3S):S90–S97. https://doi.org/10.1097/ACM.0000000000004535
Schuetz E, Obirei B, Salat D, Scholz J, Hann D, Dethleffsen K (2017) A large-scale peer teaching programme–acceptance and benefit. Z Evid Fortbild Qual Gesundhwes 125:71–79
Seeman H (2010) Preventing disruptive behavior in colleges: a campus and classroom management handbook for higher education. R&L Education, Lanham, MA
Shariati M, Kaffashi A, Ghalehbandi M, Fateh A, Ebadi M (2002) Mental health in medical students of the Iran University of Medical Sciences (IUMS). Payesh 1(3):29–37
Shirey MR (2004) Social support in the workplace: nurse leader implications. Nurs Econ 22(6)
Shrader S, Kern D, Zoller J, Blue A (2013) Interprofessional teamwork skills as predictors of clinical outcomes in a simulated healthcare setting. J Allied Health 42(1):1E–6E
Snyder J, Lee-Partridge J, Jarmoszko A, Petkova O, D'Onofrio M (2014) What is the influence of a compulsory attendance policy on absenteeism and performance? J Educ Bus 89(8):433–440
Stewart WF, Ricci JA, Chee E, Morganstein D (2003) Lost productive work time costs from health conditions in the United States: results from the American productivity audit. J Occup Environ Med 45(12):1234–1246
Task Force Two (2005) A physician human resource strategy for Canada: innovation service models in Canada database. Government of Canada, Ottawa
Thistlethwaite J, Spencer J (2008) Professionalism in medicine. Radcliffe, Abingdon, UK
Times Higher Education (2022) Class attendance plummets post-Covid. https://www.timeshighereducation.com/news/class-attendance-plummets-post-covid#:~:text=Asked%20what%20the%20typical%20level,than%20two-thirds%20of%20respondents/. Accessed 18 Oct 2024
Whitla DK, Orfield G, Silen W, Teperow C, Howard C, Reede J (2003) Educational benefits of diversity in medical school: a survey of students. Acad Med 78(5):460–466
Wikimedia Commons (2009) Maslow's hierarchy of needs. https://commons.wikimedia.org/wiki/File:Maslow%27s_Hierarchy_of_Needs.svg. Accessed 21 Oct 2024

Building Belonging in Biological Sciences

2

Kelly Edmunds

Abstract

Students in Higher Education (HE) are reporting record high levels of anxiety alongside lower levels of self-efficacy, and these are contributing to an engagement crisis. Students are struggling to engage with their studies as they manage declining mental health—poorer mental health than that reported by the wider population—and increasing numbers of students are juggling their studies alongside paid work, just to meet their basic living costs.

University life brings many new opportunities, both social and academic. For an increasing number of students, growing external demands on their time mean they are making hard choices about which education opportunities to engage with. Biosciences courses present specific opportunities to perform experiments, make use of new equipment and engage in trial-and-error as part of the scientific method—all activities which can often lead to failure, the consequences of which for a student already managing challenges with their mental health, financial situation or time pressures, can be to disengage further from their studies. One way to mitigate the impact of these factors on student engagement and success is to help students build a strong sense of belonging within their studies.

This chapter sets the context for some of the challenges facing our students, with a focus on student belonging and how this links to academic success. Drawing on examples from across different UK biosciences departments, this chapter explores how we can embed activities into our biology teaching which encourage students to experiment whilst also building a growth mindset and a stronger sense of belonging. The examples shared will cover aspects of the important transition into HE through to a range of activities that can be embedded into teaching and assessment at various points throughout the student journey.

Keywords

Student belonging · Academic belonging · Mattering · Student experience · Student support · Embedding wellbeing

2.1 Introduction

2.1.1 Belonging Within the Context of Higher Education

Belonging is the feeling that we associate with security and acceptance within a space. A strong

K. Edmunds (✉)
School of Biological Sciences, University of East Anglia, Norwich, UK
e-mail: k.edmunds@uea.ac.uk

R. Door (ed.), *Accessibility, Inclusivity and Diversity in Education and Beyond*, Biomedical Visualization 8, https://doi.org/10.1007/978-3-031-81795-3_2

sense of belonging typically requires us to feel valued, included and that we can feel these things when we bring our true authentic self to a place. And when we feel that we belong in a space we are more likely to engage both with that space and with the others in it. When we sense that we genuinely belong then we are more likely to build connections with others, to speak up and share our ideas and opinions and more likely to collaborate. When we feel that we belong, then we are more likely to thrive (Winstone et al. 2020).

Students studying within Higher Education (HE) are reporting higher levels of anxiety (Lewis and Bolton 2023), lower levels of self-efficacy (Meng and Zhang 2023) and this feeds into an engagement crisis (UPP Foundation 2022). Students are struggling to engage with their studies as they manage declining mental health—poorer mental health than that reported by the wider population—and increasing numbers of students are juggling their studies alongside paid work, just to meet their basic living costs. We know that students with below average mental health are almost twice as likely to feel that they do not belong at university and are less confident in their abilities compared to those who are reporting "better" mental health and thus likely to be experiencing fewer mental health-related challenges (Lewis and Bolton 2023; UPP Foundation 2022).

As the HE sector navigates the dynamic changes to our student cohorts that have been seen post-pandemic and during a cost-of-living crisis, the topic of student belonging has seen a surge of interest and attention (Gilani 2023, see Fig. 2.1). There have been many attempts to delve into this complex arena and understand what is needed to ensure that students feel a sense of belonging within their institution, but the consensus is that there is no one-size fits all approach that works for any given student, course or institution. Student belonging is a complex and evolving landscape. As individual students progress through their academic journey and as life within and beyond their studies changes, so too does their expectations and requirements of their institution, to offer them a place for belonging. Recent research by Wonkhe and Pearson (published as Blake et al. 2022) explored how to build belonging within HE and identified four pillars of belonging: connection, inclusion, support and autonomy. Together these four pillars encompass the very foundations which are central to students

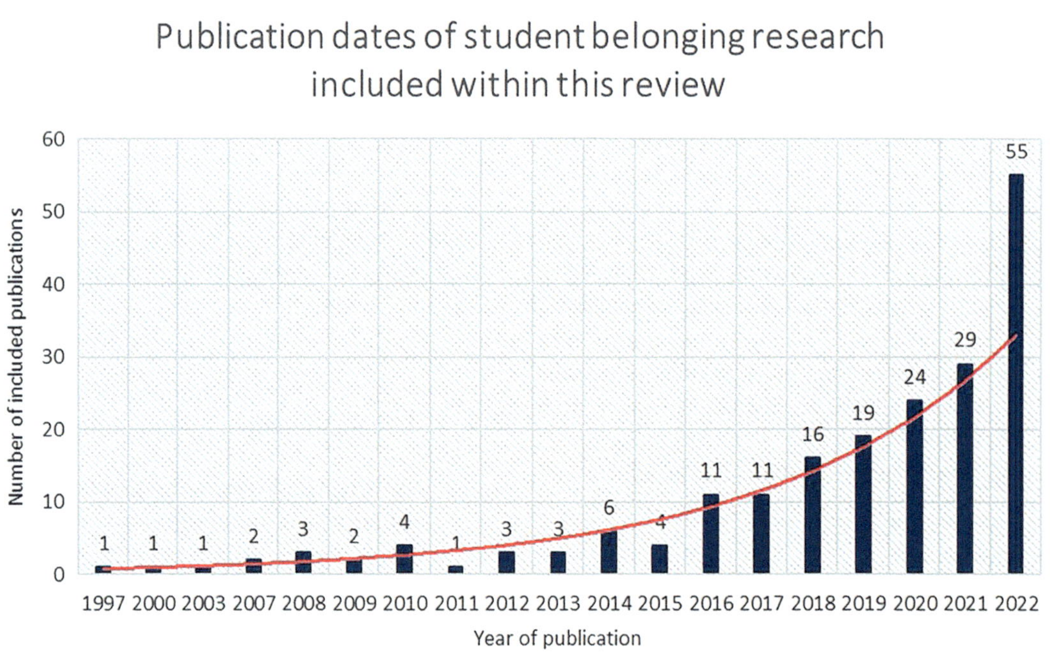

Fig. 2.1 The growth in publications from research into student belonging, used (with permission) from Gilani (2023)

feeling a sense of belonging within their HE institution. But what are we asking students to belong to and how can we nurture and support them to find their place? This chapter will begin by exploring how students develop a sense of belonging and how this can be influenced by some of the challenges facing students within HE and the relationship between these challenges. The chapter will then move on to think about factors which can affect a student's sense of belonging and how we can influence these at different levels, from the institutional level through to that of the course. The second half of the chapter shifts to the more practical aspects of supporting students to build a sense of belonging through teaching-related activities. With the inclusion of case studies from across different levels of bioscience teaching, the chapter provides advice and guidance on activities that can be embedded into a range of teaching activities that can help students to foster connections and build a support network—two of the key activities known to support the development of a strong sense of belonging. The examples presented in this chapter can be applied across a range of taught settings from larger lecture theatre activities through to smaller seminar settings and the lab environment, with examples and guidance also given for pre-arrival activities and assessments.

2.1.2 The Unique Experience of Belonging

How we foster a sense of belonging is not universal nor is it static in its retention or development as our personal circumstances and needs change over time. There is no one formula that can be applied to capture all the elements which feed into an individual student's sense of belonging. Discussions around student belonging often fall short on two assumptions, the first being that all students within HE want to feel that they belong, despite there being evidence that there are some students who actively choose not to belong (Gravett and Winstone 2019) and the second being the assumption that all students will be influenced by the same aspects of their university experience when building their sense of belonging. However, there are a multitude of factors, both within and beyond the institutional boundaries, which influence the sense of belonging that a student may attribute to their HE institution (HEI).

Key to building and maintaining a sense of belonging for students is the connection that they feel to their institution (Blake et al. 2022). These connections can be made across experiences, opportunities and interactions within a range of curricular, extra-curricular and residential contexts, including:

(i) Within the formal taught environment.
(ii) Within add-on teaching-related activities such as field courses, work placements or research days.
(iii) As part of university-based clubs and societies.
(iv) Through social opportunities with course peers, housemates or other students.
(v) Associated with university-led pastoral and support services.

Given the sheer volume and range of opportunities available for undergraduate students to undertake alongside and within their studies, identifying one formula that can predict how to ensure that any one student will feel a place of belonging within the institution is essentially mission impossible. As well as each student being unique, courses are not equal in terms of their structure and the opportunities that they afford and certainly not all students are experiencing their courses in the same way.

2.1.3 Factors Influencing Student Belonging

Student belonging is a dynamic and rapidly evolving space. Trying to understand the influence of each of the factors which impact on the sense of belonging for a student and the interplay between these both within and across cohorts is a complex task. As already mentioned, there is no one formula which can be universally applied to all students to ensure that they all feel a sense of belonging. Despite occupying the same physical space, we each have a different map of the world (Borland 2018) and how one student engages with and perceives an opportunity or interaction, another student may perceive it very differently (Palmer 2024). What we do know is that student belonging is becoming a priority area of focus for many HEIs, and the research effort and attention being given to student belonging is soaring (Gilani 2024). This means that our knowledge about factors influencing student belonging is growing rapidly but it is such a fluid and dynamic space that we will never stop learning and likely never know enough.

2.1.3.1 Influencing Belonging: Beyond the Institution

Previous research by Hoffman et al. (2002) has identified three specific factors which are known to influence a student's sense of belonging:

1. Connections with empathetic and understanding tutors
2. The perception of having peer support and
3. The perception of comfort within the physical learning space

Despite the research of Hoffman et al. taking place more than 20 years ago, the fundamental factors that are influencing a student's sense of belonging have not changed (see, e.g., Blake et al. 2022; Gilani 2024), however, how students engage with their HE studies is changing (Gravett and Ajjawi 2021). Students are managing their studies alongside increasing external demands on their time (Sutton Trust 2023) with more students than ever before engaging in paid work alongside their studies (National Union of Students 2023). Record high numbers of students from lower participation areas enrolled in HE in 2022 (UCAS 2023) during a cost-of-living crisis. Financial pressures disproportionately impact on students from lower participation areas and make them more likely than their counterparts, to withdraw from their courses (Sutton Trust 2023; Million Plus 2022). This juggle of paid work alongside academic studies forces students to make choices about prioritising their time and is resulting in many students opting to engage with their studies remotely using lecture recordings, asynchronous resources and web-based content to keep up with the taught materials.

In addition to those who are managing paid work alongside their studies, the numbers of students who are commuting to university is also increasing (Finn and Holton 2019), with this being attributed to changes to funding in the UK (Finn and Holton 2019; Khambhaita and Bhopal 2015). Living at home whilst studying or having to undertake a lengthy commute to campus contribute to a decreased sense of belonging (De Beer et al. 2009) and it has been shown that having greater parental support can lead to students experiencing a declining sense of belonging during their studies (Haussmann et al. 2009).

Students who commute to campus tend to live off campus for reasons either driven by finances or care responsibilities though there is also evidence which suggests that some students prioritise academic attainment over social integration and as such, choose to live away from campus (Grayson 2003). There are also external factors influencing belonging that relate to culture for some groups of students, for example Asian female students are less likely to move away from home to live at a university campus compared to white female students, due to cultural differences associated with living alone (Pokorny et al. 2017).

2.1.3.2 Influencing Belonging: Within the Institution

Institutions strive to create a culture that enables students to be successful so that they can embrace the opportunities to grow, learn and reach their potential, producing graduates who go on to make positive contributions to society. For this to happen, students need to feel accepted, able to be their authentic selves and valued for who they are. There are many ways in which organisations have achieved success with this from embracing staff–student partnerships (e.g. Curran and Millard 2016) when developing institutional policies and curriculum content through to providing shared study spaces for students to use (Standen 2024). Students who experience interactions within their courses that they perceive to be positive are more likely to persist with their studies (Tinto 2023). This persistence with their studies because of positive interactions demonstrates the close links between student engagement and that all important sense of belonging.

Who students experience these positive engagements with matters (Tinto 2023). Throughout their studies, students will encounter countless opportunities to build connections and develop their own network of individuals who can offer support and guidance across a range of areas, from academic and wellbeing through to social opportunities and transferable skills development. Students who actively engage with developing themselves a network containing multiple, diverse connections are more likely to benefit from social support (Eggens et al. 2008), a greater sense of belonging (Dawson 2008) and potentially greater academic success and engagement with their studies (Tinto 2023).

How students make peer and staff connections and the nature of the subsequent interactions between them also matters. There are differences between individual students and how they interact and utilise in the physical university space as well as differences in the duration of time that individual students will spend physically on campus. Students who are campus-based are more likely to feel a stronger sense of place and belonging alongside their studies than students who feel on the periphery of the student experience, such as commuting and mature students (Tinto 2023; Pokorny et al. 2017). These differences in contact opportunities with the physical institutional spaces also affect how connections are built and maintained. Student support services are typically delivered centrally at an institutional level and the role of these services and the staff who provide them is known to be important for supporting students to find a sense of belonging (Gilani 2023; Levett-Jones et al. 2008).

2.1.3.3 Visualising Belonging: Within the Institution

Many of the elements which influence a student's feelings of belongingness vary throughout their course; their modules may change, friendship groups may be fluid as students take up placement and study abroad options and flatmates may change each year but one constant that remains throughout a course for the majority of students is their institution. Examples of ways in which an institution can visually help to connect students to their institution and the student community, helping to foster and retain a sense of belonging amongst their student cohort, are shown in Fig. 2.2 below.

We know that students interact with both the physical and virtual spaces of their institution differently based on their own personal needs and circumstances. We also know the importance of these interactions for a student's sense of belonging and consequently their engagement, motivation and success on their course. The remainder of this chapter will explore academic belonging with a focus on teaching within biological sciences and ways in which we can embed elements within our teaching which are inclusive and adaptable, to promote a student's sense of belonging and student success.

2.1.3.4 Influencing Belonging: The Course Level

Academic belonging has been described as cultivating a learning environment where individuals feel able to contribute their ideas, ask questions and where students feel able to be their authentic selves, where they feel valued and where they are encouraged to succeed (Kahu et al. 2022; Versteeg et al. 2022). Those connections that have already been mentioned in this chapter apply to the course level with the building of positive, supportive connections and interactions with faculty, who deliver their teaching through inclusive, positive teaching approaches and enable students to engage through a variety of methods being among the many course-based factors that influence an individual student's sense of belonging (Kirby and Thomas 2022; Rueda et al. 2017).

The how, where and when of a student building connections and developing a sense of belonging within their courses vary by course and individual. Across an institution the student cohort may consist of a mixture of both undergraduate and postgraduate students, students

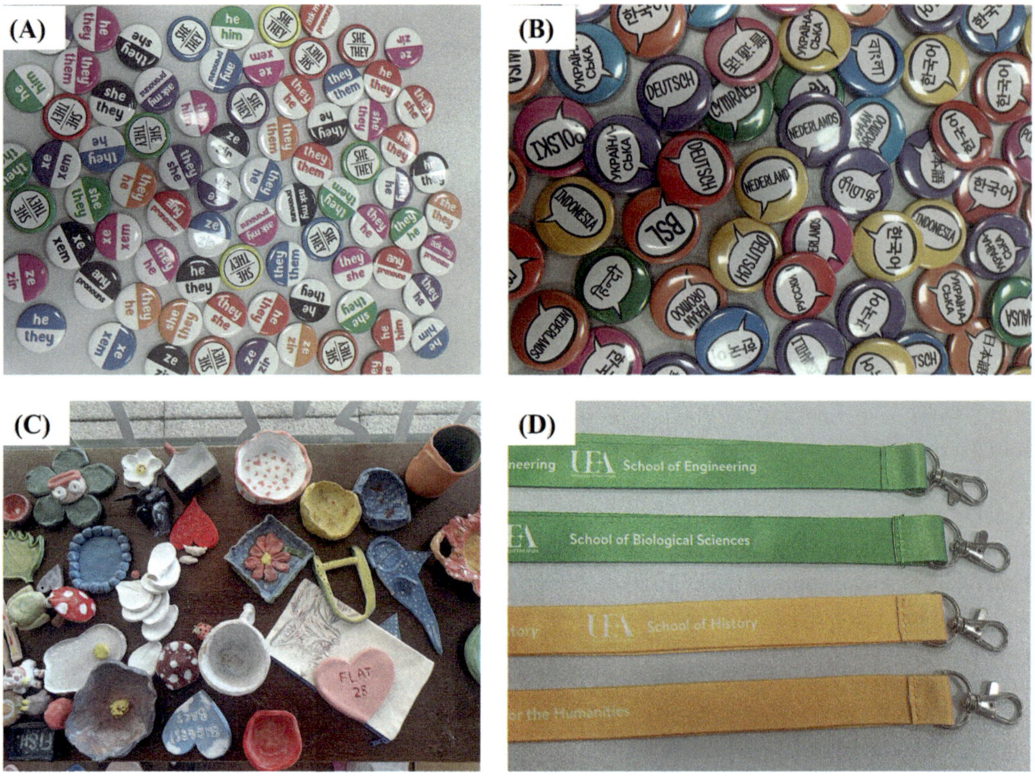

Fig. 2.2 Images showing examples of institution-wide initiatives led by the Student Information Zone at the University of East Anglia, to help students develop and retain a sense of belonging. The initiatives range from those which help students to connect with others who might be from a similar background or studying on a similar course through to those which help to personalise their campus home space. (**a**) Badges in a rainbow of colours showing a range of preferred pronouns and students select their preferred badge and can pin it to their lanyard. (**b**) Badges in a kaleidoscope of colours and languages for students to select their native or preferred language. (**c**) Objects made by students at Creative Calm Art Club, a club which runs throughout the year to encourage students to make pieces which reflect themselves and help to personalise their home space on campus. (**d**) Each student has a campus card and can carry it on a lanyard, which is colour themed by their Faculty of study and helps students to identify at a glance, other students which study within that Faculty

focusing on research and those engaging in a teaching-focused course. Even at a course level there can be differences in the opportunities and experiences that the students have based on the taught elements that they opt to undertake or engage with. For example, there may be students who are undertaking regular placements for clinical practice or for a placement year in industry alongside other students who are studying part-time or via a distance learning option. Table 2.1 attempts to illustrate some of the many differences that can be seen within the formal taught structure of courses within the biological sciences.

2.1.4 Embedding Belonging Within Teaching

A student's HE journey begins before they arrive on campus to start their course. Prior to starting a course, many students will have visited their institution for open days, applicant days or outreach events, others may live locally and be very familiar with the university campus. At the very least, all new students will have received communications from an institution prior to arriving there. All these pre-arrival factors help students in forming that initial sense of belonging and an idea of where their place might be, where they

Table 2.1 The formal components of different Higher Education courses and their potential to influence a student's sense of belonging for biological sciences courses. Majority denotes the main time allocation (>50%) of a course is typically given to that activity, minor denotes a minority (<50%) of course time allocation is allocated to that activity and nominal denotes that very little time on a course is given to that activity (<10%) and optional reflects that students can opt to engage in that activity at some point during their course

	Course-related activity			
Students by course	Research	Teaching	Placements	Field courses
Doctoral student	Majority	–	Optional	–
Masters by research	Majority	–	–	–
MRes	Majority	Nominal	Optional	–
MSc	Minority	Majority	Optional	Optional
MSci	Minority	Majority	Optional	Optional
BSc	Minority	Majority	Optional	Optional
Apprenticeships	Minority	Majority	–	–

might find their people—those that they can connect with and who they feel comfortable with—and how they envisage themselves spending their time—both social and academic—at university. But in the same way that each of our lived experiences are unique, so too are the interactions and experiences that we have and how they impact on our sense of belonging. Thus, when embedding activities within teaching with the aim of helping a student to find their place, space or people, for example, it is important not to assume that all students will get the same benefit from the same activity or experience.

Here are a few pieces of advice that might be helpful for those who are seeking inspiration for how to embed activities that promote student belonging, within teaching:

Some of the advice below has been drawn from examples shared within the Inclusive Education Framework by Hubbard and Gawthorpe (2024).

1. Engage with the students. As well as getting formal feedback from students through evaluations, speak to the students informally as well. From these informal conversations, you can find out what is working well for them on their course, what they are enjoying about their time at university and what challenges they are experiencing. A recent conversation with some of my Foundation Year students during a workshop led to finding out that they love baking, and this led to a bake sale being organised, involving nearly a dozen student bakers where they raised money for a conservation charity that was important to them.
2. Be authentic. Ensure that you are listening to and supporting your students, responding to their individual lived experiences rather than treating the cohort as one. Be empathetic to their situation and challenges and avoid assuming that a student's needs should reflect that of previous students or even yourself when you were studying.
3. Offer regular opportunities. Don't just offer one field trip or fun team quiz at the start of the year and think that's it, job done. Aim for multiple touchpoints through the teaching cycle to allow students to establish new connections or restore previous ones, as their needs change through the year.
4. Provide variety. Try to avoid offering the same approaches each week as the impact decreases over time. This does not need to be a complete rethink for you, for example, if you like to engage the students in team quizzes to test their learning can you introduce a picture round—maybe even get the students to draw their answers and get them marked by another team?
5. Don't go it alone. You don't have to do everything yourself. Many institutions have teams who are experts at running sessions for students that help develop academic skills or support wellbeing whilst also supporting students to build and consolidate their sense of belonging. Reach out to these teams and see what they can offer for your students.

2.1.4.1 Transition into Higher Education

The transition into HE is challenging for many students but can be particularly tough for students from disadvantaged backgrounds for whom the university environment may present an even starker contrast to their home environment (Pokorny et al. 2017; Morosanu et al. 2010; Krause and Coates 2008). The start of a degree programme results in many students juggling an array of shifting circumstances, such as increased academic demands and their associated workload, as well as new living arrangements and social settings (Ramler et al. 2016) and the consequential changes in their immediate support networks.

There is an argument to be made that the traditional approach to inducting students into a new HE setting is no longer fit for purpose (Walker 2024), particularly given modern advances in technology and how intertwined it has become within our day-to-day lives. There are many ways in which institutions are making use of technology to revolutionise their transition and welcome to new students, with pre-arrival platforms fast becoming a core element of the welcome package that are offered to new students (Walker 2024) and being seen to have a positive impact on a student's sense of belonging (Blake et al. 2022). An interdisciplinary team at the University of East Anglia have developed an award-winning pre-arrival module (see Case Study 2.1) that has seen success with their new undergraduate biological sciences students and has subsequently been rolled out across the institution. In Case Study 2.1, Dr. Becky Lewis explains the lessons that they have learnt on their journey to developing their award-winning pre-arrival support module from suites of resources that were poorly engaged with to a site that receives strong and sustained engagement from the students.

Case Study 2.1 Pre-arrival Platforms for Building Student Belonging Within a First-Year Biological Sciences Cohort, Contribution from Dr. Becky Lewis

The University of East Anglia (UEA) is a campus-based, dual-intensive research and teaching institution in the East of England. Each September, the UEA School of Biological Sciences welcomes several hundred new undergraduates onto their BSc degree programmes which range from Biomedicine through to Ecology and Conservation. In response to observations made of students struggling with anxiety associated with the 'unknown' element of university study, over the last few years, we have developed a pre-arrival module which we have embedded within our institutional virtual learning environment. The pre-arrival module was co-designed with student interns, informed by a comprehensive peer survey and provides an immersive introduction to academic study while alleviating anxieties associated with starting university.

This course contains a range of resources including virtual lab tours (see Fig. 2.3), interviews with academic, technical and professional services staff, virtual lab simulations (see Fig. 2.3), an introduction to the advising system and advice for navigating campus. Content is released in weekly blocks and used alongside embedded content such as discussion boards and an embedded Padlet site for students to introduce themselves to each other. Through the use of virtual socials, the pre-arrival module helps to promote early connections among students and faculty as well as to help students to get a feel for what they can expect from their initial weeks at UEA. The module leads into the institutional 'welcome week', a brief but impactful

(continued)

introduction to university life but with the content remaining accessible to students throughout their degree, providing a valuable resource they can revisit as needed. Indeed, despite deployment in the pre-arrival period, students continue to access the module well into their second semester. In a spot check half-way through the second semester of 2024, 30 students from the 2023 first-year cohort had accessed resources within the past week. Beyond that, for the previous year's site, we saw 58 students returning to engage with the module content in their first semester of their second year, over a year after the release of content. Resources available with the first week of the module had 664 hits and by the end of the first semester, there were almost 5000 hits across the resources.

2.1.4.2 Course/Curriculum Design

Belonging is closely linked to academic success and motivation (Freeman et al. 2007). Establishing a feeling of belonging within an academic environment is determined largely by people and place. For students, their sense of belonging to an academic environment is known to be heavily influenced by the connections that are made on their course—with their peers, administrative staff and academic teaching team—and the quality and frequency of the connections that are made with them (Rueda and Lowe Swift 2024). These connections then help to build a sense of mattering, of both belonging to and mattering to, a space. This idea of acceptance and mattering to an academic environment is fundamentally shaped by feeling able to connect with both the content of a course and the people associated with it (Rueda and Lowe Swift 2024; Van Gijn-Grosvenor and Huisman 2020). When students feel able to connect to the content and people linked to a course, that then strengthens their confidence, academic competence and self-efficacy—their ability and belief in themselves to be able to succeed.

There are many inter-linking elements that feed into a student's sense of belonging, within an academic, course-based context. There is clear intersectionality between the factors which influence an individual student's sense of belonging such as connections with staff and peers, having study partners, financial stress and feeling valued and safe. Known to influence a student's sense of belonging to their course are many elements which the individual student can influence and to some extent control, such as belonging to academic societies through to other elements which are beyond the influence and direct control of the student, such as building connections with peers and feeling safe within a teaching environment.

The suite of elements that feed into an individual student's sense of belonging varies between students and the strength that each element plays in the development and maintenance of that sense of belonging varies at different stages of a degree programme and across the academic year. An inclusive and institutional approach to student belonging seems to be the cornerstone of successful practice in this area (Blake et al. 2022). The majority of the 5233 students surveyed by Pearson and Wonkhe felt a stronger connection to their courses than they did to their institution (55% for course level connection compared to 39% for the institution level) (Blake et al. 2022). Indeed, the findings of the Pearson and Wonkhe belonging survey also found that one of the key barriers to belonging that students face is when there is disconnect and a lack of integration between a student's course and curriculum and the wider student experience (Blake et al. 2022). It is no surprise then to hear that institutions which are embedding belonging and student wellbeing within course content and promoting an inclusive campus culture seem to be leading the way for student experience and the sense of belonging felt amongst their student cohorts.

Fig. 2.3 Screenshots taken from the pre-arrival module developed for the first-year Biological Sciences cohort at the University of East Anglia (see Case Study 2.1). (**a**) Depicts the opening page of the pre-arrival module, dominated by a view across the UEA lake to some of the student residences and university buildings. (**b**) One of the first activities that students are invited to complete when they first enrol on this module is to virtually introduce themselves to the other students on their course and to also meet some of their teaching team for that year. (**c**) Week 2 of the pre-arrival module includes a series of resources to prepare students for the laboratory environment, including a series of lab simulations created by LearnSci. These simulations range from how to prepare to be in a lab through to simulations for using various items of lab equipment. (**d**) A screenshot taken from a 360° tour of the teaching laboratories, included with the week 2 content for the pre-arrival module

2.2 Developing Teaching Practices to Positively Influence Student Belonging

Student-centric teaching practices place students at the centre of the learning experience and are widely acknowledged as fostering a stronger student experience and sense of belonging than the more traditional teacher-centric approach (Wong 2024; Shaw et al. 2023; Kong 2021; Chatterjee and Correia 2020). Student-centred teaching spaces help to build a sense of belonging amongst the student body because students are partners in the learning process and their voices are heard (Wong 2024). For students who are from traditionally marginalised groups, they are more likely to hear comments from students of a similar background or from someone that they can recognise themselves in when teaching is taking place in a space that is proactively encouraging student contributions (Wong 2024).

Adopting elements of a student-centric teaching approach can be relatively straightforward. The fundamental principles are to make sure that students are given time and space to engage in discussions about their learning so that they feel able to contribute and that their contributions are valued. Helping students to feel valued from early in their studies is one of the most impactful ways of helping to develop a level of trust between them as a learner and you as their teacher. Here are a few suggestions of ways to help bring a student-centric approach to biosciences teaching:

- Begin the academic year by engaging your students in discussions about what excites them about their chosen subject. Why are they pursuing studies in biology? What area of biology are they excited to learn about? How do they see their future biology career developing? Discussions such as these take no preparation time but often result in interesting spin-off discussions and can result in subjects that can feed into future taught sessions. Perhaps most importantly, by asking students what matters to them shows that you are interested in them as individuals and starts to build that all important trust between learner and teacher.
- Ask students to provide suggestions of potential assessment topics, for example "What are the grand challenges facing biologists today?" Then allow students to choose one of the discussion topics as the subject for a future presentation/seminar discussion/written assignment.
- Meet students where they are, not where you think they should be. At the start of the academic year, ask students what skills (this can be lab skills, soft skills, analytical skills, etc.) they feel they lack confidence in or that they want to develop. And then support them to access opportunities where they can develop those skills either through formal taught sessions, provision of virtual resources such as lab simulation videos or through signposting them to support being offered within the institution, such as drop-in essay writing or data analysis sessions.
- Rather than dictating the subject of seminar discussions, offer a range of subjects and let the students decide which best suits their current interests. Better yet, invite students to contribute suggestions for the subjects that are offered!
- Ask students to co-create course reading lists by making suggestions of books or articles that they have read that they have found to be particularly interesting, engaging or thought-provoking.

In addition to the above suggestions, providing opportunities for students to engage in experiential learning opportunities can be transformational for students in terms of both their sense of belonging to their place of learning and the community within it but also in helping them to see how their current course is going to reach their future goals. Experiential learning is a successful, widely used teaching approach that supports students to engage in active learning through real-world experiences and being able to engage in group work or experiential learning is known to positively influence student belonging (Gilani 2023; Peace 2023). Within the field of biology, which can often feel like a subject full of theory, much of which is intangible to our students, having the opportunity to align biological theory with real-world experiences can be very powerful. Opportunities for building experiential learning into biosciences subjects can include multi-week (or month!) activities such as industrial placements, shorter timescale events such as field trips or even shorter activities such as client-focused projects where students work to a brief from a client or assessments which mimic workplace tasks such as producing executive summaries, a report on a series of diagnostic tests or a literature review on the latest knowledge within a specific area.

2.2.1 Building Connections

2.2.1.1 Facilitating Peer–Peer Connections

Building and maintaining authentic connections with staff and other students can have a significant influence on the success of an individual. Previous work looking at the retention of second year students found that the retention of a student's friends had as much effect on a student remaining on their course as any other background variable that was tested (Eckles and Stradley 2011). Helping students to make meaningful connections with others on their course is one of the top three most impactful activities that we can do to help with a student's sense of belonging and their retention on their course (Blake et al. 2022).

Throughout their studies, biosciences students engage in a range of teaching events that take place across a range of settings from large lecture theatres with hundreds of students through to smaller seminar groups of a dozen students or fewer and some may even be fortunate enough to venture on overseas field courses to tropical forests. The variety of physical learning environments that a student may experience throughout their degree (or in some cases, within a week!) and the corresponding variation in the class size can make it challenging for students to connect with their peers and teaching faculty. Some students may feel a greater sense of belonging within a large group setting where they feel like they are a member of the anonymous majority whereas other students will feel a greater sense of belonging when they are able to actively contribute to discussions and be visible to their peers. There will be no one-size-fits-all teaching approach that is able to cater to the needs of every student within every taught event but there are teaching strategies that can be used to help create a more inclusive, holistic and supportive learning environment where students feel valued, supported and able to thrive. Examples of approaches which can help students to develop peer–peer connections and to widen their support network include:

- Embed low-stakes teamwork activities into teaching from the first week to enable students to interact with a range of students before they have established their course friendships (see Case Study 2.2 for a more detailed example of this). Course friendships can be an invaluable source of support for students, but they can also isolate students from others in their cohort and hinder them from developing a wider support network.
- Early in their studies, set a challenge for student teams within a course. This could be discipline specific such as a quiz-style picture round of different cell images that students need to identify, or it can be unrelated to biology and something like a treasure hunt around campus collecting answers to specific questions.
- Outside of teaching, group advisee meetings can be an effective way for students to meet other students outside of their course friendships, but who are studying the same modules. Facilitated in the right way, these meetings can lead to a sharing of challenges being found within a module, e.g. "I am finding it so hard to find sources to use in the microbiology essay!" and the other students then sharing advice or that they are having the same problem too.
- Field courses are always great for helping students to connect with each other. And they don't need to be expensive or time consuming to have impact. A consistent favourite amongst students is a 90-minute fungal foray where

students explore campus armed with nothing more than a fungi ID book, their smartphone and a link to a class Padlet for sharing photos. We have a few members of staff who join in and move about the groups of students helping them to ID their finds and we encourage students to share their photos on the class Padlet (see Fig. 2.4), which then becomes a resource that we can refer back to in later teaching sessions. This field trip event is a favourite amongst the students. Whilst the students are in different physical locations during their fungal foray, they are connected to each other via the Padlet digital platform and can interact with each other and explore each other's finds—a really simple way of connecting these digital native students with both their learning and their physical environment.

Whereas Case Study 2.1 described a pre-arrival module which incorporates elements to support the making of peer–peer connections before students arrive on campus, Case Study 2.2 describes activities that I have used at the University of East Anglia to support students with building their own peer support networks from their first week on campus.

Case Study 2.2 Using Expertise from the Arts to Build Belonging in the Sciences, Based on Examples Given in Edmunds and Fridd (2023)

Keen to facilitate students making peer connections from their first day at university, we utilise expertise from performing arts facilitators to provide interactive student workshops for those studying on our Biological Sciences courses. The overwhelmingly positive feedback from students during the first cohort (2022/23) led to workshop expansion during 2023/24 and we have now run 8 variants of this workshop with different groups of students with >95% of students agreeing that all new students should be offered this workshop.

Our approach is built on the core principles of mystery, fun and permission to fail. We schedule the workshop for all students during the early part of their induction week and intentionally give it a vague name, i.e. "Your Biology Studies" to entice any students who might avoid attending if they knew they would be working with strangers for two hours!

Workshops begin with the facilitators briefly introducing themselves before moving to the first activity. Activity 1 is always one that requires limited peer–peer discussion, will likely be familiar to the students and which includes a strong element of play, e.g. Grandma's footsteps or a variation of catch. Each activity can be levelled up to a more complex activity by adding rules such as students needing to take a seat before reaching "Grandma" or varying the order/direction/sequence in which the ball must travel around the room. Activity 2 then levels up to incorporate elements of teamwork, e.g. working in pairs to reach "Grandma" without being seen to move or by introducing a pair-based challenge such as building a tower from a deck of cards. For Activity 3 (and 4 if time allows) we move to team-based activities, usually with an element of challenge or competition. A popular team activity has students in groups of ~6 and their challenge is to arrange themselves (and props if any are available) into a random object shouted out by the facilitator, for example, "As a group, arrange yourselves to become a microscope!" Other objects that work well for this activity include a washing machine, racing car, camera, cell and a toaster.

Our workshops are well received by students with a typical scenario being the start of the session being students sitting on their own, not engaging with each other. Within 10 min, the volume, chatter and laughter in the room increase visibly and by the end of the session students are leaving the room chatting together in small groups, having made connections with their peers.

Fig. 2.4 Screenshot taken from the Padlet site used to accompany the Fungal Foray undertaken by first year students studying within the Biological Sciences courses offered at the University of East Anglia. The use of the Padlet for offering a virtual photo album-type repository for students to share their finds and interact with each other's discoveries provides a unique meshing together of a student's physical space, digital technology and their learning within biology

2.2.1.2 Facilitating Staff–Peer Connections

For many students, making connections with a member of staff who they find to be approachable and supportive can be a transformative experience. If the student can make that connection early in their studies and to maintain that connection throughout their course, the constancy of that connection can be grounding and help to strengthen a feeling of community and belonging (Blake et al. 2022).

When communicating with students via digital formats, e.g. by email or via a virtual learning environment, students respond most positively to communications that are sent from a familiar person and that are tailored to them as an individual rather than addressed to a whole cohort (Blake et al. 2022). Whilst this might not be possible for modules which have a large cohort, it can be possible for smaller groups of students such as members of a seminar or problem-based learning (PBL) group or to a pastoral group of advisees. Communications sent to larger groups of students can still present opportunities for building a connection by engaging with students about guidelines of such communications, i.e. time of day for sending them, preferences for bullet-pointed lists of immediate actions for the students, etc.

2.2.2 Embedding Wellbeing Within Taught Content

With a growing number of students experiencing mental health challenges alongside their studies, finding ways to increase awareness of and access to wellbeing support is essential. Despite significant investment in resources, many institutions are struggling to meet the demands of their student body for access to wellbeing support. Within taught elements of a course there are both indirect and direct ways of embedding wellbeing support. Indirect approaches include making use of captions/subtitles for all presentations (see Fig. 2.5 for how to do this for PowerPoint presentations) which forms a key part of inclusive teaching practice, reduces cognitive load for students and once again, recognises the experiences and preferences of our students, as digital natives—two recent surveys have found that when viewing TV shows, significantly more young adults choose to use subtitles compared to any other age group (YouGov UK 2023; Stagetext 2021).

There are also many approaches that can be built into teaching to both foster and strengthen the staff–student connection. One approach that has been seen to help students see that they matter to the staff teaching them is the use of wellbeing check-in slides at the start of workshop sessions (see, e.g., Edmunds and Gulliver 2021). These anonymous check-in slides (Fig. 2.6) can be used to both enable the staff to understand how well the student cohort feel they are coping at a given moment in time as well as provide an opportunity to signpost to support that may help and to spark discussions around wellbeing and managing the various demands on their time.

Using taught sessions to signpost students to the support available or even delivering wellbeing sessions within taught sessions can be an efficient use of staff time whilst also ensuring a wider reach across the student body. For many sessions related to this, expertise and support can also be sought from professional services teams within an institution. Examples of wellbeing-related topics that have relevance within the content of a bioscience curriculum include:

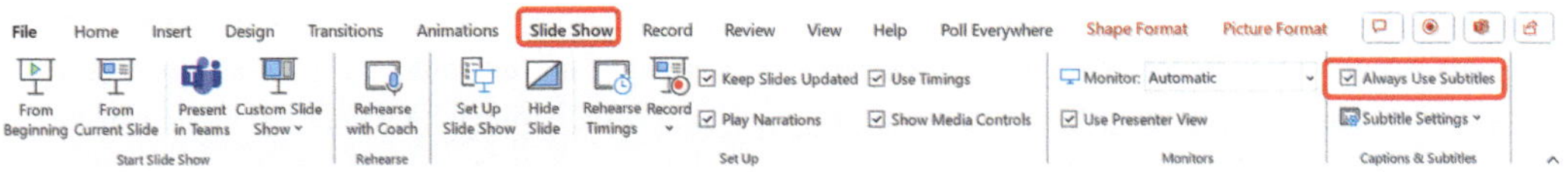

Fig. 2.5 Screenshot showing how to opt to have subtitles automatically generated for PowerPoint presentations, to demonstrate one element of inclusive teaching practice

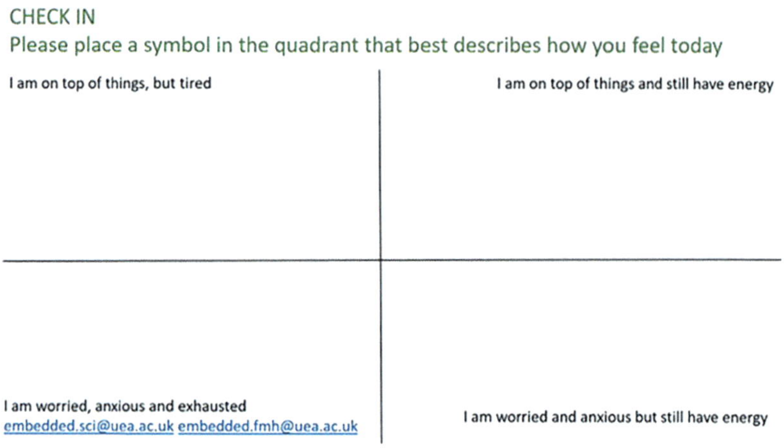

Fig. 2.6 An example of a check-in slide that can be adopted for use within taught sessions, to gauge how well the student cohort are coping with their studies, at that point in time (taken from Edmunds and Gulliver 2021)

- The physiological effects of stress on the body.
- The influence of sleep on the physiological and cognitive processes within the body.
- The cognitive impacts of dehydration on the body.
- The wellbeing benefits of time spent in nature.
- External influences on our circadian rhythm and how our body responds.

As previously mentioned in this chapter, studying biosciences and engaging in scientific experiments is likely to lead to moments of failure. Indeed, across all academic subjects where assessment is involved, students are likely to experience moments that they perceive as failure. Experiencing failure can result in feelings of disappointment, embarrassment, shame and can exacerbate many mental health conditions. Consequently, when supporting the learning of new students, there is often a desire to protect them from failure. But when experienced within a safe, secure and supportive environment, failure can result in personal growth, active learning, strengthening connections between peers to create a stronger sense of belonging and the development of a growth mindset; all key skills that will help students to thrive. To avoid exacerbating any potential mental health problems being experienced by students, activities which can lead to feelings of failure need to be well thought-out, supported and designed to ensure that the positives outweigh any potential negatives. Staff at the University of East Anglia have been doing this by creating space for low-stakes failure with the aim of fostering growth mindset developing and scientific experimentation within laboratory practical sessions on their Science Foundation Year courses, with promising results (see Case Study 2.3).

Case Study 2.3 Encouraging Low-Stakes Failure and Experimentation to Build Learner Agency and Resilience, Contribution from Dr. Helen Leggett

The Foundation Year programme within the Faculty of Science at the University of East Anglia (UEA) teaches students progressing onto courses across a range of science, technology, engineering, mathematics and medicine (STEMM) subjects. The largest modules on the programme are for biological sciences, with cohorts of >200 students primarily studying across biosciences, medicine and pharmacy courses.

In semester 1 of the 2021/22 academic year, we noticed that students in our cohort were not developing peer networks, were lacking learner agency and struggling to make decisions which enabled them to fill in the gaps within a set of simple instructions and this was particularly noticeable when they were in the laboratory. Compared to

(continued)

previous intakes they required more precise, explicit instructions and without these their default response was inaction or "panic". It was clear that the students were not able to relate their scientific theory to a "real-life" scenario and "getting something wrong" led to higher levels of anxiety than we could recall from previous years. Given the disruption to education for most of these students, this was unsurprising, but students need to learn how to work through these feelings if they are to thrive in their degree and beyond. We developed a plan to help our students to build their confidence and resilience and to do so subtly and with support.

We developed a low-stakes experiment with the primary aim of activating and centring the development of soft skills, with the acquisition of discipline knowledge and skills a secondary aim. Specifically, we had a simple aim: to get the students working together to devise their own rigorous experiment and once they had the results, to reflect on how it had gone. We did this through providing students with:

1. An equipment list
2. An experimental aim—"Determine the antibacterial properties of essential oils against *Escherichia coli*"
3. And a simplified methodology of "performing a disc diffusion assay".

Students were given time to plan their experimental approach and in groups of four, they conducted their experiment. After reviewing their results in a reflective workshop, the students asked for a second chance to "do better" and so the following semester, they had an opportunity to redesign and repeat the experiment. Once again, we led workshops to help them reflect on the whole process. In subsequent laboratory practicals, the increase in confidence, active participation and decision-making by the students was obvious.

In response to sector demand for examples of how to effectively embed wellbeing into practice, a recent Quality Assurance Agency (QAA)-funded study sought to collate examples of best practice for embedding wellbeing into HE teaching and developed these into a toolkit (Lister and Allman 2024). As part of this project, a thematic literature review was conducted, and the results used to develop a five-point definition to guide the embedding of wellbeing into practice (Lister and Allman 2024). The five points proposed by Lister and Allman (2024) suggest that embedded wellbeing activities should be:

1. Modelled and situated within practice
2. Tailored to meet the needs of the students
3. Embedded and widespread within institutional culture
4. Be delivered holistically and in partnership and
5. Represented within strategy and resourced appropriately so as to ensure ongoing practice

Embedding wellbeing activities into teaching can go a long way to helping students to feel that they are valued, seen and matter. Those teaching within biosciences are unlikely to also have professional training and qualifications in mental health support but there are ways that we can build elements into our teaching—such as delivering sessions or essay topics on physiological impacts on the body—or draw on the expertise of wellbeing professionals within an institution to help deliver these sessions. As with everything suggested in this chapter, none of this work needs to be done alone and indeed we can often have the greatest impact when we are working in collaboration with our colleagues and the students.

2.3 Discussion

Belonging is not static for a student. Just because a student feels that they belong at one point in their studies does not mean that they will always feel that same sense of belonging. Whilst student belonging is currently experiencing a boom in research effort and institutional attention, the idea of unbelonging lags somewhat behind. As a sector, we are learning rapidly about factors and interventions which can help a student to develop a sense of belonging and as a sector, we are rapidly developing an arsenal of activities and approaches that can help to foster that sense of belonging and there are growing numbers of opportunities to share and disseminate the successes (and failures!) of those approaches. But where we can do better is with understanding how we can help students who have found themselves in a phase of un-belonging, having previously experienced a strong sense of belonging. What is it that leads a student to a shift in their sense of belonging towards un-belonging? How can we support students who are in that feeling of un-belonging and can we help steer them to a place where they feel a sense of belonging again?

As with all students, those studying within the biosciences may well feel a sense of belonging beyond their course—perhaps to a specific sports club or society, maybe even to the institution. Many of the suggestions and case studies within this chapter have application across disciplines, e.g. co-creating content and assessments with students, using language and engaging with students so that they feel that they matter to us and that they are valued within our learning spaces. But bioscience students do have unique experiences within their courses and several of these lend themselves to being excellent opportunities for students to develop a stronger sense of belonging. Time spent trying to develop expertise with new lab-based equipment without ruining an entire experiment or trying to fathom complex formulae and intricate biochemical processes are two excellent examples of when students can bond over the shared experience of trying to avoid failure and the joy of (hopefully) being successful and develop those crucial peer–peer connections! Field trips and time spent away from a formal, indoor learning environment offer incredible, authentic bonding experiences for students both with their peers and with staff. Seminars or workshops where students are trying to master new analytical skills or to interpret that valuable lab-derived data can yield opportunities to facilitate and encourage peer discussions between groups of students who might not otherwise be interacting with each other.

To conclude, there are many individuals across many institutions doing many wonderful, creative and innovative things to help their students to develop a stronger sense of belonging. Connecting these different approaches is key—the sum is greater than the parts. Belonging work has most impact when delivered with a joined up, connected approach. This chapter has been written with HE practitioners in mind. It is hoped that the chapter will provide ideas and spark inspiration to those who want to build belonging into their practice and for those who are already engaged in building student belonging, perhaps this chapter will give them a boost or reassurance of the value of their efforts. Embedding belonging into biosciences teaching has impact, it makes a difference, and it is valued.

References

Blake S, Capper G, Jackson A (2022) Building belonging in higher education: recommendations for developing an integrated institutional approach. Wonkhe and Pearson. https://wonkhe.com/wp-content/wonkhe-uploads/2022/10/Building-Belonging-October-2022.pdf

Borland J (2018) Coaching in the great unknown: why not knowing matters more than you know. Independent Publishing Network

Chatterjee R, Correia A-P (2020) Online students' attitudes toward collaborative learning and sense of community. Am J Dist Educ 34(1):53–68. https://doi.org/10.1080/08923647.2020.17034792

Curran R, Millard L (2016) A partnership approach to developing student capacity to engage and staff capacity to be engaging: opportunities for academic developers. Int J Acad Dev 21(1):67–78. https://doi.org/10.1080/1360144X.2015.1120212

Dawson, A.-M., (2008). Transnational flows and generational disjunctures: Conceptions of "homeland" among Melbourne Greek Cypriots. In Lee, H. (Ed) Ties to homeland: Second generation transnationalism (pp. 72–91). Newcastle: Cambridge Scholars Publishing

de Beer J, Smith U, Jansen C (2009) Situated' in a separated campus—students' sense of belonging and academic performance: a case study of the experiences of students during a higher education merger. Educ Change 13(1):167–194. https://doi.org/10.1080/16823200902944922

Eckles JE, Stradley EG (2011) A social network analysis of student retention using archival data. Soc Psychol Educ 15(2):165–180. https://doi.org/10.1007/s11218-011-9173-z

Edmunds K, Fridd L (2023) Break the ice early to build student belonging. Times Higher Education Campus. Published 02 August 2023. https://www.timeshighereducation.com/campus/break-ice-early-build-student-belonging. Accessed 02 Aug 2023

Edmunds K, Gulliver B (2021) Building an oasis of calm in a sea of chaos: supporting students during Covid-19 disruption. J Found Year Netw 4:13–23. https://jfyn.co.uk/index.php/ukfyn/article/view/65/59

Eggens, L., M. Van der Werf, and R. Bosker. (2008). The Influence of Personal Networks and Social Support on Study Attainment of Students in University Education. Higher Education 55 (5): 553–573. https://doi.org/10.1007/s10734-007-9074-4

Finn K, Holton M (2019) Everyday mobile belonging: theorising higher education student mobilities. Bloomsbury, London

Freeman TM, Anderman LH, Jensen JM (2007) Sense of belonging in college freshmen at the classroom and campus levels. J Exp Educ 75(3):203–220

Gilani D (2023) Student belonging: critiquing simplistic notions of its factors and links to student success. Figshare, Presentation. https://doi.org/10.6084/m9.figshare.24166746.v1

Gilani D (2024) Time to belong—exploring the dynamic nature of students' sense of belonging. Presentation for the student belonging conference at the University of East Anglia. https://youtu.be/rbU20x_YIX0. Accessed 20 May 2024

Gravett K, Ajjawi R (2021) Belonging as situated practice. Stud High Educ 47(7):1386–1396. https://doi.org/10.1080/03075079.2021.1894118

Gravett K, Winstone N (2019) Storying students' becomings into and through higher education. Stud High Educ 46(8). https://doi.org/10.1080/03075079.2019.1695112

Grayson JP (2003) The consequences of early adjustment to university. High Educ 46(4):411–429

Haussmann L, Ye F, Ward Schofield J, Woods R (2009) Sense of belonging and persistence in white and African American first-year students. Res High Educ 50(7):649–669. https://doi.org/10.1007/s11162-009-9137-8

Hoffman M, Richmond J, Morrow J, Salomone K (2002) Investigating sense of belonging in first year college students. J Coll Stud Retent Res Theory Pract 4(3):227–256. https://doi.org/10.2190/DRYC-CXQ9-JQ8V-HT4V

Hubbard K, Gawthorpe P (2024) Community and belonging within the inclusive education framework. https://www.inclusiveeducationframework.info/category/community-and-belonging/

Kahu ER, Ashley N, Picton C (2022) Exploring the complexity of first-year student belonging in higher education: familiarity, interpersonal, and academic belonging. Stud Success 13(2):10–20. https://doi.org/10.5204/ssj.2264

Khambhaita P, Bhopal K (2015) Home or away? The significance of ethnicity, class and attainment in the housing choices of female university students. Race Ethnicity Educ 18(4):535–566. https://doi.org/10.1080/13613324.2021.759927

Kirby LA, Thomas CL (2022) High-impact teaching practices foster a greater sense of belonging in the college classroom. J Furth High Educ 46(3):368–381. https://doi.org/10.1080/0309877X.2021.1950659

Kong Y (2021) The role of experiential learning on students' motivation and classroom engagement. Front Psychol 12:771272. https://doi.org/10.3389/fpsyg.2021.771272

Krause, K., & Coates, H. (2008). Students' engagement in first-year university. Assessment & Evaluation in Higher Education, 33 (5): 493-505. https://doi.org/10.1080/02602930701698892

Levett-Jones T, Lathlean J, Higgins I, McMillan M (2008) Staff-student relationships and their impact on nursing students' belongingness and learning. J Adv Nurs 65(2):316–324. https://doi.org/10.1111/j.1365-2648.2008.04865.x

Lewis J, Bolton P (2023) Student mental health in England: statistics, policy, and guidance. House of Commons Library Research Briefing No. 8593. https://researchbriefings.files.parliament.uk/documents/CBP-8593/CBP-8593.pdf

Lister K, Allman Z (2024) Embedding mental wellbeing in the curriculum: a collaborative definition and suite of examples in practice. Front Educ 8:1157614. https://doi.org/10.3389/feduc.2023.1157614

Meng Q, Zhang Q (2023) The influence of academic self-efficacy on university students' academic performance: the mediating effect of academic engagement. Sustainability 15:5767. https://doi.org/10.3390/su15075767

Million Plus (2022) Learning with the lights off-students and the cost of living crisis. https://doi.org/https://www.millionplus.ac.uk/documents/Learning_with_the_lights_off_-_students_and_the_cost_of_living_crisis.pdf

Morosanu L, Handley K, O'Donovan B (2010) Seeking support: researching first-year students' experiences of coping with academic life. High Educ Res Dev 29(6):665–678

National Union of Students (2023) New data reveals the hidden student job crisis, Published 25 September 2023. https://www.nus.org.uk/new_data_reveals_the_hidden_student_job_crisis

Palmer E (2024) Balancing belonging with a diverse student community. Poster presented at the Student Belonging Annual Conference at the University of East Anglia on 20/05/2024

Peace M (2023) Deeper pedagogies are key to cracking open the black box of experiential learning. Wonkhe, Published 14 May 2023. https://wonkhe.com/blogs/deeper-pedagogies-are-key-to-cracking-open-the-black-box-of-experiential-learning/

Pokorny H, Holley D, Kane S (2017) Commuting, transitions and belonging: the experiences of students living at home in their first year at university. High Educ 74:543–5458. https://doi.org/10.1007/s10734-016-0063-3

Ramler, T. R., Tennison, L. R., Lynch, J., & Murphy, P. (2016). Mindfulness and the college transition: The efficacy of an adapted mindfulness-based stress reduction intervention in fostering adjustment among first-year university students. Mindfulness (7): 179–188. https://doi.org/10.1007/s12671-015-0398-3

Rueda E, Lowe Swift C (2024) Introduction within academic belonging in higher education—fostering student connection, competence, and confidence. Routledge, New York, pp 1–12. https://doi.org/10.4324/9781003443735-1

Rueda E, Ballard T, Gonzalez K, Gutierrez A, Herrera J, Magdaleno L, Majarali V, Martinez J, Molina, Walker A (2017) In search of belonging: first generation, low-income students navigating financial, bureaucratic, and academic experiences at Vassar (Faculty Research Report No. 112). Vassar College. https://digitalwindow.vassar.edu/faculty_researc_reports/112/

Shaw C, Bhardwaj R, Condon K, NeJame L, Martin S., Rich J, Janson N, Bryant G, Fox K (2023) Listening to learners. https://tytonpartners.com/app/uploads/2023/09/Tyton-Partners-Listening-to-Learners-2023-1.pdf

StageText (2021) Captions and subtitles in the UK. Research conducted by Sapio on behalf of StageText and findings. https://sapioresearch.com/success-stories/stagetext/

Standen G (2024). How our study together programme promotes belonging and improves well-being. Times Higher Education Campus, 02 February 2024. https://www.timeshighereducation.com/campus/how-our-study-together-programme-promotes-belonging-and-improves-wellbeing

Sutton Trust (2023) Cost of living and university students, Published 26 January 2023. Cost of Living and University Students - Sutton Trust

Tinto V (2023) Reflections: rethinking engagement and student persistence. Stud Success 14(2):1–7. https://doi.org/10.5204/ssj.3016

UCAS (2023) Statistical releases—daily clearing analysis 2023. https://doi.org/https://www.ucas.com/data-and-analysis/undergraduate-statistics-and-reports/statistical-releases-daily-clearing-analysis-2023

UPP Foundation (2022) A student futures manifesto. UPP Foundation Student Futures Commission. https://upp-foundation.org/student-futures-commission/news/a-student-futures-manifesto/

van Gijn-Grosvenor, E. L., & Huisman, P. (2020). A sense of belonging among Australian university students. Higher Education Research & Development, 39(2), 376-389. https://doi.org/10.1080/07294360.2019.1666256

Versteeg M, Kappe RF, Knuiman C (2022) Predicting student engagement: the role of academic belonging, social interaction, and resilience during COVID-19 emergency remote teaching. Front Public Health 10:1–14. https://www.ncbi.nlm.nih.gov/pmc/articles/PMC8971557/

Walker, N. (2024). Introduction. In W. Garnham & N. Walker (Eds.), Supporting the Student Journey into Higher Education: How Pre-Arrival Platforms Can Enhance Widening Participation in Supporting the Student Journey into Higher Education (1, pp. 1-6). Routledge. https://doi.org/10.4324/9781003427575-1

Winstone N, Balloo K, Gravett K, Jacobs D, Keen H (2020) Who stands to benefit? Wellbeing, belonging and challenges to equity in engagement in extra-curricular activities at university. Act Learn High Educ 23(2). https://doi.org/10.1177/1469787420908209

Wong M (2024) Inclusion may not Lead to belonging: the case for student-centred classrooms. Coll Teach 1–3. https://doi.org/10.1080/87567555.2024.2307884

YouGov UK (2023) When watching TV shows or movies in your native language, do you generally prefer to have the subtitles on or off? YouGov UK poll conducted 24/02/2023. https://yougov.co.uk/topics/entertainment/survey-results/daily/2023/02/24/9a34f/3

3 Unlocking Undergraduate Students' Potential: Developing Play-Based Learning Strategies for Lifelong Learning in STEMM Disciplines

Noelia Dominguez Falcon

Abstract

Play-based learning is an engaging educational approach that uses play and interactive activities as a method for learning. It encourages exploration, creativity, and hands-on experiences to develop effective learning, as well as fostering teamwork. Although traditionally associated with early childhood education, play-based learning has been recently posed as a transformative pedagogical approach for fostering creativity, collaboration, and critical thinking among higher stages of education, particularly university students. This approach is especially beneficial for Science, Technology, Engineering, Mathematics and Medicine (STEMM) careers, where experimentation and exploration in a low-pressure environment boost creativity and problem-solving skills, while strengthening teamwork and communication abilities. Herein, we illustrate the incorporation of play elements in two different sessions within our Biomedicine course. Ultimately, we are aiming to inspire educators to integrate play in their sessions and advocate for a paradigm shift that embraces play as essential to effective higher education pedagogy in STEMM fields.

N. Dominguez Falcon (✉)
School of Biological Sciences, University of East Anglia, Norwich, UK
e-mail: N.Dominguez-Falcon@uea.ac.uk

Keywords

Play-based learning · Gamification · Active learning · Critical thinking · STEMM · Student engagement

List of Abbreviations

FHEQ	Framework for Higher Education Qualifications
HE	Higher Education
IBMS	Institute of Biomedical Science
RA	Reasonable Adjustments
STEMM	Science, Technology, Engineering, Mathematics and Medicine
TBL	Team-Based Learning

3.1 Play-Based Learning: Foundations, Educator Roles, and Lifelong Impact

3.1.1 Revealing the Dynamics of Learning and the Role of Educators

Learning is, in simple terms, a dynamic and multifaceted journey characterised by the acquisition, integration, and application of knowledge, skills, and understanding. It is a highly

R. Door (ed.), *Accessibility, Inclusivity and Diversity in Education and Beyond*, Biomedical Visualization 8, https://doi.org/10.1007/978-3-031-81795-3_3

individualistic experience, involving a complex interplay of cognitive, affective, and behavioural components, influenced by individual experiences, social interactions, and environmental factors (Newton and Miah 2017).

Additionally, learning is at the core of all professional careers, regardless of its specific nature. However, one must note that not all learning approaches are adequate for fully leveraging an individual's intellectual capacities. Effective learning, as such, is described as the most efficient and durable learning strategy. This is because effective learning involves critical thinking and problem-solving (Rea et al. 2022).

Interestingly, effective learning poses a challenge to an opportunistic strategic learning process, which can be common for passive learners. Generally, unmotivated students tend to learn when convenient, and once they use that knowledge for an assessment item, is discarded as it does not serve a purpose anymore. Therefore, learning should shift to an effective spectrum, so the student's learning serves as a tool to acquire and apply knowledge, instead of being assessment-centred, ultimately fostering personal growth and intellectual development (Biggs and Tang 2011).

Nonetheless, effective learning is not a solo effort, and it will most likely be achieved in conjunction with the people who are standing on the other side of the lecture room: the educators.

Educators are not only pivotal to ensure students are actively learning, but their role is key in making the most out of the potential of their students. It is known that students who undergo effective learning are more likely to thrive in their academic curriculum, especially with a supportive educator. Studies describe that an engaging educator aids students in feeling like they have their place within their university and/or department, they have a purpose, and their wellbeing is looked after, on top of receiving a good education for their future endeavours (Ahn and Davis 2023). Interestingly, studies carried out with psychology undergraduate students revealed how a supportive lecturer gave rise to (1) well-behaved students, (2) efficient communication between students and lecturer, (3) a stronger community of learning, and (4) a climate of higher academic achievement, overall improving classroom belonging (Kirby and Thomas 2022).

However, this is not an easy task. Inevitably, the world keeps changing, and so the future generations and their struggles. Remarkably, educators have been challenged in recent years by a reduced attention span among students. Rapid technological advancements, coupled with the prevalence of an online world and multitasking behaviours, have unfortunately contributed to shorter and more fragmented attention spans. Constant exposure to stimuli from smartphones, social media, and online platforms has conditioned students to seek instant gratification and quick information consumption, which rivals with a "traditional" lecture. Students can often fall into a pattern of passively listening to a 50-min speech that can fail in involving or engaging them (May et al. 2023). In fact, studies support that if students do not feel present in the lecture room and are not properly stimulated, they can easily disengage from the content, and their attention span can last as little as 10–15 min in a traditional lecture, according to 2024 statistics (Gitnux 2024).

It is because of these reasons that in recent years, the importance of incorporating engaging and interactive learning experiences, such as gamification, multimedia presentations, and experiential activities has been rapidly evolving, with the ultimate goal of enhancing student focus and participation. The backbone of such activities should be to create a more student-centred environment; hence, students would feel like they belong to the session and/or module via active participation in one or more activities designed by the educator (Miguel Sicart 2014; Whitton and Moseley 2014).

3.1.2 The Pedagogical Roots of Play-Based Learning

There are many strategies educators can use to promote active learning and overall improve student engagement and satisfaction. These could entail problem-based learning applied to real-case scenarios, using innovative technology that involves quizzes or other interactive apps, in-class activities that offer students a break within lectures, peer review, continuous assessments, etc.

However, there is one strategy that has emerged in recent decades as a transformative pedagogical approach that goes beyond traditional methods of education, and it is commonly known as play-based learning.

Play-based learning is, in essence, an educational approach that advocates for the integration of playful activities into an educational setting. Such activities can involve exploration, games, and/or tactile activities (Siviy 2016). Those elements not only are designed in a way that students have an active participation in their learning, but educators can also design them to emphasise relevant content for prospective assessments. Ultimately, play-based learning is a teaching strategy that harnesses the students' ability to explore, experiment, and discover, offering students the opportunity to develop critical thinking skills, problem-solving capacities, and resilience (James and Nerantzi 2019).

From the outside, learning through play might be seen like a simple concept. However, play-based learning has profound roots in many pedagogical studies. The pedagogical roots and theoretical foundations of play-based learning are initially drawn from studies on child development, written by influential scholars such as Lev Vygotsky (Veraksa et al. 2022), Jean Piaget (Nicolopoulou 2010), and Maria Montessori (Marshall 2017).

Lev Vygotsky was a phycologist and developmental theorist, and his work mostly emphasised the relevance of social interaction for child development. Vygotsky believed that children's cognitive growth is strongly influenced by interactions with individuals that they perceive as more knowledgeable, such as parents or teachers. In fact, Vygotsky stressed that the way such interaction happens is important, and play has a key role in an individual's development. According to his work, play provides a unique environment in which children can take roles while following instructions, making them think and work through ideas and/or concepts while actively learning. The overall social interaction among children significantly helps them to develop not only an internal language, but also the ability to think for themselves and communicate that to others (Veraksa et al. 2022).

The psychologist Jean Piaget likewise described play as pivotal for the development of children's intelligence. Piaget constructed a theory that defended play as an ideal way of encouraging cognitive and language development, especially during child maturation. In addition, his theory also highlighted the relevance of hands-on experimentation as a mechanism for assimilating and acquiring new knowledge (Nicolopoulou 2010).

Dr. Maria Montessori, a physician and educator, was acclaimed for her educational methods and for theories in child development. Throughout her work with young and disadvantaged children, she forged a style of teaching based on play, defending that playing with real-like objects makes children participate actively in their day-to-day actions and/or tasks, as they constantly interact with their surroundings. Interestingly, Montessori observed that if children are exposed to activities that capture their spontaneous interest, they can focus on one or several tasks for extended periods of time (Marshall 2017).

It is clear by the work of Vygotsky, Piaget, and Montessori (and many others not mentioned in this chapter) that play promotes deeper understanding and meaningful learning experiences via open-ended activities, imaginative exploration, and intrinsic motivation. Therefore, play underpins the capacity of children to actively engage, experiment, and discover.

3.1.3 Transition to Adulthood: Reintegrating Play-Based Learning in Young Adults

It is worth mentioning that the described theories of Vygotsky, Piaget, and Montessori were founded exclusively on child development, and although they do describe play as a natural and essential aspect of their development, one might wonder about the significance of play during the transition from child to adult learner.

Indeed, there is an inherent emotional turmoil that arises during the transition from childhood to adulthood. Transitioning from childhood to adulthood is a complex and individualised process that typically requires significant time investment in career development, building relationships, and possibly starting families (Ho 2022). This period often comes with added stress due to societal pressures such as job-seeking, financial responsibilities, and career obligations. Therefore, and given the mentioned pressures and the daily stress of a hectic lifestyle, it is important to encourage (and teach) adults to develop their emotional intelligence within their educational and professional environments. This essential life skill enables adults to understand, express, and regulate their emotions, helping them to build resilience and manage life's challenges (Van Vleet and Feeney 2015).

Strikingly, the further we go into education; the less play there seems to be. Why does an engaging activity that initially contributed to our cognitive development disappear from higher educational and professional environments, precisely when we are still evolving as individuals?

Unfortunately, there is no straightforward answer to this question, and literature that refers to the relevance of play in adulthood is scarce. Nonetheless, a common theme in psychology studies is that adults can perceive play as trivial and frivolous (Van Vleet and Feeney 2015). This might be because both young people and adults are expected to be productive and professional, with play often viewed as a hindrance to these expectations, leading to its neglect. This neglect can sadly result in a disregard for emotional intelligence, as environments that promote relaxation and connection with oneself and others become scarce.

The absence of environments conducive to mental and physical relaxation can significantly impact young adults, who represent the future generations of adults. According to the UK census in 2021, 29.1% of all people in England and Wales were under 25 years old (GOV.UK 2021). Although it is challenging to pinpoint individuals struggling with mental health, it is estimated that at least 25% of young adults are facing depression, marking a seven-fold increase from pre-COVID-19 levels (Owens et al. 2022). Particularly, university students often experience diminished wellbeing and sleep quality, placing them at a vulnerable stage for mental illness onset. Studies conducted in the UK have indeed reported increased rates of anxiety, mood swings, and self-harm among university students (Owens et al. 2022).

Given these challenges, there is a pressing need for studies that demonstrate activities promoting emotional intelligence, awareness, and positive emotions. Ho conducted a study involving full-time undergraduates aged 19–25, exploring the impact of play-based activities such as sports, artistic hobbies, and leisure pursuits on emotional traits and resilience (Ho 2022). Through questionnaires, the study revealed a positive correlation between resilience and engagement in play, attributed to the relaxation and enjoyment derived from such activities. Participants reported feeling better equipped to tackle life's challenges and more prepared for their professional life, suggesting that the benefits of play are deeply embedded in those who actively participate. Ho's (2022) research thus contributes to our understanding of how play fosters knowledge and skill acquisition, leading to enjoyment and relaxation.

Considering the aforementioned factors, it is natural to contemplate whether Higher Education (HE) environments, such as universities, represent the optimal junction in the transition from youth to adulthood. Given the numerous benefits associated with play-based learning, it is only natural to wonder: Can universities and other institutions incorporate playful activities to educate young adults within environments that foster enjoyment, positivity, connection, and belonging, thereby facilitating effective learning? Could play-based learning help for the integration of young adults into the professional world? Could play-based learning aid educators in guiding young individuals towards becoming better-prepared adults, aligning their cognitive development with the skills acquired during childhood?

3.2 The Transformative Potential of Play-Based Learning Within STEMM Programmes

In today's world, there is a pressing need for young adults to become active citizens, particularly after completing HE (Rodríguez-Ferrer et al. 2023). Consequently, students are even more aware of the quality of education they receive, understanding its importance for their future (May et al. 2023). HE, therefore, plays a crucial role in integrating students into society, enhancing their employability, and supporting their emotional wellbeing.

To achieve these outcomes, the curriculum must meet specific learning objectives and benchmarks to adequately prepare students for real-world challenges. As previously mentioned, traditional lecture-based, teacher-centred pedagogy is no longer the most effective method for engaging students and fostering the skills needed to become active members of society. Thus, there is a growing need for more interactive teaching methods.

The shift towards interactive and online-centred pedagogies has particularly been accelerated by the COVID-19 pandemic. With the shift to online learning and social distancing, universities quickly adapted by offering online lectures and multimedia resources to engage students and assess their progress (Khaldi et al. 2023). However, the abundance of quizzes, tests, and similar tools often diminished student interest (Dost et al. 2020). This decline may have been due to the general anxiety and uncertainty during the pandemic and the challenges educators faced in rapidly transitioning their materials to an online format.

This is where play-based learning plays a crucial role for students and educators. With the return of face-to-face teaching, play-based learning has significant potential to reinvigorate student attention. It offers the added benefits of fostering connections between students and educators, enhancing engagement in the classroom (Alhammad and Moreno 2018).

Although all degree programmes are worth discussing, this section of the chapter is going to focus on the potential of play-based learning for Science, Technology, Engineering, Mathematics, and Medicine (STEMM) degrees. Science-related degrees have, indeed, unique demands. This is because STEMM programmes should aim to train students not only as active researchers who practice truthful science, but also to develop critical thinking, efficient communication, problem-solving skills, and to contribute with valuable knowledge to the scientific community (Imad et al. 2023). In fact, these are some of the key premises of the Institute of Biomedical Science (IBMS) accreditation for degrees (QAA Benchmark statements) (May et al. 2023), which emphasises the need for STEMM programmes to deliver such learning outcomes.

The efficacy of play-based learning within STEMM programmes has been tested in several institutions, although quantitative evidence within the literature is scarce. Nonetheless, there are some key studies which are worth mentioning.

Educators at the University of Bristol recently published a study in which they evaluate the efficacy of Resimion, an online platform (Resimion.com) that allowed students from a Biomedical Science programme to work on problem- or scenario-based activities (May et al. 2023). For example, students were presented with patient case studies and had to make decisions about the hypothetical future of the patient by selecting appropriate laboratory experiments as biomedical scientists. Resimion was also used as an assessment tool and was configured to accommodate students with Reasonable Adjustments (RA). The study, which involved 769 individuals studying BSc (Hons) Biomedical Science during the period of September 2020 to June 2023 at the University of the West of England, demonstrated that Resimion was largely successful in promoting student engagement, as evidenced by positive student feedback, an excerpt of which is provided in Table 3.1.

Expanding on this play element, Garden (2022) relied on 3D resources, such as Lego® building blocks. This study described how educators used Lego® to assist with exam revision, enhancing student engagement and reinforcing

Table 3.1 Qualitative feedback from students of their experiences of using Resimion. Students that used Resimion were asked for their feedback when using the tool during their assessments. An excerpt of such qualitative feedback is shown. Adapted from May et al. (2023)

Theme/question	Feedback
Application of knowledge	• I really enjoyed the coursework being split into three Resimions, really helped understand the lecture more. I believe it should be kept as it gives a fair chance to students to do well. • The coursework is great! It does not take up a huge amount of time but really solidifies the content taught in lectures and compliments the practical.
Interactivity/practical application	• I really enjoyed the case study aspect rather than a written assignment because you engage practically with knowledge you have been given. • I really enjoyed the coursework. Very interactive and allowed me to bring out all I've learnt. Great for memory training too. • Very helpful, because you implement knowledge based on possible real-life scenarios.

their learning after a year-long Cell Biology module (Garden 2022). Students received cards with specific topics from the module, which were distributed among groups. With the cards in hand, students can choose one topic that they feel the most familiar with, and their task is to represent it using the Lego® blocks in 10 min. Examples of topics used in this study were "Metabolism", "Cell division", and "Cell signalling". The study reported the overall success of the activity, as evidenced by a post-activity questionnaire that collected quantifiable data. Students were asked to answer each question by providing a score from 1 to 7, where 1 is a "Strongly disagree" feeling, and 7 is a "Strongly agree" feeling. In total, 21 students submitted the questionnaire.

As seen in Fig. 3.1, 20 students scored the question "The session encouraged me to be creative in thinking about biology" with either a 6 or a 7, which means that they strongly agree with the sentiment that this activity enhanced their creativity when thinking about scientific concepts learnt in the module.

Similarly, and as seen in Fig. 3.2, 14 students scored the question "This session improved my understanding about cell biology" with either a 5 or a 6, which means that they generally agree with the fact that this activity improved their understanding in specific biology concepts.

Overall, the results from Garden (2022) clearly indicated that the use of Lego® as an interactive activity enhanced student's active learning and supported cognitive and affective engagement, which in turn increased their understanding of the topic.

Other studies also relied on tactile activities to increase students' understanding of complex topics. Terrell and colleagues published a study in 2021 advocating for the use of card-sorting games to simplify intricate subjects found in traditionally challenging modules or programmes like Biochemistry (Terrell et al. 2021). For instance, one task called "Proteins—a collection game" involved students working in small groups with a set of cards containing protein-related components, such as proteins, amino acids, and peptides. During each round, students had to collect cards from one of these three categories by continuously flipping their cards into the centre of the table. The group that first recognised three cards belonging to the same category (for example, three cards that displayed an amino acid) would collect the cards and win that round. The group with the most correct sets at the end of the game won overall. The authors noted that the game was particularly challenging because the cards could feature either cartoon representations of biochemistry concepts or symbolic representations, such as peptide sequences. An example of these cards is shown in Fig. 3.3.

Despite the challenges, the authors reported that students were able to reflect and self-assess in a low-stakes environment while connecting with their peers. This is especially beneficial for complex disciplines within STEMM, where students often feel reluctant to engage with difficult

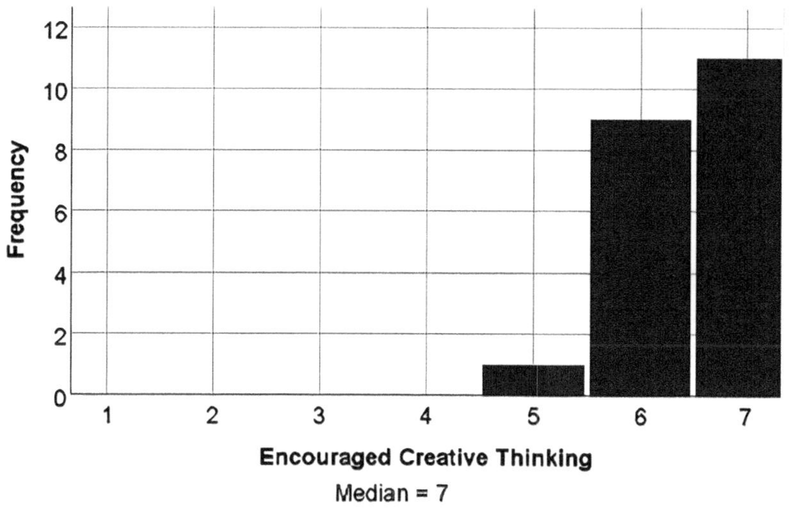

Fig. 3.1 Results from the Encouraged Creative Thinking evaluation questionnaire given to students that participated in a Lego® Serious Play activity. Students were given a questionnaire to reflect their opinions after they engaged with the Lego® Serious Play activity that was part of a revision session to consolidate the learning from a Cell Biology year-long module. This figure represents the answers to the question "The session encouraged me to be creative in thinking about biology". The numbers on the X axis represent the score the students used to answer the question, where 1 is a "Strongly disagree" feeling, and 7 is a "Strongly agree" feeling. The numbers on the Y axis represent the frequency of students who answered the questions with a particular score. Figure taken from Garden (2022)

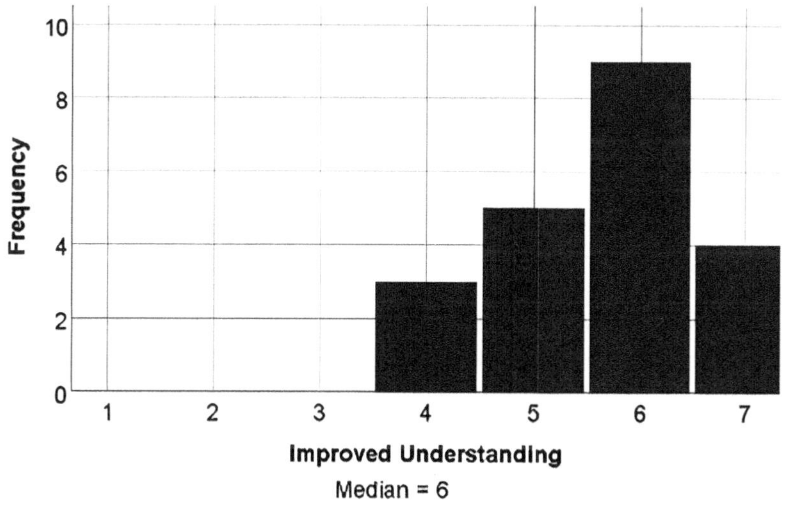

Fig. 3.2 Results from the Improved Understanding evaluation questionnaire given to students that participated in a Lego® Serious Play activity. Students were given a questionnaire to reflect their opinions after they engaged with the Lego® Serious Play activity that was part of a revision session to consolidate the learning from a Cell Biology year-long module. This figure represents the answers to the question "This session improved my understanding about cell biology". The numbers on the X axis represent the score the students used to answer the question, where 1 is a "Strongly disagree" feeling, and 7 is a "Strongly agree" feeling. The numbers on the Y axis represent the frequency of students who answered the questions with a particular score. Figure taken from Garden (2022)

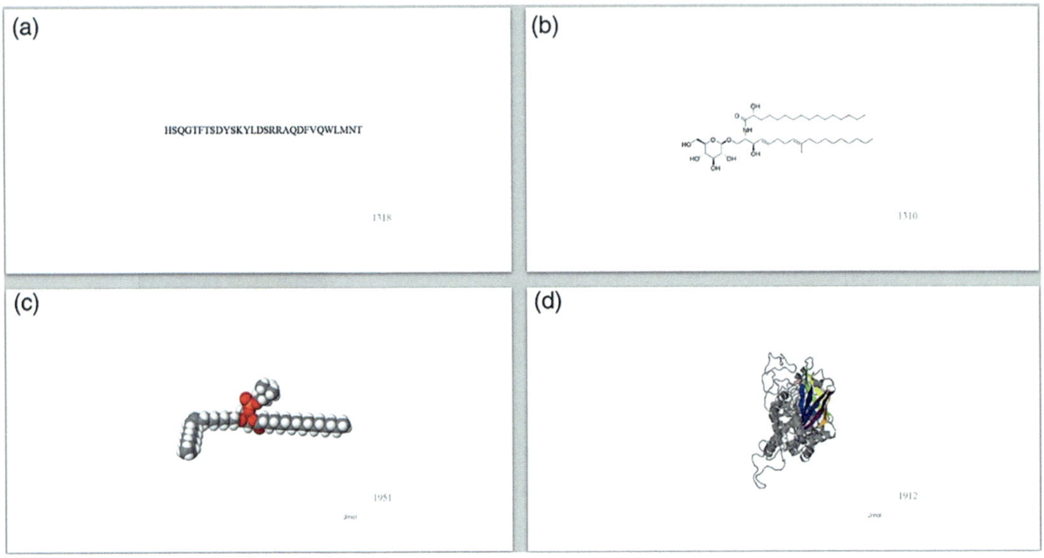

Fig. 3.3 Examples of symbolic and cartoon-based cards from the biochemistry card deck. Students were given a deck of cards that included biochemistry concepts, represented either as a (**a**, **b**) symbolic sequence, chemical structure, or similar; or a (**c**, **d**) cartoon representation of a molecule or protein structure. Figure taken from Terrell et al. (2021)

concepts, making revision and exam preparation a mammoth task.

Overall, these studies clearly demonstrate that play-based learning has significant potential to promote motivation-based learning. It simultaneously fosters positive experiences for students, enhances their decision-making, autonomy, reflection, and understanding of complex concepts (Koivisto et al. 2016; Njie-Carr et al. 2017). Furthermore, the diverse array of tools designed and applied in these studies highlights the successful integration of game-based activities within a STEMM curriculum across various modules and subjects, showcasing the versatility and applicability of play-based learning.

This was precisely what inspired us to develop play-based learning activities in our institution. As educators, we wanted to develop activities that hopefully improve the learning process of our students, facilitating content comprehension, and promoting educator–student interaction. The following section of the chapter will detail the methodology for two activities that demonstrated a good level of interaction, so other educators can get subsequently inspired and apply them to their own subjects and modules.

3.3 Aims and Objectives of the Present Study

The play-based learning strategies presented in this section of the chapter have been applied within our BSc Biomedicine course at the University of East Anglia (UEA). Biomedicine is an exciting degree that has a growing interest among young scientists. It encompasses diverse subjects such as Biochemistry, Cell and Molecular Biology, Genetics, and Human Physiology. Given the high level of expertise that is associated with these topics, students often struggle to grasp intricate and specialised content fully. Additionally, students may fail to integrate their knowledge effectively throughout their degree and into their future careers.

Therefore, bearing in mind the pedagogical foundations discussed earlier in this chapter, and the pressing need for developing more game-based activities in HE environments, we want to share two play-based learning strategies that have been implemented in our Biomedicine course. The first strategy was implemented during a Framework for Higher Education Qualifications (FHEQ) Level 5 exam revision session for the Human Physiology module. This approach aimed to help students review and consolidate their knowledge from the entire module interactively. The second strategy was applied during a series of four workshops as part of a FHEQ Level 6 Cell Biology and Mechanisms of Disease module. This involved the creation of a board game to facilitate active learning and collaboration among students. The activities were designed with several objectives:

1. *Enhance engagement:* These strategies offer a break from traditional lectures and exam revision sessions, aiming to increase student motivation and participation.
2. *Improve content retention:* By providing an enjoyable learning experience, these strategies help students retain and apply information more effectively.
3. *Foster critical thinking:* The strategies encourage students to apply their knowledge in practical scenarios, enhancing critical thinking and problem-solving skills.
4. *Promote collaboration and communication:* These activities promote teamwork and communication, essential skills for success in HE and beyond.
5. *Reduce student anxiety:* As earlier mentioned, incorporating play into learning can reduce stress and anxiety, particularly during exam revision, creating a more positive learning environment.
6. *Provide a holistic understanding:* These strategies support a deeper understanding of complex concepts by encouraging exploration and reflection in a relaxed setting, making the topics more accessible and interesting.

Overall, the developed play-based learning activities aim to foster a culture of experimentation and innovation, empowering students to navigate ambiguity and embrace complexity. While these practical examples are designed to meet the learning outcomes of specific modules, educators can adapt these strategies to suit their own requirements.

3.4 Strategy 1. BINGO! How to Unleash Undergraduate Exam Performance: A Systematic Play-Based Learning Strategy for Effective Exam Revision Sessions

3.4.1 Introduction

Traditionally, at the end of a semester-long module, educators might schedule additional sessions for students to review lecture content in preparation for their exams. In modules with more than three lecturers, there is often one session scheduled per educator or pair of educators, allowing students to revisit each theme or content block. During these sessions, the lecturer typically provides highlights, bullet points, and key aspects of their lectures to make the session manageable and help students grasp the essential concepts they need to review.

For educators, the goal of revision sessions is to present key aspects of the subject or module while ensuring students effectively learn and retain the concepts for their exams. For students, in addition to naturally wanting to perform well in the exam, they want to feel that they have had a good opportunity to review the content.

However, these sessions usually last 1 or 2 hours, which may be too brief to cover an entire semester's content. The time factor is crucial when planning a revision session, as condensing a considerable number of lectures into a 50-min session might not be the most effective revision strategy. With access to recorded lectures, students can arguably do this themselves.

Several reports in the literature discuss the effectiveness of technology, such as quizzes (Tong and Chow 2013), mobile interactions

(Hashim et al. 2011) and even podcasts (Evans 2008) for exam preparation. These studies indicate that students feel they revise better when involved in practice questions and critical thinking. Additionally, some studies show that students who attended revision sessions incorporating interactive methods performed better in exams compared to those who did not attend (Hashim et al. 2011).

Therefore, while exam revision sessions might be considered an "extra" at the end of the module, and their importance might be underestimated, it is evident that the quality and quantity of the revision session, the resources used, and the support from the teacher are crucial considerations when planning these sessions.

3.4.2 Description of the Activity

As previously explained, this activity is part of our FHEQ Level 5 Human Physiology module, offered during the autumn semester with typically 100–115 students enrolled. We closely follow Vander's Human Physiology textbook outline (Widmaier et al. 2023) for lecture content, resulting in 2–5 lectures per organ/tissue system. In total, we deliver approximately 20 lectures from September to December.

Initially, we planned three exam revision sessions, each with two lecturers. However, due to the vast material covered in this module, we found that traditional revision sessions were impractical. From past experience, lecturers often rush through these sessions, mostly because they are trying to highlight too much content, and also because we need to allocate equal time to each lecturer if more than one educator is scheduled in one session. We needed to come up with an activity that gathers the entire content of the module without overwhelming students with extensive sessions.

We therefore designed an interactive activity that is based in a popular game: Bingo. Bingo is a game where players mark numbers on a card as they are randomly called out by a caller, aiming to complete a specific pattern such as a row, column, or diagonal. The first player to achieve this pattern shouts "Line", and the first player to complete the full card shouts "Bingo" to win the game.

For our session, we created a 4x4 Bingo card that contains one word per square or box. These words will be the answer to questions that the lecturer (caller) will shout. The questions will be extracted from the lectures of the module. Students must find the answers on their cards without referring to notes, promoting immediate concept recall and active participation. This approach also fosters Team-Based Learning (TBL), encouraging students to collaborate and engage with the material, enhancing their learning, performance, and communication skills (Akl et al. 2012).

The following section will explain the methodology for both preparing and conducting the session. This strategy is particularly beneficial for modules with extensive content to review, and we encourage educators to adapt it to their own material.

3.4.3 Methodology

3.4.3.1 Preparation of Questions and Answers

The initial step in preparing for this activity is to draft questions and their corresponding answers. Given the adaptable nature of this activity across various modules and content, each educator can tailor their questions to suit their specific formats. However, we suggest adhering to several guidelines:

(a) *Generate two questions per lecture.* It is advisable for the organiser to request the teaching team to generate two or three questions per lecture, particularly if the module comprises more than 20 lectures.
(b) *Formulate questions with concise answers.* Regardless of the quantity, questions should produce responses consisting of one or two words maximum. This facilitates fitting the answers within the Bingo grid and simplifies the process for students to locate the correct answer.

(c) *Craft questions aligning with key module concepts.* Questions should be strategically devised to prompt active recall of fundamental concepts crucial for understanding the module. This aids students in identifying key concepts for revision during exam preparation.

Once all questions are prepared, they can be compiled with their answers into a Word document or similar for printing. Once printed, questions can be cut into individual pieces of paper, which can then be placed into a bag or box. This method ensures that during the session, questions are randomly drawn from the container, ensuring equitable distribution. We have included an example of questions we prepared for the session and their corresponding answers in Table 3.2.

3.4.3.2 Preparation of the Bingo Cards

Once the questions have been prepared, the activity organiser must proceed to incorporate the answers into Bingo cards. We strongly recommend using a free online Bingo card generator for this purpose. By typing "Free Bingo card generator" into Google, educators will find numerous options to choose from according to their preference.

In essence, all one or two-word answers are compiled into a list and entered into the generator, which then arranges them within the boxes of the grids. When designing the cards, we suggest considering two main factors:

(a) *Decide the grid size.* Online generators typically offer the option to select the grid's size. We recommend opting for a 4x4 grid, as we have found it ensures randomness of answers without overwhelming students in locating them.

(b) *Ensure randomness of the generated Bingo cards.* During the session, each student or team should receive a unique Bingo grid. It is essential that each card is different as it adds an exciting and competitive element among students to win the game or a prize. For our session, we generated 33 different cards.

Once all the grids are prepared, they can be printed and stored until the session commences. Below (in Fig. 3.4), we have included an example of two different grids used in our session to demonstrate the randomness of the cards, their layout, and the types of answers included.

3.4.3.3 Running the Session

Once all the materials are prepared, it is time to commence the session. We advise to encourage students to arrive early for the session, especially if it is 1 h long. We decided to run a 1-h long session to keep a fast-paced activity and due to timetable constraints, but there is scope for developing a 2-h long session.

If organisers wish to gauge attendance in advance to ensure adequate number of cards and potential prizes are available, we suggest creating a Microsoft Forms survey or similar, to register students prior to the session. This way, the number of expected attendees can be determined. However, organisers should also consider the possibility of last-minute dropouts or students opting not to attend. We decided to not gauge attendance prior to the event.

Before proceeding, it is important to address the inclusivity of the session. We believe that the

Table 3.2 Example of questions and answers generated for the FHEQ Level 5 Human Physiology exam revision session

Question (to be called during the session)	Answer (to be located within the Bingo card)
Which fibrillar collagen produces an elastic network and is characterised by a low molecular weight?	Collagen III
What is the name of the structure that allows for rapid depolarisation of the muscle membrane deep within muscle fibre cells?	T tubules
What is the name of the structural component of striated muscle cells that connects the sarcomere of the muscle to the cell membrane?	Costamere

The thyroid gland	Acetylcholine receptors	Elastic filaments	SA Node
Sarcoplasmic Reticulum	Infundibulum	Osteoclast	Alveoli
Bone	Sarcomere	Hydroxyapatite	SERCA pumps
Ca 2+ release	Voltage-gated Na+ channels	Epitenon	Median Eminence

Elastic filaments	Neuromuscular junction	Bicarbonate	ADH
Pleural cavity	T tubules	Ca 2+ release	SERCA pumps
The thyroid gland	Thin filaments	Troponin-Tropomyosin complex	Collagen III
Median Eminence	Hydroxyapatite	Acetylcholine	Melatonin

Fig. 3.4 Two Bingo cards generated for the FHEQ Level 5 Human Physiology exam revision session. Each team of students that wanted to participate in the session received an individual/team Bingo card, which was randomly produced with a free of charge Bingo card generator

session can be conducted either synchronously or asynchronously to accommodate all students. For those who prefer to participate online, educators can live stream the session. Bingo cards can be made available online in advance, allowing students to choose one before the session begins. Online participants can mark their answers and follow along as if they were in the physical classroom.

Real-time verification of Lines and Bingo for online students can be challenging unless an educator monitors an online chat to track their progress. If the latter is not possible, all questions and answers can be uploaded online after the session, enabling students to self-assess their performance.

For an in-person session, we recommend allocating minimum three members of staff to the session. This will ensure that question calling, answer checking, and overall running of the session happen smoothly.

The running of the session can be tailored to suit the educator's preferences and the availability of staff and students. We have compiled a sequence of events illustrating how the session unfolds as seen in Fig. 3.5 .

As illustrated in Fig. 3.5, the session begins with organising the students. The session can accommodate students who prefer to work individually or in teams, though we recommend team collaboration. Working in teams promotes communication and interaction among peers, which enhances collaborative problem-solving and strengthens understanding and retention of the material. For team setups, we suggest groups of no more than four students to balance the contribution of individual answers.

Once students are seated in their respective groups, a staff member distributes the Bingo cards. To minimise paper waste, these cards could be laminated for reuse in future sessions. As the cards are handed out, staff should explain the session's format. Educators need to remind students to mark the correct answer on their cards in response to the questions called out. It is important to note that sometimes the correct answer may not be on their cards, and they will need to wait for the next question. The rules for achieving a Line and Bingo should be clearly established. In our session, we accepted vertical and horizontal lines, but not diagonal ones. Bingo is achieved with a full card, and only one Bingo winner is allowed. For Lines, a maximum of two per group or individual is permitted. These rules can be adjusted to fit the educator's preferences.

Questions are then drawn randomly from a container to ensure variety. We recommend reading each question twice and allowing about one

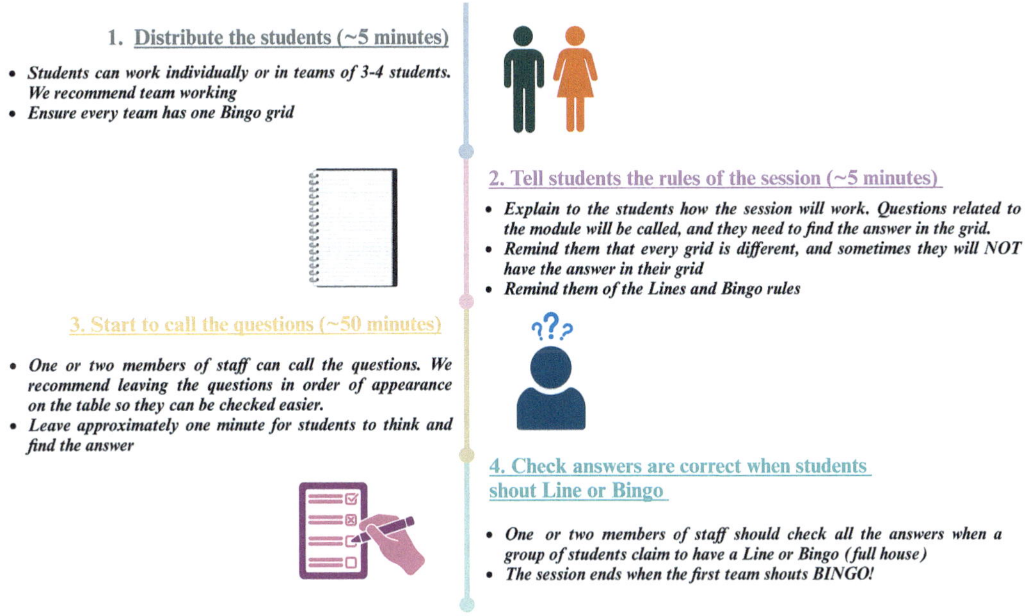

Fig. 3.5 Timeline of events for the running of the FHEQ Level 5 Human Physiology Bingo session. Created with Biorender.com

to one and a half minutes between questions to maintain a fast pace while giving students enough time to discuss and find the answers. Answers can be noted by staff on a separate piece of paper. When a group claims to have a Line, a staff member verifies the answers at their desk. Alternatively, questions and answers could be displayed on a traditional screen projector to support students with additional needs and ensure inclusivity.

The session continues with questions being called until a group of students shouts Bingo. Based on our experience, it takes around 30 min for students to gather enough correct answers to call for Lines, with Bingo typically being achieved towards the end of the session.

For added motivation, we provided prizes for the students. Prizes not only enhance the excitement and engagement of the session but also serve as a reward for participation. We have given out sweets, chocolates, stationery, and textbooks, but the choice of prizes (if chosen at all) is up to the educator. Prizes are also a good way to reward students who may not have won but came close or achieved a Line.

After the session concludes, the remaining task is to make all materials available online. The Bingo cards, along with the questions and answers, can be made available so students can repeat the session as often as they like at their convenience. Additionally, if the session is recorded, students can watch it and participate as if they were playing the game live. The questions provided in this game also allow students to train for multiple-choice question formats, which are common in FHEQ Level 5 exams.

3.5 Strategy 2. LabQuest: How to Help Students to Problem-Solve their Scientific Essays Through Interactive Workshops

3.5.1 Introduction

One persistent concern for students throughout their degrees is writing essays. Whether as coursework or part of an exam, essays often pose significant challenges for students. Unlike A levels and other educational stages, the essays demanded in HE are more complex and require substantial research skills to support an argument or claim (Winter 2003). Additionally, students often struggle with essays due to a lack of structuring skills, insufficient understanding of the subject, inadequate preparation, and feeling overwhelmed by the numerous academic references available (Harackiewicz and Priniski 2018).

Paradoxically, essay writing is a valuable skill to develop during university, regardless of the degree programme. Writing about a topic involves a cognitive process where students must analyse and evaluate information, creating a new perspective through their own words. This process fosters active learning, critical thinking, and long-term retention and understanding of the material (Nickerson et al. 1985). Consequently, essay writing enhances employability by developing writing and research skills, critical thinking, the ability to summarise and present different perspectives, as well as organisational skills (Torrance et al. 2000).

Therefore, it is evident that as educators, we should allocate sufficient resources to teach our students how to write essays. Although writing is a skill that takes time to master, there are certain elements that can enhance students' active engagement with their essays, allowing them to maximise their understanding of the assigned topic.

One such element for ensuring essay success is fostering the student's interest in the topic they are writing about. Typically, educators assign a fixed topic for essays to ensure equal opportunities among students and to accommodate various writing styles and ways of expressing ideas. However, even with a single set topic, educators should consider whether the topic can be approached from different angles or perspectives. If so, could students be allowed to choose their own "subtopic" freely? Could this be the technique needed for students to actively choose what they want to write about, thus researching with greater motivation and enthusiasm?

If the answer is yes, we are looking at a powerful motivational process. Several studies have shown that interest in a topic can significantly enhance learning and lead to greater academic success (Harackiewicz et al. 2016). Therefore, it is worth considering a shift in the paradigm for essay topics. Allowing students to actively choose what they write about can lead to more thorough research and deeper learning, as well as potentially higher grades. Educators, on the other hand, might find it more engaging to read and grade diverse essays and perspectives, rather than repeatedly evaluating similar standard essays.

3.5.2 Description of the Activity

As previously explained, this activity is part of our FHEQ Level 6 Cell Biology and Mechanisms of Disease module, offered during the spring semester. As an optional module, it typically enrols 55–60 students. The module includes traditional lectures, but also workshops that are led by the lecturers who teach in the course. Each lecturer chooses the workshop topics based on their expertise, generally focusing on novel biomedical approaches to diseases or pathologies that lack targeted therapies and that are currently managed with one-size-fits-all treatments. These specialised workshops enable students to deepen their understanding of cutting-edge biomedical techniques used in research labs and thus expand on the lecture material.

The workshops span 4 weeks, with one workshop per week. Students are randomly assigned to workshop leaders at the start of the module, ensuring an equal distribution of participants per lecturer. At the end of the workshop series, students submit an essay on their workshop topic, contributing to their summative module mark. Lecturers have flexibility in running these workshops; however, they should include problem-solving activities to enhance engagement. These activities might be quizzes, preparatory questions for the next session, peer discussions, etc., to ensure the workshops are interactive rather than just additional lectures.

For our workshop, titled "Investigation of Cell Therapies for Treating Tendinopathies", we focus on studying tendinopathies. Tendinopathies, or pathologies of the tendon tissue, have no successful treatment so far, and they would largely benefit from biomedical research. In particular, cell therapies are very pertinent to study, since tendons require a large number of cells to regenerate damaged tissue, and many host cells die after the injury, hence requiring the exogenous injection of cells.

However, cell therapies for tendon regeneration are difficult to develop, since there are multiple variables to investigate, such as the type of cells to use, or if the cells will be stimulated with biomolecules to potentially express target genes, among others. Therefore, students enrolled in these workshops need to come up with a combination of variables that should be investigated in the laboratory, learning about each variable's role in tissue regeneration and the disease mechanisms, along with state-of-the-art biomedical research and its human applications.

As we recognise the complexity and vastness of this topic, we developed an interactive activity; a board game, where students navigate through a board to obtain the variables that will be part of their essay. The variables are taken from a fixed set, decided by the workshop leader. By reducing the number of variables to four, this play-based learning approach simplifies the essay writing process and makes the workshops more accessible.

The game is called LabQuest, and it is based on a reward system. Reward-based games encourage skill practice in a safe environment, foster peer collaboration, and improve the retention of complex topics by providing rewards at the game's end, creating positive expectations (Koster 2013).

In LabQuest, students act as biomedical scientists working in a laboratory to investigate variables for the development of a cell-based therapy for tendon tissue regeneration. They navigate a board to collect four different variables from the fixed set, which they use to craft their essay. The game incorporates elements of surprise, challenges, and problem-solving to stimulate student engagement and attention, offering a relaxing break from traditional sessions. Students can play alone or in teams, at their discretion.

The following section will outline the methodology for preparing and running the game. This game is particularly beneficial for modules with small groups of 6–10 students, and we encourage educators to adapt it to their own topics.

3.5.3 Methodology

3.5.3.1 Designing the Board and the Game Layout

The first step in this activity is to design the board and the game layout. The following guidelines have been effective for us when preparing the game and are tailored to our workshop content. However, we encourage educators to adapt them to fit their own topics.

(a) *Determine the number of variable elements.* First, determine the number of variables to be included in the essay. If the essay requires students to describe a series of independent elements, such as collecting four variables in our case, decide on the exact number for them to research and write about. Given the 1500-word limit for our students' essays, we set the number of variables at four. For each of these four variables, we further included a series of elements within each. For instance, within the variable "Cells", we decided on four types of cells that will be interesting for students to know and shuffled the card deck. This approach introduces more randomness to the game, allowing each student to potentially finish with a unique set of variables.

(b) *Plan student navigation through the board.* For this game, we wanted students to arbitrarily collect variables from a fixed set, adding an element of uncertainty and competition as students strive to collect all their variables for their essays. We colour-coded the four variables and distributed them randomly around the board. We also adopted a classic turn-based game pattern where students roll the dice to advance through different categories.

(c) *Design the board layout*: The layout can be customised by the educator. We designed a classic path-type layout with laboratory-based illustrations. We also included a "cut-through" path allowing students to skip half the board, adding an element of surprise. We used the online platform Biorender (Biorender.com) to design the board.

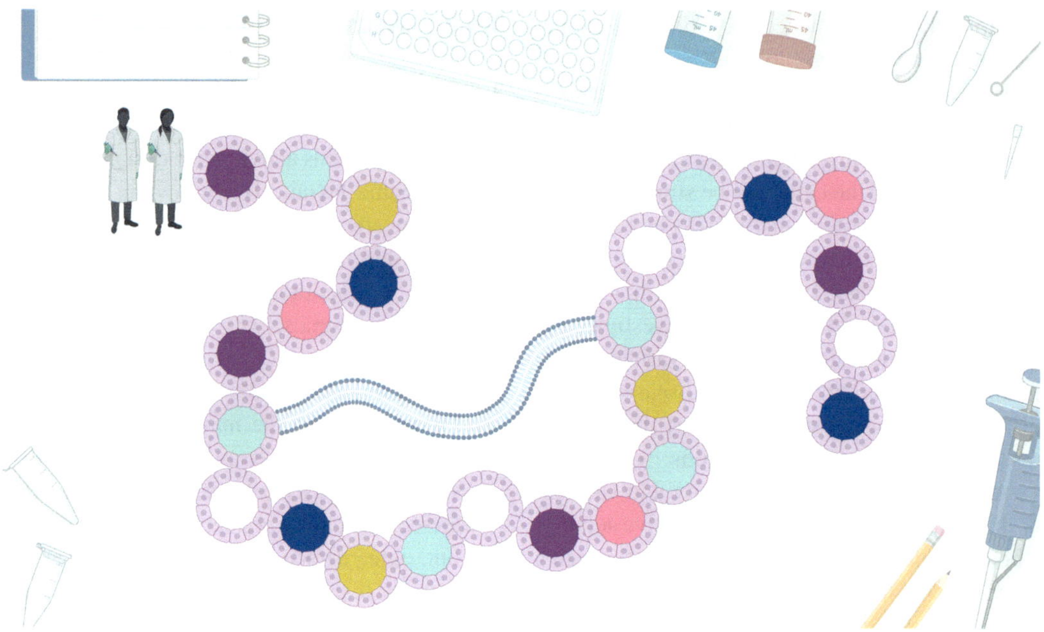

Fig. 3.6 The board of the LabQuest game used for the FHEQ Level 6 Cell Biology and Mechanisms of Disease workshops. The board follows a traditional linear pattern with colour-coded areas that represent the variables students need to collect to write their essay. Created with Biorender.com

The resulting board can be seen in Fig. 3.6. The board can be printed to the desired size. We recommend sizes larger than A3.

Regarding the rhythm of the game, all students start at the cartoon of the scientist, rolling the dice to advance through the board (We purchased wood tokens to give to each student and standard dice that are inexpensive and can be found in craft shops). Each turn, students will look at the colour-coded box they land on and collect the corresponding variable from the deck. The objective is to gather one element contained within each of the four different colours. However, students may end up with duplicates of some colours. To aid in collecting all necessary variables to write their essay, we introduced three elements of surprise:

1. *When a student picks up a variable for the first time, we explain it.* At the game's start, all variables are new to students. We have prepared a deck of slides with bullet points that explain the main aspects and importance of each variable for their tissue system research. Students take notes, deepening their understanding as they progress. This encourages self-reflection and critical thinking by fostering curiosity about the details and roles of each variable in tissue repair.
2. *We included "action cards".* Since students might end up with multiple variables of the same colour, we added an extra colour on the board that requires drawing cards from a special deck. These cards allow students to exchange or purchase variables, making it easier to collect all four colours and increasing competition among students. This will be detailed further in the next section.
3. *We introduced a "cut-through" path.* This path, represented on the board by the cells' phospholipid membrane cartoon, allows students landing on it to skip half of the board. This adds an element of competition and urgency to complete their variable collection.

3.5.3.2 Creating the Variables/Cards

Once the board game and layout are finalised, it is time to create the variables or independent elements for students to discuss or compare in their essays. We encourage educators to design them for their own content and to follow specific learning outcomes.

As previously mentioned, we decided to limit the number of variables for students to investigate to four, each colour coded. Each colour/variable corresponds to an independent card deck containing all the sub-elements from that variable. Additionally, we included two extra colours as elements of surprise. An image of the independent card decks and a key for students to not lose track can be seen in Fig. 3.7a, b, respectively. All the colour-coded elements of the board and the explanation behind them can be seen in Table 3.3.

As shown in Table 3.3, we included action cards to aid students who might have duplicates of some variables or are missing others entirely. The purpose of these cards is to facilitate the exchange or purchase of variables and encourage collaboration, simulating the work of biomedical scientists in a laboratory. These cards enhance students' curiosity as they may want to obtain different variables after learning from the educator's slides.

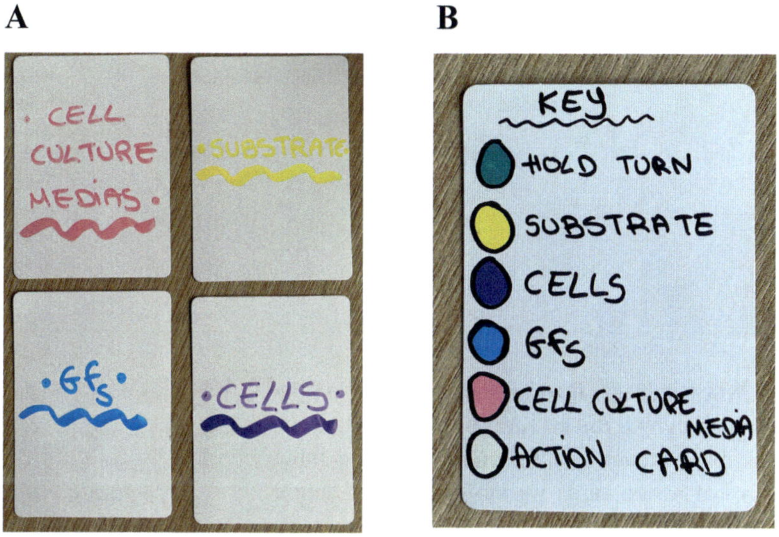

Fig. 3.7 The independent card decks to draw variables from, as well as their colour-coded key created for the LabQuest game used for the FHEQ Level 6 Cell Biology and Mechanisms of Disease workshops

Table 3.3 The type of variables depicted on the board of the LabQuest game used for the FHEQ Level 6 Cell Biology and Mechanisms of Disease workshops

Colour	Variable that corresponds to the colour	What does the variable investigate?	What does the variable contain?
Purple	Type of cell	Cell types that have been used for tendon regeneration	• Tendon-derived stem cells • Bone-marrow-derived stem cells • Adipose-derived stem cells • Perivascular stem cells
Yellow	Type of substrate/cell culture surface	Tissue culture surfaces that have been used to study tendon regeneration	• Two-dimensional surfaces • Collagen-coated surfaces
Blue	Type of growth factor	Growth factors with tenogenic properties	• Transforming growth factor β family • Bone-morphogenetic proteins family
Pink	Type of cell culture media	Tissue culture medias that have been used to incubate cells to study tendon regeneration	• Medias with foetal-bovine serum • Medias without foetal-bovine serum
Green	Hold a turn	N/A	N/A
White	Draw an action card	N/A	N/A

Table 3.4 The action cards created for the LabQuest game used for the FHEQ Level 6 Cell Biology and Mechanisms of Disease workshops

Type of action card	Main aim behind the card	What are the possibilities?
Funding	To purchase variables. Beneficial for those teams who are missing variables	1. Students can purchase one variable from any deck 2. Students wait until another team draws the same card, hence allowing them to purchase variables from each other 3. Dummy card, they return the card to the deck and continue the game
Time	To hold turns. Beneficial to strive for competition and obtaining all the variables.	1. Missing one turn 2. Missing two turns 3. Missing three turns
Team	To collaborate with other teams. Beneficial for students to discuss and exchange variables and enhance critical thinking	1. Discuss with the adjacent team what variables they have so far, to help each other if they are hesitant about what variables to purchase/ collect 2. Take one variable from one of the teams 3. Discuss with a team that is stuck in some of the variables, given them the possibility to exchange them

When students land on the white box on the board, they will roll an additional dice with three possible outcomes, corresponding to the three types of action cards we designed. Each action card contains three different possibilities to increase the game's randomness. These options are detailed in Table 3.4. The physical cards and the custom-made dice are illustrated in Fig. 3.8, while the card designs created using Biorender are shown in Fig. 3.9.

A

B

Fig. 3.8 A detail of the action cards and the custom dice created for the LabQuest game used for the FHEQ Level 6 Cell Biology and Mechanisms of Disease workshops

3.5.3.3 Running the Workshops

As previously mentioned, our module's workshops are scheduled to run once per week, each lasting 2 h. To balance content and play-based learning activities, we decided that the game will be conducted over two sessions. The workshop schedule is as follows.

The first session is an introductory session, in which the students will be introduced to the workshop topic in detail, along with an explanation of the game and its basic rules. Educators should clearly outline the game's objectives and the importance of collecting the four types of variables for the essay. It should also be communicated, either during the session or via an assessment brief, that while the essays will vary, they will all be assessed using the same rubric.

During the second and third session, students play the board game. When the second session starts, students are encouraged to sit randomly to promote interaction. Given that

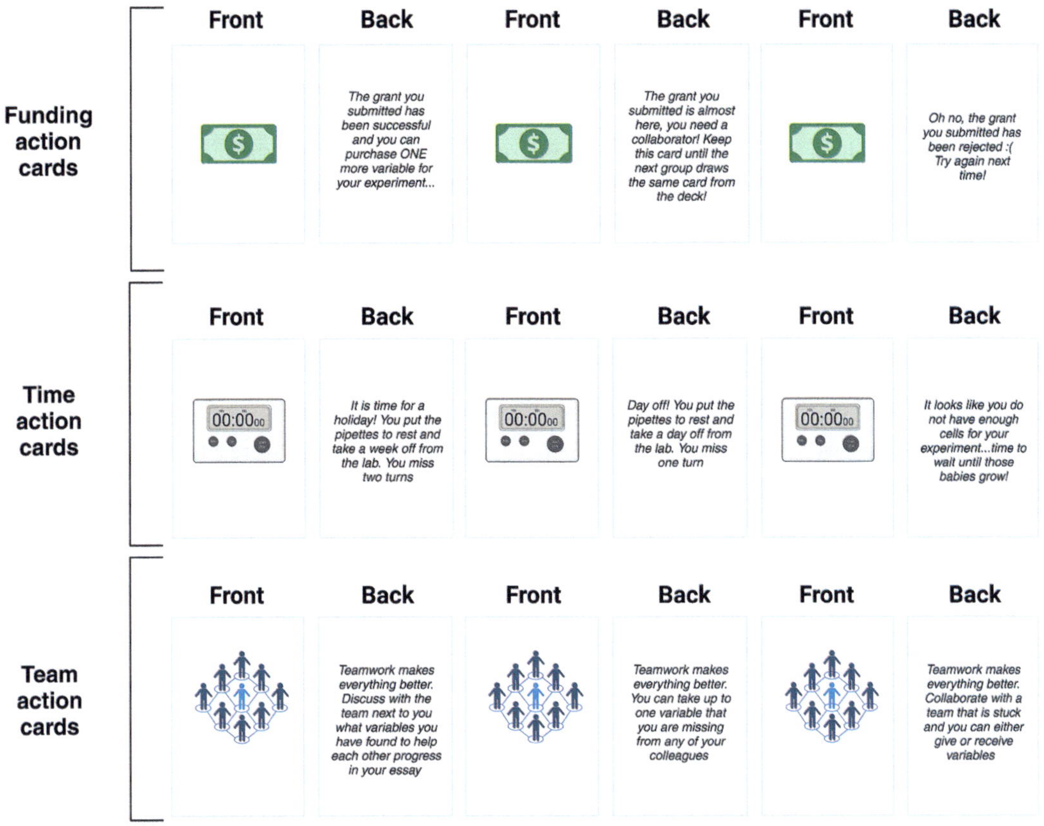

Fig. 3.9 A detail of the three different possibilities students can draw from the action cards created for the LabQuest game used for the FHEQ Level 6 Cell Biology and Mechanisms of Disease workshops. Created with Biorender.com

this module is optional and available to students from different degree programmes, this diversity can enhance discussions. Students can work in pairs, but individual work is recommended to ensure each essay is unique. The game will proceed as described in earlier sections. Splitting the game into two sessions ensures a balanced approach, preventing any aspect from being rushed. To facilitate a smooth transition between sessions, we noted the students' seating arrangements and the variables they obtained. Students are encouraged to research their variables in their own time to begin gathering literature for their essay.

Finally, in the fourth session, we provided a summary of the workshop and reminded students of the essay requirements. This is an excellent opportunity for students to ask questions about their topics.

3.6 Reflection and Next Steps

The final section of the chapter will discuss the two play-based learning strategies presented, mainly focusing on the overall development of the session and areas for improvement. As this was our first implementation of play-based learning strategies in our course, we did not collect measurable feedback from students. Our priority was, for the moment, creating and developing the activities and gathering qualitative evaluations of students' reactions. Thus, the following discussion is based on qualitative observations during the activity and supported by module feedback evaluation forms where these activities were mentioned.

The first play-based learning strategy, the Bingo session, was a great success overall and ran smoothly. Positive aspects include the students' willingness to participate and their high engagement with the activity. They arrived on time, curious about what would happen. A key highlight was observing how much they remembered from their lectures without looking at their notes. Students had to quickly reflect on the questions and think rapidly about the content to find answers on the Bingo grid, increasing their awareness of key concepts for future assessments.

Additionally, we observed excellent teamwork among students. They naturally sat with peers and frequently discussed possible answers, stimulating each other's thinking to review the content. Students were visibly excited when close to completing the grid, and there was a relaxed atmosphere in the room. This demonstrates the effectiveness of the activity, especially since typical lecture-based revision sessions are less engaging. About 70 students attended, indicating that this activity works well with large groups.

We did obtain some informal feedback via email after the session run. One student mentioned "I really enjoyed Bingo this morning and the care and time that went into it was evident. Thank you so much for your time, it was so helpful and quite reassuring to know that I haven't forgotten everything. It also pointed out some areas I need to go over!". This demonstrates the success of the session and how the aims and objectives were met, which is extremely rewarding. However, we encountered a few minor issues. Some students preferred engaging in the activity alone, while others chose larger groups, leading to an uneven distribution. Although this did not hinder the activity, we feel that solo students missed out on peer-to-peer interaction. Nonetheless, we acknowledge the value and need for some students to work independently, as the goal is interactive content revision. Additionally, some students decided not to attend or dropped out last minute. While this did not affect the session since materials were available online, those students missed out on the communication and relaxation factors provided by the activity.

The second activity, LabQuest, also ran smoothly. Students were engaged and curious about the game. A major highlight was observing how they interacted with the game's variables. For instance, in the second session, when most students had an almost complete set of variables, they became excited when an action card for exchanging variables appeared. Students cleverly chose new sets of variables, either purchasing or exchanging them with peers. This demonstrated their engagement with the variables and their understanding of the workshop topic, leading to the "crafting" of essays based on their own critical thinking. Overall, the sessions were relaxed and engaging, with positive student participation.

Nonetheless, there were a few challenges with this activity. Educators must ensure that all possible combinations of variables have been studied and, most importantly, published. The game is designed to promote student initiative and engagement with written assessments, so there must be literature available on each included variable. We encountered one student who struggled to find published research for their chosen combination of variables, necessitating an adaptation of their essay content. This can be a significant limitation for less-explored fields, so we suggest reducing the number of variables in relatively new areas.

Another issue was that two students attended only one game session, and another missed all the sessions. While this did not affect the progress of other students, the nature of the activity requires regular attendance. Therefore, there needs to be a way for absent students to access the same level of freedom for their essays as their peers. We temporarily solved this by having one-to-one discussions with such students so they could access a mentoring session prior to writing their essay.

As mentioned, we did not conduct a quantitative analysis of student satisfaction and engagement with the activities. Similarly, it is difficult to gauge the impact of play-based learning activities on students' exam or assessment performance, as we lack a comparative baseline of students who took the same assessments without participating in play-based learning activities.

Addressing this is a key objective for future years. With an understanding of how students generally received the activities, our next step is to develop surveys and feedback forms. We aim to ask students how the activities impacted their learning, how prepared they felt for assessments, and whether the activities fostered a sense of belonging and communication among peers. This feedback will help us identify both positive and negative aspects of the activities and provide an opportunity for reflection and improvement.

Despite the lack of quantitative data, we received numerous positive comments about the implemented strategies. In module evaluation forms, students highlighted factors such as:

(a) How the activities helped them revise content.
(b) The confidence they gained before exams by identifying topics they needed to review and those they already knew.
(c) The usefulness of the activities as a break from traditional lectures and the relaxed atmosphere of the sessions.
(d) The importance of having the autonomy to decide what they wanted to write about.
(e) Their appreciation for the organisation of the activities by the staff.

This demonstrates the versatility of the presented activities for STEMM careers, and we deeply encourage educators to be inspired and implement their own. Our first implementation of play-based learning strategies in our course has provided valuable insights into both the benefits and areas for improvement of such activities. Through qualitative observations and feedback from module evaluation forms, we have identified several positive outcomes, including increased student engagement, improved teamwork, and enhanced content retention. The activities were particularly effective in creating a relaxed and interactive learning environment, which contrasts favourably with traditional lecture-based sessions.

Our main aim was to ensure that play-based learning strategies foster a sense of belonging, communication, and effective preparation for assessments. Despite the challenges, the overwhelmingly positive student reaction to such activities underpins the potential of play-based learning strategies to enrich the educational experience from our students.

In conclusion, we believe that play-based learning can effectively create dynamic and stimulating learning environments that cater to the diverse needs and learning preferences of university-level students. As long as the strategies are aligned with the learning outcomes and educators have student learning in mind, we firmly believe that play is and will be a key aspect of HE practices.

References

Ahn MY, Davis HH (2023) Students' sense of belonging and their socio-economic status in higher education: a quantitative approach. Teach High Educ 28:136–149. https://doi.org/10.1080/13562517.2020.1778664

Akl IB, Ghaddar F, Sabra R et al (2012) Teaching clinical pharmacology using team-based learning: a comparison between third- and fourth-year medical students. J Clin Pharmacol 52:1806–1814. https://doi.org/10.1177/0091270011428986

Alhammad MM, Moreno AM (2018) Gamification in software engineering education: a systematic mapping. J Syst Softw 141:131–150. https://doi.org/10.1016/j.jss.2018.03.065

Biggs JB, Tang C (2011) Teaching for quality learning at university, 4th edn. Open University Press

Dost S, Hossain A, Shehab M et al (2020) Perceptions of medical students towards online teaching during the COVID-19 pandemic: a national cross-sectional survey of 2721 UK medical students. BMJ Open 10. https://doi.org/10.1136/bmjopen-2020-042378

Evans C (2008) The effectiveness of m-learning in the form of podcast revision lectures in higher education. Comput Educ 50:491–498. https://doi.org/10.1016/j.compedu.2007.09.016

Garden CLP (2022) Lego Serious Play: building engagement with cell biology. Biochem Mol Biol Educ 50:216–228. https://doi.org/10.1002/bmb.21608

Gitnux (2024) Student's attention span statistics and trends in 2024

GOV.UK (2021) UK Census, 2021

Harackiewicz JM, Priniski SJ (2018) Improving student outcomes in higher education: the science of targeted intervention. Annu Rev Psychol 69:409–435. https://doi.org/10.1146/annurev-psych-122216-011725

Harackiewicz JM, Smith JL, Priniski SJ (2016) Interest matters: the importance of promoting interest in education. Policy Insights Behav Brain Sci 3:220–227. https://doi.org/10.1177/2372732216655542

Hashim AS, Ahmad WFW, Ahmad R (2011) Mobile learning course content application as a revision tool:

the effectiveness and usability. In: 2011 international conference on pattern analysis and intelligence robotics, pp 184–187

Ho WWY (2022) Influence of play on positive psychological development in emerging adulthood: a serial mediation model. Front Psychol 13. https://doi.org/10.3389/fpsyg.2022.1057557

Imad M, Reder M, Rose M (2023) Recasting the agreements to re-humanize STEM education. Front Educ (Lausanne) 8. https://doi.org/10.3389/feduc.2023.1193477

James A, Nerantzi C (2019) The power of play in HE: creativity in tertiary learning. Palgrave Macmillan

Khaldi A, Bouzidi R, Nader F (2023) Gamification of e-learning in higher education: a systematic literature review. Smart Learn Environ 10:10. https://doi.org/10.1186/s40561-023-00227-z

Kirby LAJ, Thomas CL (2022) High-impact teaching practices foster a greater sense of belonging in the college classroom. J Furth High Educ 46:368–381. https://doi.org/10.1080/0309877X.2021.1950659

Koivisto J-M, Multisilta J, Niemi H et al (2016) Learning by playing: a cross-sectional descriptive study of nursing students' experiences of learning clinical reasoning. Nurse Educ Today 45:22–28. https://doi.org/10.1016/j.nedt.2016.06.009

Koster R (2013) Theory of fun for game design, 2nd edn. O'Reilly Media

Marshall C (2017) Montessori education: a review of the evidence base. NPJ Sci Learn 2:11. https://doi.org/10.1038/s41539-017-0012-7

May JE, Anderson E, Clark D, Hull J (2023) Gamification in biomedical science education: the successful implementation of Resimion, a scenario-based learning tool. Br J Biomed Sci 80. https://doi.org/10.3389/bjbs.2023.11756

Newton PM, Miah M (2017) Evidence-based higher education—is the learning styles 'myth' important? Front Psychol 8. https://doi.org/10.3389/fpsyg.2017.00444

Nickerson RS, Perkins DN, Smith EE (1985) The teaching of thinking, 1st edn. T&F Books

Nicolopoulou A (2010) Play, cognitive development, and the social world: Piaget, Vygotsky, and beyond. Hum Dev 36:1–23. https://doi.org/10.1159/000277285

Njie-Carr VPS, Ludeman E, Lee MC et al (2017) An integrative review of flipped classroom teaching models in nursing education. J Prof Nurs 33:133–144. https://doi.org/10.1016/j.profnurs.2016.07.001

Owens M, Townsend E, Hall E et al (2022) Mental health and wellbeing in young people in the UK during lockdown (COVID-19). Int J Environ Res Public Health 19. https://doi.org/10.3390/ijerph19031132

Rea SD, Wang L, Muenks K, Yan VX (2022) Students can (mostly) recognize effective learning, so why do they not do it? J Intell 10. https://doi.org/10.3390/jintelligence10040127

Rodríguez-Ferrer JM, Manzano-León A, Aguilar-Parra JM (2023) Game-based learning and service-learning to teach inclusive education in higher education. Int J Environ Res Public Health 20. https://doi.org/10.3390/ijerph20043285

Sicart M (2014) Play matters. MIT Press

Siviy SM (2016) A brain motivated to play: insights into the neurobiology of playfulness. Behaviour 153:819–844. https://doi.org/10.1163/1568539X-00003349

Terrell CR, Nickodem K, Bates A et al (2021) Game-based activities targeting visual literacy skills to increase understanding of biomolecule structure and function concepts in undergraduate biochemistry. Biochem Mol Biol Educ 49:94–107. https://doi.org/10.1002/bmb.21398

Tong VCH, Chow DSL (2013) A study of student participation and nonparticipation in prelecture electronic surveys. Br J Educ Technol 44:869–880. https://doi.org/10.1111/j.1467-8535.2012.01374.x

Torrance M, Thomas GV, Robinson EJ (2000) Individual differences in undergraduate essay-writing strategies: a longitudinal study. High Educ (Dordr) 39:181–200

Van Vleet M, Feeney BC (2015) Play behavior and playfulness in adulthood. Soc Personal Psychol Compass 9:630–643. https://doi.org/10.1111/spc3.12205

Veraksa N, Pramling Samuelsson I, Colliver Y (2022) Editorial: early child development in play and education: a cultural-historical paradigm. Front Psychol 13. https://doi.org/10.3389/fpsyg.2022.968473

Whitton N, Moseley A (2014) Deconstructing engagement: rethinking involvement in learning. Simul Gaming 45:433–449. https://doi.org/10.1177/1046878114554755

Widmaier EP, Raff H, Strang KT (2023) Vander's human physiology ISE, 16th edn. McGraw Hill

Winter R (2003) Contextualizing the patchwork text: addressing problems of coursework assessment in higher education. Innov Educ Teach Int 40:112–122. https://doi.org/10.1080/1470329031000088978

4 Making Psychology Accessible to Students in Medical Education

Karen M. Adams

Abstract

Many medical students find psychology challenging and this can have a negative impact on their engagement with the subject. This chapter will explore some of the key barriers that medical students might encounter that may explain why they find psychology difficult or inaccessible. In particular it will consider how the lack of familiarity and exposure that students have with the subject may play a pivotal role. Psychology is very much an interdisciplinary subject, and this is a strength that can be exploited by linking the subject to concepts that medical students are likely to be more familiar with. This linking should help students to understand what psychology is and to appreciate how and why it holds such value in medical education. Some practical examples are presented to illustrate its value, for example introducing students to the concept of uncertainty which is prevalent in medicine yet frequently an unexpected and uncomfortable concept for them. It is important that students have opportunities to actively apply and discuss their psychological knowledge rather than simply learning the theoretical concepts. Ideas to promote active engagement such as authentic and relatable learning material, opportunities for experience sharing and including practical activities they can engage with can all promote the practical and applied nature of psychology and increase medical students' engagement with the subject by making it more accessible to them.

Keywords

Behavioural science · Medical education · Applied psychology

4.1 Introduction

Engaging medical students in behavioural and social science subjects as part of their medical degree can be a challenge. It has been reported that psychology in medical education can be perceived as 'soft and fluffy' (Gallagher et al. 2015) and 'nice to know' but not 'need[ed] to know' (de Visser 2009) and this can be reflected in the students' attitudes towards it.

One of the key contributory factors to this is likely to be their prior exposure—or lack of it—to these subjects (Dikomitis et al. 2022). Standard entry requirements for students progressing

Supplementary Information The online version contains supplementary material available at https://doi.org/10.1007/978-3-031-81795-3_4.

K. M. Adams (✉)
School of Medicine, Keele University, Keele, UK
e-mail: k.m.adams@keele.ac.uk

R. Door (ed.), *Accessibility, Inclusivity and Diversity in Education and Beyond*, Biomedical Visualization 8, https://doi.org/10.1007/978-3-031-81795-3_4

directly from school for all UK medical schools are a minimum of three A levels (or equivalent) which must, with only one exception currently, include chemistry and/or biology. Although there is some variation and flexibility, many medical schools specify two A levels must be in a 'lab-based science'. There are a small number of medical schools (currently) who do accept psychology as a second science subject (Medical Schools Council 2024). Graduate students would normally require a minimum of a 2:1 degree (there is variation between medical schools as to whether the degree should be related to science), usually in addition to three A levels that include chemistry and/or biology.

Based on these entry requirements, it is unsurprising that most students enter medical education from a biological and physical sciences background rather than behavioural (or social) science. Bioscience and behavioural science generally adopt rather different philosophies. This is likely to affect what students believe and value knowledge to be, as well as how they consider they can achieve understanding of this. Bioscience subjects taught at A level tend towards a positivist epistemological perspective. This assumes all knowledge is achieved through objective observation and measurement. If these measurements are consistently replicated, the evidence can be used to construct theories that are assumed to provide a correct answer and can be used to predict and control different phenomena. Psychology might be considered as more interdisciplinary than other science subjects; as such, it adopts a range of philosophical views and epistemological approaches. As well as embracing positivism, it employs interpretivism which believes that knowledge is subjective and open to interpretation. It acknowledges that experience is likely to differ between individuals, but also that this experience may be understood differently depending on the perspective of observers. Taking this perspective means that there is unlikely to be a single correct theory that can explain all instances of a particular phenomenon; instead, there are a range of factors that need to be considered when attempting to predict and control different phenomena. Other epistemological approaches are also encountered in psychology; for example, critical theory believes that social interactions and power dynamics are embedded in culture and society, and that these work to mould the knowledge and understanding that individuals gain. Pragmatism takes a more practical stance in considering how knowledge should be created for it to be usefully applied to the real world.

Most students have not been exposed to the range of epistemological approaches prior to their entering medical school and this is likely to have an influence in how they (expect to) learn medicine. The bioscience background does not necessarily prepare them for ambiguities and 'grey' areas that are more commonly encountered within psychology (Armitage et al. 2024) or to take a critical perspective and weigh up conflicting evidence which are necessary skills for future clinicians.

Curiously, while the body of knowledge studies in A levels such as biology and chemistry is assumed to be the product of scientific method, there is no explicit coverage of research methods or the generation and testing of hypotheses. The body of knowledge is presented as more-or-less undisputed fact, with little exploration of how it is known. The practical component of these qualifications focuses on use of apparatus and techniques, presentation of data and drawing conclusions from single experiments (AQA 2024a), albeit with scope for consideration of conflicting evidence. In contrast, psychology A level includes specific content on research methods that includes the variety of types of experiment and appropriate methods of data analysis for testing of hypotheses (AQA 2024b), all of which are essential for the study of any kind of science beyond school level.

The medical curriculum covers a huge and diverse range of subjects that is necessary to meet the graduate outcomes required by the General Medical Council (GMC). The early years in medical education are generally pre-clinical. Although they tend to focus on equipping students with the basic science foundation they will continue to build on, the patient remains pivotal to all their learning. Students are informed at the very beginning of their studies that as doctors they will treat people and not the disease or condition, and this brings psychology and the study of mind and behaviour to the very heart of medicine. People are diverse and although there may be some commonalities between how they think, behave, and feel, no two individuals are identical and even individuals are likely to differ over time and across situations. This means it can be difficult to make predictions, for example, of how patients might cope with symptoms, adhere to treatment, and engage in health risks. There is rarely a single 'correct' answer to these issues, so medical students must learn and apply, sometimes competing, psychological theories and models to take this diversity into account. Thus, perhaps for the first time, they will need to manage uncertainty in their learning.

Psychology is not just of value to students to apply to their patients in their future clinical role. It covers a vast range of areas that can help them with their own professional development including gaining an understanding of how we learn, how teams can work and break down, critical thinking, and research design, methods and statistics. They should also be introduced to how it enables them to understand their own behaviour, thoughts, and feelings (Robiner et al. 2021). In the UK, the Behavioural and Social Sciences Teaching in medicine (BeSST) network has developed a curriculum of psychology topics that are considered a priority for inclusion in medical education (BeSST 2010). This provides good guidance for *what* should be covered but, in terms of *how* psychology should be taught, there are many ways and at present there is no specific recommended approach. However, it is essential that psychology is delivered in a way that makes it as accessible as possible for all learners in medical education, including those who may not even know what it is, to inspire engagement and facilitate their understanding.

This chapter will consider some ideas and principles that educators may find useful to consider for delivering psychology to medical students. It is recognised that there are likely to be constraints on the practicalities of how some of these may be adopted, due to how different medical schools construct and deliver their curriculum. However, it is hoped that they may be adapted or, indeed, inspire new ideas that will keep psychology accessible and engaging to learners, particularly those that may not have previous exposure to the subject.

4.2 What Methods Can We Use to Make Psychology Accessible?

4.2.1 Clear Understanding and Application of Psychology

Many medical students are unaware of the range of subjects covered in psychology and think of it narrowly and as only being relevant for patients with mental health concerns or in relation to counselling. Psychology is the study of mind and behaviour; the British Psychological Society (BPS) states that it applies sound scientific methodology to understand why people act the way that they do, so we can then use that knowledge and understanding to address real-world issues (British Psychological Society 2024). The GMC (2018) Outcomes for graduates document also details how newly qualified doctors must be able to apply psychological principles. Although they provide seven clear and key points that cover a wide range of areas to apply psychological principles (see Fig. 4.1), the sub-heading for this GMC outcome (23) may be perceived as misleading due to its focus on identifying, managing, and referring patients with mental health conditions.

Ensuring medical students understand what psychology is enables them to have a better understanding of its value and relevance, but it is also important to ensure they have the ability to apply this knowledge in a meaningful way (Gallagher et al. 2015). It is essential that students are signposted to appropriate learning resources for psychology. Many students do not feel the need to read beyond lectures for psychology and there can be a perception that it is just 'common sense' (de Visser 2009). Psychology should be clearly represented in the medical cur-

Applying psychological principles

Newly qualified doctors must explain and illustrate by professional experience the principles for the identification, safe management and referral of patients with mental health conditions.
They must be able to:

- Describe and illustrate from examples the spectrum of normal human behaviour at an individual level
- Integrate psychological concepts of health, illness and disease into patient care and apply theoretical frameworks of psychology to explain the varied responses of individuals, groups and societies to disease
- Explain the relationship between psychological and medical conditions and how psychological factors impact on risk and treatment outcome
- Describe the impact of patients' behaviours on treatment and care and how these are influenced by psychological factors
- Describe how patients adapt to major life changes, such as bereavements, and the adjustments that might occur in these situations
- Identify appropriate strategies for managing patients with substance misuse or risk of self-harm or suicide
- Explain how psychological aspects of behaviour, such as response to error, can influence behaviour in the workplace in a way that can affect health and safety and apply this understanding to their personal behaviour and those of colleagues

Fig. 4.1 GMC Outcomes for graduates (2018) applying psychological principles

riculum, and it is valuable to demonstrate to students why the taught sessions are relevant to them, either now in their academic learning or in their future clinical practice. As the GMC graduate outcomes (2018) state, it is important that students can understand and apply psychological principles not only to their patients, but also to their own and colleagues' thoughts, feelings, and behaviour. All of these contribute to the patient experience in some way; for example, good team dynamics ensure effective and efficient interdisciplinary meetings which impact on the management and level of care for patients.

It is recommended that students receive an introductory session early in the undergraduate medical programme explaining the topics that psychology covers, where and how they will be taught psychology, and how it will be assessed. Introducing students to the core curriculum (BeSST 2010) and GMC graduate outcomes (2018) at this early stage can also help lay the foundation for their psychology learning and manage their expectations of the necessity to view psychology with the same importance as bioscience and clinical learning. This promotes the importance of taking psychology seriously and ideally the session itself should be engaging and make students curious and interested so that they want to engage with it, rather than feeling they must. Relating some key psychological concepts to their situation can be a useful strategy in this introduction. So, an interactive lecture could discuss how they feel about starting to study medicine, how they think they need to behave and whether they consider this creates a dilemma when they are taking part in some of the University welcome week social events, and how they feel when they are starting to make friends and form groups. Discussing these issues and making links to some of the psychological factors and concepts that may contribute to these can help students identify the wider relevance of psychology. These issues can then be related to specific health care scenarios. A frequent comment from students is that psychology is 'common sense' (de Visser 2009), so it is worth including a discussion on this in the introductory lecture. Providing a number of examples of some common-sense beliefs and what these might be based on, examples of decisions or behaviours that they themselves have perhaps demonstrated that are not necessarily 'common sense' and discussions as to why they might have engaged in these, can result in students recognising that 'common sense' is a nebulous concept that is not an effective explanation or strategy.

Continuing to support students to apply their knowledge should be modelled by educators. Whilst it is important to provide theoretical frameworks for psychological concepts, students should be provided with clear, concrete examples that they can relate to (Gallagher et al. 2015). These may be examples that align to weekly problem-based learning (PBL) cases or scenarios from other sessions. Cues to different behaviours, emotions, or thoughts should be included in scenarios so that they become more 'life-like' and can be perceived as 'proxy-patients'. This allows students to notice these and consider how they can best understand them. Alongside these scenarios, educators may deliver lectures to support the students' learning using elements from the scenarios or PBL case. These applied examples can highlight to students the relevance and importance of psychology in being able to understand and appreciate the thoughts and behaviours of patients, families, or healthcare staff. A specific example of this may be a PBL case based on an individual who has been involved in an accident resulting in acute injuries. The case may describe the patient's reaction immediately after the injury and their responses a few weeks later. It may also include details on how the patient's family or any attending medics respond. A supporting lecture could be based on the psychological impact of injury and propose models that might explain individuals' reactions and draw on examples from the case regarding the patient's behaviour, emotional state and beliefs about what has happened to them, and how these might be predictors for how the patient will adjust to and manage with the injury. It might also consider the wider impact of the accident on medics, witnesses, and family members and consider how their responses and reactions towards the patient are likely to have an influence on how the patient perceives what has happened to them.

A key challenge of integrating psychology into PBL is that facilitators are generally experts in a particular bioscience or clinical field, so the majority are unlikely to have a sound working knowledge of psychology. This can lead to them feeling uncomfortable prompting students to discuss the subject or pressing them into providing more depth or applying their understanding to explain examples of behaviour illustrated in the PBL scenario. Facilitators can be supported with this by the provision of tutor notes that include clear and basic explanations of the concepts that students are expected to cover and possible prompts and questions that tutors can use to extend discussions that appear vague or surface level. PBL tutor briefings are also a valuable opportunity for psychologists to give further explanations and which allow facilitators to discuss, ask questions, or seek clarification on any aspects they feel uncomfortable about.

It is essential that educators use specific applied examples with students, so they demonstrate and model what students are expected to do. It is not just learning the theoretical psychological concepts being taught, but being able to apply them in a meaningful way to situations that students are likely to encounter both now and in their future.

4.2.2 Introduction of Uncertainty

Medical students will face uncertainty from the very beginning of their education; this will likely increase as they progress through the course and their future clinical practice. Uncertainty itself is difficult to define, but it is suggested that it arises from a lack of knowledge and understanding (Lee et al. 2020). This can be influenced by personal characteristics such as an individual's affect and experience. The responses to these encountered uncertainties can be cognitive, emotional, and behavioural (see Fig. 4.2). Students who have been mostly exposed to the positivist approach before commencing their medical undergraduate degree may not have the experience or world view that allows them to feel comfortable with uncertainty and ambiguity and they may still perceive that there should be a single correct answer for most of the issues that they encounter. This is reinforced by the assessment strategy adopted by most medical schools that relies on single best answer questions. Using this type of question may perpetuate the idea of there always being a correct answer in medicine, with no room for uncertainty or alternatives. Furthermore, students are rewarded (with marks) when they identify the correct answer: thus reinforcing a perception that there *is* always a correct answer. One way in which student uncertainty manifests is in their requests for solutions to class quizzes or tasks (Witt et al. 2022) or for a list of 'official' learning outcomes for PBL. This dem-

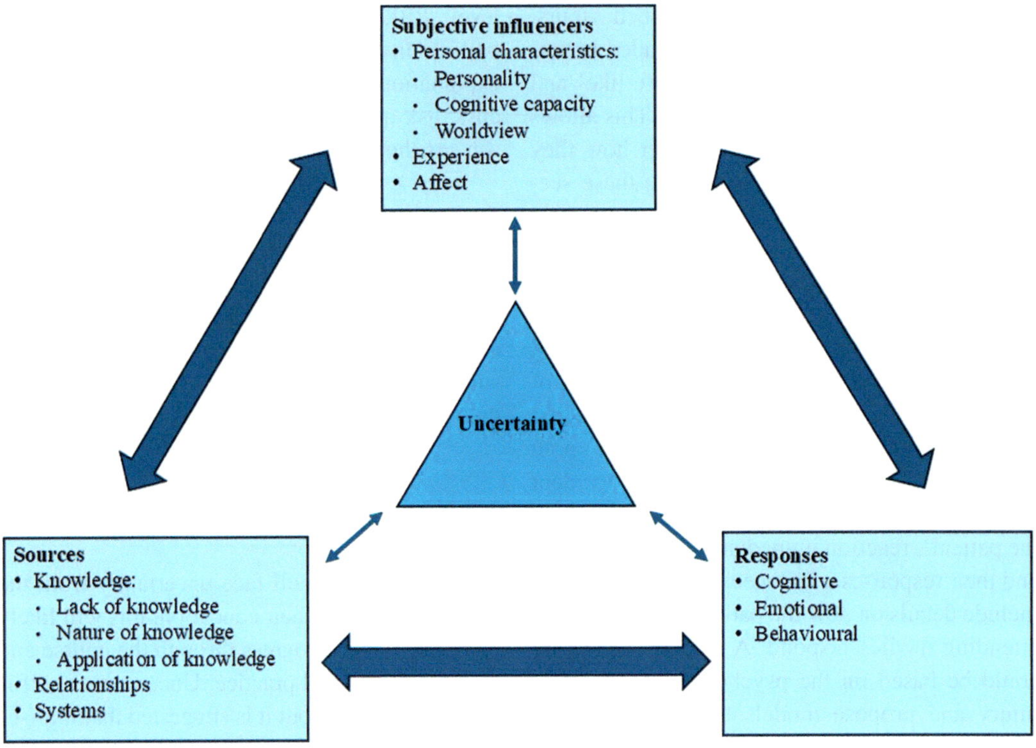

Fig. 4.2 Adapted from Lee et al. (2020) Framework for uncertainty in clinical practice. Subjective influencers of uncertainty, sources of uncertainty, and responses to uncertainty are the three key factors associated with uncertainty, each being interdependent with bi-directional influence

onstrates their desire to focus on getting the right outcome rather than engaging in the learning process that will enable them to acquire knowledge, understanding, and reasoning skills. A further factor that may contribute is likely to stem from medical students needing to be high performers to be accepted to medical school. They are likely to have been rewarded academically for 'correct' answers and perhaps less likely to have experienced failure or made mistakes.

Psychology is well placed to introduce uncertainty, with its focus on understanding behaviour, and can therefore play a key role in supporting students to better prepare for and tolerate ambiguities. Thinking that we live in a predictable environment where things happen for a reason rather than unexpectedly provides us with a sense of reassurance and safety. Our perception of living in an ordered and predictable world is supported by mental schemas, or cognitive frameworks, that provide internal representations by organising and interpreting our experiences and knowledge (Piaget and Cook 1952). Mental schemas are then drawn on to help us manage in new or unfamiliar situations. An example of a mental schema is how we categorise people into different personality types based on their known traits and characteristics. This helps us to understand how we might expect them to behave and react in different situations. These expectations then enable individuals to plan a response to these predicted behaviours and responses. However, it is always easy to think of examples where individuals have behaved unexpectedly, or 'out of character', and the feelings of disempowerment or helplessness that can be experienced by those around them when this happens. A different type of example is when someone encounters an unfamiliar situation and has limited prior knowledge or past experiences to draw on to help them work out what they need to do. In these situations, they may utilise different thinking styles (see Fig. 4.3). They could take an analytical approach that draws on knowledge or understanding that they think may be helpful (system 2 thinking), or make rapid, intuitive decisions (system 1 thinking) that are more instinctual and based on a 'gut feeling' (Kahneman 2011). The key is to initially introduce students to the concept of uncertainty and its prevalence in medical education and practice so they can start to accept it as something that they will encounter and should expect.

Specific strategies using psychology can be helpful in aiding students prepare for uncertainty, whilst also highlighting the usefulness of psychology to students as a learning tool. One area of psychology teaching that may introduce students to the idea of uncertainty is considering how 'risky' decisions are made; for example, the likelihood of engaging in health risk behaviours where individuals may be exposed to some sort of danger. Using examples that students can identify with enables them to understand some of the factors that impact on decision-making for these choices that are likely to involve a level of uncertainty. This might be choosing to participate in adrenaline sports that have a high risk of injury, having unprotected sex, or not revising for a forthcoming assessment. All these situations carry an element of risk, although the perception of the level of risk is highly subjective and likely to differ between individuals. Asking students to rate the level of uncertainty that each situation carries and presenting these ratings back to them is an effective way to demonstrate this subjectivity. This can be extended by asking them to consider why they might think that and what information might have influenced their decisions (e.g. experience or knowledge of the situation, social norms). The type of thinking that students employed is a useful area to explore with students too, as this can lead to flaws in decision-making when there is no clear answer. Exposing students to different scenarios where they are likely to make biased decisions engages students and provides clear and concrete demonstrations that they can then discuss. This can be achieved by providing students with differing information beforehand:

- Framing—Half the students see information presented in a positive way (80% of attendees decided to engage in the bungee jumping team building event), whereas half the students see it framed in a negative way (20% of attendees decided not to engage in the bungee jumping team building event).
- Stereotypes—Students see a scenario involving a patient presenting with specific signs and symptoms but are given differing stereotypical information about the patient's background or appearance to see how this affects a list of proposed differential diagnoses.

The resultant discussions from these kinds of activities can really help students to use their knowledge of psychological concepts to explore the different types of uncertainty that they are likely to encounter, to raise their awareness of some of the pitfalls they could fall into, and also to consider strategies that they might find helpful.

Thinking styles	
System 1 (intuitive)	System 2 (analytical)
• Automated mode of thinking that requires no conscious effort • Based on experience, gut feelings • Happens in routine or time constrained situations • Prone to error	• Deliberate, rational, logical judgement • Processing of information from past learning and experience • Slower, requires significant cognitive effort • Less prone to error

Fig. 4.3 Illustrating system 1 and system 2 thinking styles (Kahneman 2011). System 1 thinking is a rapid, automatic process that has no conscious deliberation, system 2 thinking is slower with conscious thought, such as weighing up pros and cons

4.2.3 Application of Psychological Theory to Student Learning and Practice

Psychology has made a huge contribution to learning theories. Early behaviourist theories proposed learning happened through associations and reinforcements. Cognitivist and constructivist perspectives considered how our thought processes and strategies seek to find solutions to presented problems and that knowledge is constructed into mental schemas by learners because of these cognitive processes. Our social environment also provides learning opportunities: we can provide support and scaffolding to aid another's understanding, as well as observing and imitating the behaviour of other individuals and groups in our community. Other learning theories highlight the importance of self-learning and personal reflection, which may result in identifying beliefs, assumptions, and attitudes that we hold or may wish to change (Stewart 2012). Many of these theories will feature in different areas of the psychology curriculum for medical students as they are relevant to, for example, how best to structure consultations with patients, understanding patient adherence, understanding child development, and understanding why patients may remember or misremember information. Whilst explaining these concepts, it can also be helpful to make the link to how these theories and models may help students in their own learning practice and collaboration with their peers and future colleagues.

Practical examples of how this might be achieved include when teaching about the psychology of memory, and how to increase the effectiveness of something being 'memorable'. Whilst this might be in the context of giving information to patients, it can also be explicitly linked to students' learning. For instance, advising them to 'chunk' information together to create more meaningful units of knowledge, using different sense modalities (reading, making notes, reflecting, practising skills, giving and receiving feedback) to ensure deeper processing of information that will enhance how it is encoded, forming study groups, and constructing and engaging in 'mock' exams to practise recalling information in a meaningful way. Explaining to students why methods based on active processing of new incoming information, such as relating it to their existing knowledge, linking it to other areas and thinking critically about how it might differ to their previous understanding, will help students see the value of this rather than reverting back to more passive strategies that rely on repetition and memory. By increasing students' understanding of memory, they can see it as part of a wider context as they can link a sound rationale for why they should structure information to enhance patient understanding to their own current studying behaviour.

4.2.4 Authentic Case Studies/ Scenarios

It is crucial that students realise that a person cannot be understood without considering the wider communities and backgrounds that they are embedded in. It is not possible to extract the wider social context from individuals, and students should be exposed to this concept from the beginning of their medical education so they can recognise the complexity of human behaviour. Learning scenarios must reflect this complexity and be authentic and representative of diverse communities and backgrounds. This ensures that the individuals and groups portrayed in the scenarios, including their behaviours, are perceived as realistic rather than contrived. This also enables students to recognise how portrayed individuals represent the real world and to relate to the people that are described in the scenarios. These scenarios then may be perceived as part of the community of practice (Lave and Wenger 1991) in which students can learn as they become situated within the environment depicted in the scenarios. If the scenarios are inauthentic, students may not take them seriously and either fail to understand the behaviours, attitudes, etc., that are being portrayed or consider them too simplistically (MacLeod 2011). Consequently, they may disengage from the intended psychology learning. More importantly, students may feel marginalised or offended if representations of people and/or groups are non-inclusive, stereotypical, or portrayed inaccurately, or they feel that the psychology being taught does not apply to them. As such, care must be taken to ensure that all of the student population are represented in the PBL case scenarios (Brondani et al. 2024).

Well-designed written scenarios are constructed and reviewed by a team of subject specialists from different disciplines. However, this can focus on the subject content and meaning rather than the wider context of how the individuals are presented. The wider context is what creates the authenticity within the scenario and is not necessarily based on the medical condition. Examples of the context might be including details on the beliefs that may emerge from different religions, the values held by different cultural groups, and healthcare concerns experienced by specific LGBTQ+ groups. Often, the authors of the scenarios are not as diverse as the student community so, although the intention is always to make scenarios authentic, there is a danger of them being stereotypical if none of the contributors are from the community being described. Most undergraduate cohorts are very diverse, and it can be helpful to engage students in the development of psychology (and wider) teaching material. This ensures that any individuals or communities represented in the scenarios are introduced in a genuine way with authentic beliefs, values, and behaviours that avoids stereotypes. The lived experience that students can contribute is not necessarily something that the academic/clinical staff would have knowledge of. Inviting students to be collaborators in the curriculum (Harden and Harden 2023) also benefits the students by reinforcing their sense of being valued and respected. It is important to recognise that written scenarios can only represent the views and experiences of those involved in the writing: any lived experience may be specific to those individuals and not necessarily be representative of wider groups, even within the same community. However, students appreciate the limitations of the scenarios whilst recognising the time and effort that has been put into making them as representative as possible, thus it helps students to feel included and valued members of the School.

A similar benefit to the authentic representation described above can be achieved by inviting expert patients to be involved in the preparation of material: having access to the lived experience of individuals with certain conditions, and how they manage these, really brings the scenarios to life. This leads to students being more inclined to treat individuals depicted within the scenarios as 'pseudo patients' who they can try to understand and apply their theoretical knowledge to.

An example of how such authenticity has been successfully achieved is a problem-based learning case on sickle cell anaemia that was co-developed by staff and students (Adams et al. 2023). The students were involved in creating and writing a new PBL case (with associated supporting notes for tutors) and they also designed and delivered an accompanying Q and A session for the students, where they explained their lived experience. The student cohort were able to electronically submit questions to preserve anonymity or ask them directly during the face-to-face session. The student collaborators gained a wider understanding and appreciation of curricular development and were able to present their work at national conferences (Ashaye et al. 2023; Okagbare et al. 2023). The result was that the student cohort were explicitly aware that the teaching material was based on clear applied examples of psychology in the real world. The engagement with the learning was excellent and it received exceptionally positive feedback in the student evaluation. An incidental advantage of this was that students felt empowered to approach faculty staff with their own ideas and constructive feedback on specific teaching materials, as well as identifying gaps and enhancements that could be made in the wider curriculum to ensure its accurate and realistic representation of society.

4.2.5 Experience Sharing and Discussion Opportunities

Medical students generally have the same underlying aim of becoming a good doctor and can have a rather idealised and narrow view of how they think they can achieve this. However, it is important for them to realise that there are a range of different cognitions (attitudes, values, and beliefs) towards health, recovery, and management. Guided discussions with students in small groups can uncover a large diversity of attitudes, beliefs, and values that are important discussion points and can be thought-provoking to other students, enabling them to gain an appreciation of wider perspectives. Experiential learning (Kolb 1984) suggests four key stages to learning: experiencing, reflecting, thinking, and acting (see Fig. 4.4). Small group experiential learning sessions can be very much based on these principles. Facilitators present scenarios and encourage all students in the group to actively participate by contributing their ideas and experiences, to fully understand different viewpoints. This sharing of experience should result in students reflecting on the different ideas that have been voiced and considering how these might complement or challenge their own experiences. This should lead to a more analytical approach where they may question their assumptions and standpoint, thus recognising any need to change their existing schemas of what they might have thought was appropriate or 'correct' in different situations. Ultimately, this will extend their understanding of the particular topic by allowing them to recognise the wider context and culture that contribute to knowledge and thus incorporate these thoughts or behaviours into their future practice and/or communications.

A specific example of a topic that would benefit from having an experiential learning session is complementary and alternative medicine (CAM). Many medical students are initially very anti CAM, making assumptions that the biomedical and, indeed, Western approach to medicine is superior. They may believe that anything that deviates from this will be ineffective or even detrimental to health. Having a group discussion where individuals are first questioned about their existing knowledge and opinions is a useful starting point. Facilitators can then ask students to

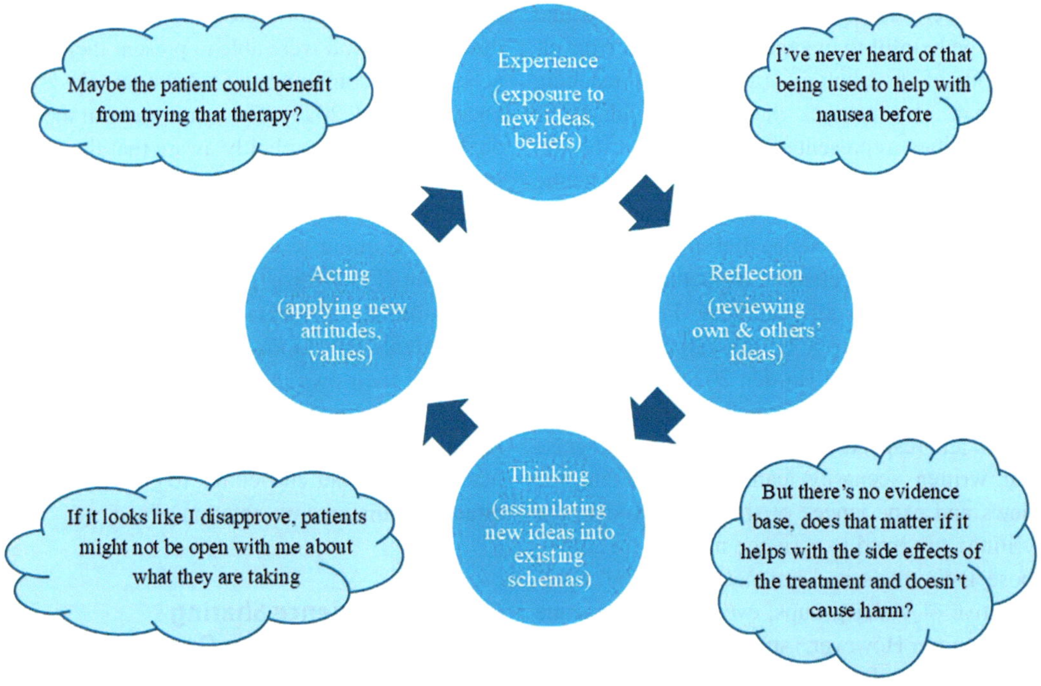

Fig. 4.4 An illustration of Kolb's experiential learning cycle with an example applied to students thinking about complementary medicine

talk about their experiences of CAM, providing prompts about various aspects of this that they might not have previously considered as being CAM (e.g. scents to help them sleep and over the counter remedies that help with stress). Often students recognise that their family members have made use of CAM and discussions can start to get very animated as they may recognise that they use it without having explicitly realised or having planned to. This can result in transformative learning moments (Mezirow 1997) as students realise their attitudes and beliefs have been challenged. At this point, facilitators may prompt students to reflect more deeply on this and start to recognise the importance of cultural beliefs and different practices in providing health care, be these traditional medicines or differing attitudes towards what makes an effective therapeutic relationship. It is useful at this point to provide students with different scenarios that they can start to apply their new thinking to, and consider how they as a future doctor might approach questioning patients about CAM and, indeed, how they would respond and advise them (see Fig. 4.3). The goal must always be that they demonstrate respect and are non-judgemental. Any perceived or encountered negative reactions by doctors are likely to impact on how open and trusting patients are and could lead to them not disclosing potentially incompatible treatments; students generally demonstrate a good grasp of this when applying their 'new' knowledge to the scenarios provided in sessions.

Experiential sessions can be more effective and feel safer for students to contribute to if they are in an established group and are familiar with others. Inevitably, when students share experiences, they will not always be positive and at times may be extremely personal, which can lead to upset. As such, it is important that educators ensure a safe space that supports students so that they feel comfortable and can trust that information they share will be respected (Kisfalvi and Oliver 2015). At the beginning of sessions, it is useful to remind students of the expectations to ensure an effective learning session. Whilst it is acceptable for students to be curious about the values and beliefs other people may hold, any questions and comments they make to get a better understanding should always be non-judgemental and respectful. Furthermore, contributions of experiences to the group should always be kept confidential within that session and not shared more widely. Educators should remain alert to any feelings of discomfort or upset and act quickly to address these. Examples of action include directly challenging the source of any comments or questions that may be perceived as upsetting. This should be done sensitively, ensuring the person making the comment does not feel targeted but perhaps questioning why they think that and what potential issues might those views/comments have on patients. Students who may become upset should be allowed time out and an opportunity to debrief with the educator afterwards, with any signposting to appropriate support provided.

4.2.6 Explicit Linking

The benefits of a spiral curriculum were first proposed by Bruner (1960). This is widely adopted in medical education and is an effective way of introducing students to topics so that they can grasp the foundational knowledge and then build on this in their subsequent learning to extend their thinking and understanding. Often, students are taught about the 'normal' working functions and structures first, then revisit this to consider how these might be affected by different pathologies; the next encounter with these might be observing these in the clinical environment and then, finally, they will be applying their knowledge in their clinical practice (Harden and Stamper 1999). Relating this to psychology, the first encounter might involve presenting concepts, models, or theories to students and explaining these with clear examples. Initially students may be tempted to use rote learning to try and memorise the information by frequent revisiting of the same material. This item specific processing (Hunt and Einstein 1981) approach may improve students' memory of the concept by making it clearer and more focused, but it does not necessarily result in their full understanding (see Fig. 4.5a or see Supplementary Video 4.1). It is therefore important that subsequent encounters build on this to encourage more relational processing. For example, describing an individual or situation where this is relevant and asking the students to apply the concept, model, or theory that they have previously been taught. They can then follow this up by observing examples of these behaviours in patients that they might see on placement in the clinical environment. The final stage would be their ability to use this knowledge and understanding to guide their interactions with patients, for example the type and level of information they provide to patients who are pre-

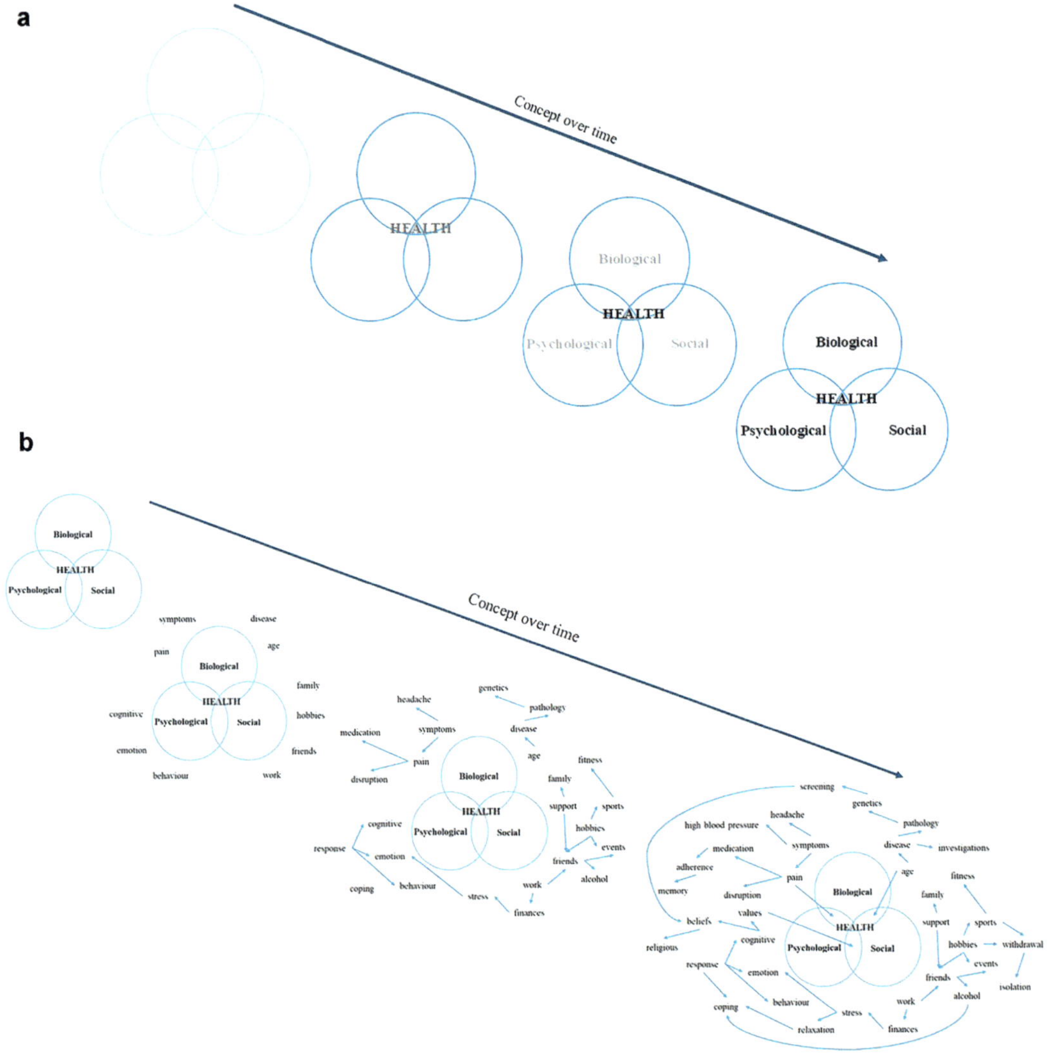

Fig. 4.5 (**a**, **b**) Highlights the difference between item specific and relational processing. (**a**) Replicates repetition, whereby the specific item is looked at repeatedly and builds up a clear and focused 'image' of the specific concept that is encoded in memory. (**b**) Replicates relational processing, whereby the information on the specific concept is processed and linked to other relevant concepts or ideas. These links enable consolidation of knowledge and are increasingly synthesised with other information. Consequently, it provides a wider context that allows greater understanding and ability to apply knowledge. Models are based on the biopsychosocial model (Engel 1977). See also Supplementary Videos 4.1 and 4.2

paring for a particular investigation or assisting patients to explore different coping strategies they may find helpful. This revisiting results in students having a deeper and progressive understanding of concepts which they should be increasingly able to relate to other areas and apply to both similar and different scenarios (see Fig. 4.5b or Supplementary Video 4.2). To achieve this building of knowledge, it is important that the principles of delivering a spiral curriculum are followed; in particular, subsequent teaching should become increasingly more detailed and/or complex rather than just repetition. Repetition frequently leads to students attempting to memorise facts, which is not an effective learning style; whilst repetition may make the particular concept itself clearer, it does not enable deep understanding, linking, synthesis, and application of knowledge.

There are several practical ways to support students to make links; one example is to include explicit links in teaching materials about where they have encountered relevant information/concepts previously and where they can expect to cover it again in the future. These links do not necessarily need to be specific to just psychological topics and concepts (vertical linking) and can also relate to other subjects or disciplines (horizontal linking). (Cross-disciplinary links will be covered more in the next section). Encouraging students to make links aids their analytical thinking, as they have to consider how and why things relate and what this adds to their current understanding.

Linking lectures explicitly to PBL cases also supports effective learning for students. When teaching students about patients' coping styles, cues should be included in the PBL case scenarios that include information allowing students to apply relevant models that are discussed in the lectures. This may include information on a patient's behavioural, emotional, and cognitive responses as well as situational factors such as the cause of the illness/injury, the healthcare provided, and the patient's family/social situation. This encourages students to think more widely about how patients might perceive and respond to different situations based on the variable factors. For example, if an individual with little prior experience of injury or illness has an accident resulting in an acute injury, they may perceive it as more of a threat than someone who might have been injured previously and made a good recovery. Their perceptions of the injury may also be affected by the reactions of onlookers, attending medics, or family members. All these factors are linked to how the individual represents and responds to it in terms of the coping styles that they may adopt. Including cues to these personal and situational factors supports students to recognise there is not likely to be a single and easily predictable coping strategy that all patients should be recommended to adopt.

4.2.7 Integrating Psychology with Other Subject Areas

Medicine is a complex subject that requires knowledge and understanding of a wide range of subjects and disciplines including psychology, anatomy, physiology, and sociology. Ensuring students adopt an interdisciplinary approach is essential for them to produce the synthesis of knowledge that they will need. The majority of UK medical schools use PBL in delivering their curriculum, and internationally over 500 academic institutions have adopted PBL or variations of it (Servant-Miklos et al. 2019). This is often delivered in a hybrid way, so whilst PBL cases are integrative and include a variety of subject knowledge that students must piece together—thus encouraging them to take the multi-disciplinary approach—the supporting sessions, e.g. lectures and lab sessions, are usually subject specific and not integrative. This conflict can send students an incorrect message, so that they learn the subjects in isolation rather than considering how they link together. An illustrative analogy of this is they may learn what the individual jigsaw pieces look like and what the information on them represents, but they don't put the pieces together to see how they interlink to provide the overall picture of the patient and the context within which they live (see Fig. 4.6 or Supplementary Video 4.3). Thus, all the psychol-

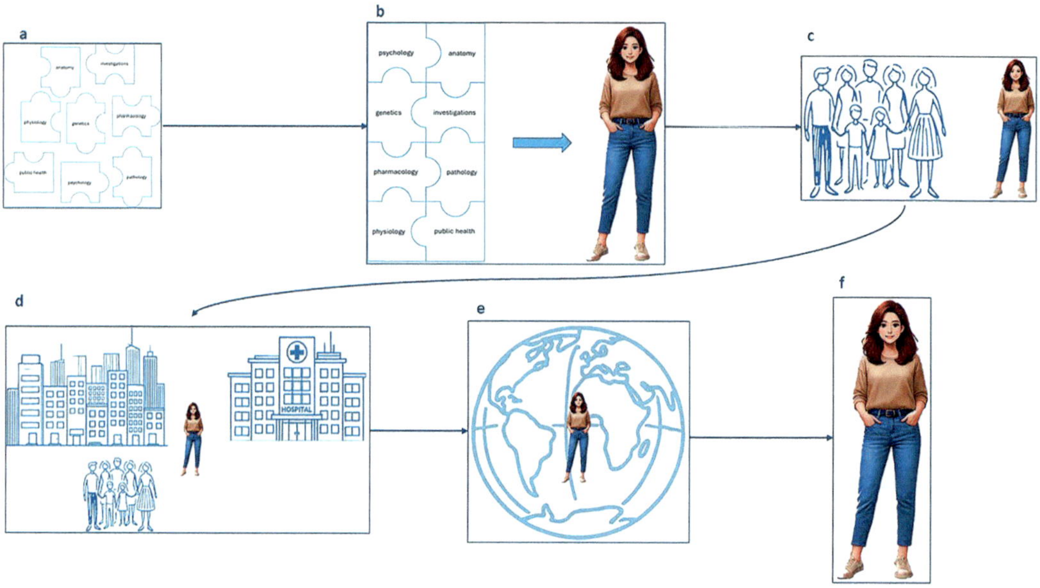

Fig. 4.6 Demonstrates the importance of taking a patient-centred approach. The different subjects should not just be learned in isolation, but pieced together to form the links and interactions that make up the person. The patient is always part of a wider context that includes the family, and local and global society, yet they always remain at the centre. Encouraging this approach ensures students learn the whole story—molecular, cellular, organic, systemic, person, family, society, population, and global. Images created by author using OpenAI ChatGPT (2023). See also Supplementary Video 4.3

ogy jigsaw pieces become standalone with no perspective of how they fit with the wider context.

One strategy that is useful in overcoming this disconnect is team or co-teaching, where colleagues from different disciplines share the same physical space to deliver teaching. This is something that has been practised with basic and clinical science practitioners (e.g. Willey et al. 2018) but can be equally valuable between different basic science disciplines. Having two (or more) teachers with different subject expertise jointly present a session provides a clearly integrated learning environment which models to students how the subjects relate to each other. Students are then positioned to more easily link and synthesise the different pieces of information together, so they form into meaningful chunks rather than separate elements (Dehnad et al. 2021). Some topics are clearly multi-factorial and would particularly benefit from a co-teaching approach by bio- and behavioural science staff, so students can understand the complex interplay of factors.

A specific example of how co-teaching has been effective in the undergraduate curriculum is a lecture on ingestive behaviour and development of food preferences delivered by a psychologist and a physiologist (Adams and Harper 2021). Trying to understand eating behaviour from a single perspective would be far too reductionist, as the biochemical and psychological factors that impact on eating are very much interdependent and impossible to disentangle. One issue that was discussed during the session was comfort eating behaviour. This was considered as emerging from an individual's beliefs, values, and the symbolic significance they placed on different food types, alongside biological considerations of neurotransmitters and the reward pathway. It is essential that students are exposed to the complex interplay of these different perspectives, so they understand the need to look at and integrate the wider context.

Whilst co-teaching psychology with other bioscience subjects can be extremely effective for student learning (Dehnad et al. 2021), it is important that it is done carefully to avoid distracting and confusing students. Session planning is likely to be time consuming and it is essential that co-teachers work together to develop a session that is coordinated and clearly organised. Both educators must have an appreciation and respect of the other's subject and ensure they present balanced rather than competitive viewpoints. Working collaboratively in this way supports students in building up and assimilating the information into something that is meaningful and makes sense to them. If careful collaboration and coordination takes place, co-teaching has clear advantages for helping students understand the relationship between behavioural and biosciences compared to more conventional teaching approaches.

4.2.8 Inclusion of Practical Activities

Students' perception of psychology can be that it is a theoretical subject with no practical activities that they can engage with, unlike some of the bioscience subjects where they may be associated with laboratory practical classes. This can make psychology appear a 'dry' subject that students struggle to engage in. Students can also think it is a straightforward, 'non-scientific' (Gallagher et al. 2015) subject that they don't need to allocate additional independent reading or research to. This can result in more 'surface level' learning which does not utilise the higher cognitive skills required for consolidation, elaboration, and synthesis.

There are numerous ways that psychology can be brought to life for students by including practical activities in lectures and workshops that students can engage with. These don't just help engage students: interactions with, and results from, the activities can provide useful learning resources in themselves and provide valuable learning points. Asking students to take part in simple memory recall experiments can provide data that can be useful to evidence some of the concepts taught in more traditional lectures. Recall experiments work by providing a verbal list of words to students who are subsequently asked to recall them. The results from this can provide clear demonstrations of key issues to students, for example:

- *Primacy-recency effect* whereby items presented at the beginning and end of lists are likely to be more memorable based on the way they are encoded. This is proposed due to the initial words presented in the list having time to be 'rehearsed' and processed, and the later words in the list still being available in short-term memory (Atkinson and Shiffrin 1968).
- *'Concreteness' of the information* to be remembered, abstract words can be harder to encode due to more limited semantic knowledge that can contribute to the encoding process and subsequently recall (Walker and Hulme 1999).
- *Relatedness of words*, if they are connected to a certain theme or category, then chunking can occur, or mental schemas activated that increases the number of items that is recalled (Poirier and Saint-Aubin 1995).
- False memories whereby information that was not presented is incorrectly recalled, this is often as a result of words in the list being related to a particular topic or category as described in the previous point (Roediger and McDermott 1995).

Participating in a word recall experiment enables students to have a direct experience and better appreciation for how we encode and recall information as well as some of the factors that impact on this. Understanding this then gives them a framework that they can apply to other areas, for example their personal learning and revision strategies, as well as understanding effective structures for giving information to patients during consultations.

Understanding biases is a crucial topic for medical students: it is vital that they understand why these may arise and how they could affect the level of care that they provide for their future patients. This can be included in the curriculum by having active sessions where students can engage in implicit association tests (IAT) such as those developed and hosted by Project Implicit

(2011). Taking part in IAT enables students to discover their own potential for bias in a range of different ways. We ran a session where students were asked to complete the Weight IAT (Project Implicit 2011) during a workshop and then discuss their thoughts around the findings, the test itself, and whether having awareness of potential biases could lead to them being overridden (Adams et al. 2020). One of the key intended learning points from the session was to prompt students to reflect on the personal unconscious biases they might hold. This was not done in a negative way as to be something they needed to feel worried about, but instead to consider why we all have biases and how these are likely to be a result of our previous experiences. The discussions that were generated covered a range of topics that related explicitly to unconscious bias and how it can lead to stigma and negatively affect health care (Tomiyama et al. 2018). Other valuable discussions also emerged, for example about experimental research design and methodology. Students were then asked to complete a critical analysis of a range of personal social media accounts (either of a celebrity or their own) to consider how biases might emerge and how pervasive these might be in society. As part of this, they considered the context of posts and images, the types of comments that were made, the likely demographics of the audience, and the impact of algorithms in tailoring individuals' news feeds. Taking this active approach 'forced' students to take an analytical approach to the task, not just about the specific subject of weight bias but also the role that the wider context has in shaping this. A final task required students to apply their knowledge and understanding to healthcare by considering the different ways in which unconscious bias might impact on patients accessing healthcare, as well as their future practice and communications with patients. To demonstrate their discussions, students created a poster aimed at a specific target audience (of their choice), which described focused strategies on improving body image and/or weight stigma to try and counteract the potential negative impacts they had identified. Posters were submitted to the session lead who analysed the key themes from the strategies and presented these back to students in a subsequent 'wrap up' session. Identified key themes revolved around 'keeping it real' campaigns in social media, a focus on abilities rather than appearance, and providing non-judgemental and supportive environments.

4.3 Conclusion

This chapter has considered some of the key barriers that medical students may experience in engaging with psychology; notably, this is likely to be principally due to a previous lack of exposure to the subject. The interdisciplinary nature of psychology means that it is well placed to position itself with other subjects that students may feel more familiar with, and this is certainly a benefit that can be exploited. The specific practical strategies and examples that have been discussed here include providing students with a clear understanding of psychology and how it can be applied, using it to introduce the concept of uncertainty which is prevalent in medicine, applying psychological theory to their learning and practice, providing authentic and relatable case studies and scenarios, enabling experience sharing and discussion opportunities, making explicit links for students to build on their existing knowledge and assimilate new information, integrating psychology with other subject areas, and including practical activities they can engage with to demonstrate the practical and applied nature of psychology. Although the specific examples provided here are aimed at psychology, they could be equally applied to how other subjects, such as sociology, public health or law and ethics could be made more accessible to students.

Acknowledgments The author would like to thank colleagues at Keele Medical School who embrace a collaborative and supportive teaching environment; this enables successful delivery of accessible and engaging psychology content. Dr. Natalie Cope and Dr. Gordon Dent provided valuable suggestions on an earlier draft of this copy, but particular thanks go to Dr. Penny List who provided continuous encouragement, recommendations, and improvements throughout the writing of this chapter.

References

Adams K, Harper A (2021, September 21) Embedding psychology and physiology in the undergraduate medical curriculum: Lecture co-presentation and integrated learning content. [Presentation]. Behavioural & Social Sciences Teaching (BeSST) in Medicine annual conference, online

Adams K, Harper A, List P, Whiteman S (2020, September 7–9) Uncovering medical student biases and challenging attitudes towards body image. In: AMEE The virtual conference

Adams KM, Ashaye A, Okagbare F Skidmore A (2023, August 27) Staff and students co-producing multidisciplinary teaching resources with an emphasis on BSS [Presentation]. Behavioural & Social Sciences Teaching (BeSST) in Medicine annual conference, University of Edinburgh

AQA (2024a) AS and A-level biology specification. https://filestore.aqa.org.uk/resources/biology/specifications/AQA-7401-7402-SP-2015.PDF. Accessed 24 July 2024

AQA (2024b) AS and A-level psychology specification. https://filestore.aqa.org.uk/resources/psychology/specifications/AQA-7181-7182-SP-2015.PDF. Accessed 24 July 2024

Armitage JS, Nash SP, Hanson JT (2024) It's "not black and white": students' perceptions of "altruism" content in preclinical medical education. Med Teach 46(1):126–131

Ashaye A, Okagbare F, Adams K, Skidmore A (2023, March 11) Decolonising the curriculum with an emphasis on sickle cell awareness [Poster]. Midlands Racial Equality in Medicine (MREM) annual conference, University of Leicester

Atkinson RC, Shiffrin RM (1968) Human memory: a proposed system and its control processes. In: Spence KW, Spence JT (eds) The psychology of learning and motivation, vol 2. Academic Press, New York

BeSST (2010) A core curriculum for psychology in undergraduate medical education. https://www.besst.info/publications. Accessed 5 July 2024

British Psychological Society (2024) What is psychology? https://www.bps.org.uk/what-psychology. Accessed 6 July 2024

Brondani M, Barlow G, Liu S, Kalsi P, Koonar A, Chen J, Murphy P, Broadbent J, Brondani B (2024) Problem-based learning curriculum disconnect on diversity, equitable representation, and inclusion. PLoS One 19(6):e0298843

Bruner JS (1960) The process of education. Harvard University Press, Cambridge, MA

de Visser R (2009) Psychology in medical curricula: "need to know" or "nice to know"? Eur Health Psychol 11:20–23

Dehnad A, Jalali M, Shahabi S, Mojgani P, Bigdeli S (2021) Students' view on supportive co-teaching in medical sciences: a systematic review. BMC Educ 21(522)

Dikomitis L, Wenning B, Ghobrial A, Adams KM (2022) Embedding behavioral and social sciences across the medical curriculum: (auto) ethnographic insights from medical schools in the United Kingdom

Engel G (1977) The need for a new medical model: a challenge for biomedicine. Science 196(4286):129–136

Gallagher S, Wallace S, Nathan Y, McGrath D (2015) 'Soft and fluffy': medical students' attitudes towards psychology in medical education. J Health Psychol 20(1):91–101

General Medical Council (2018) Outcomes for graduates. https://www.gmc-uk.org/education/standards-guidance-and-curricula/standards-and-outcomes/outcomes-for-graduates/outcomes-for-graduates. Accessed 5 July 2024

Harden J, Harden RM (2023) The changing role of medical students. Elsevier Health Sciences

Harden RM, Stamper N (1999) What is a spiral curriculum? Med Teach 21(2):141–143

Hunt RR, Einstein GO (1981) Relational and item-specific information in memory. J Verbal Learn Verbal Behav 20(5):497–514

Kahneman D (2011) Thinking, fast and slow. Farrar, Straus & Giroux, New York

Kisfalvi V, Oliver D (2015) Creating and maintaining a safe space in experiential learning. J Manag Educ 39:1–28

Kolb DA (1984) Experiential learning: experience as the source of learning and development. Prentice-Hall, Englewood Cliffs, NJ

Lave J, Wenger E (1991) Situated learning: legitimate peripheral participation. Cambridge University Press, New York

Lee C, Hall K, Anakin M, Pinnock R (2020) Towards a new understanding of uncertainty in medical education. J Eval Clin Pract 27:1194–1204

MacLeod A (2011) Six ways problem-based learning cases can sabotage patient-centred medical education. Acad Med 86(7):818–825

Medical Schools Council (2024) Entry requirements for UK medical schools 2024 entry. https://www.medschools.ac.uk/media/3060/uk-medical-school-entry-requirements-2024-entry.pdf. Accessed 5 July 2024

Mezirow J (1997) Transformative learning: theory to practice. In: Cranton P (ed) New directions for adult and continuing education, vol 74. Jossey-Bass, San Francisco, pp 5–12

Okagbare F, Ashaye A, Adams K, Skidmore A, Harper A (2023, March 11) Decolonizing the medical curriculum - importance of student input in improving teaching materials [Workshop]. Midlands Racial Equality in Medicine (MREM) annual conference, University of Leicester

OpenAI (2023) ChatGPT (Oct 31 version) [image generator]. https://chat.openai.com/

Piaget J, Cook MT (1952) The origins of intelligence in children. International University Press, New York
Poirier M, Saint-Aubin J (1995) Memory for related and unrelated words: further evidence on the influence of semantic factors in immediate serial recall. Q J Exp Psychol A 48(2):384–404
Project Implicit (2011) Project implicit. https://implicit.harvard.edu/implicit/selectatest.html. Accessed 12 July 2024
Robiner WN, Hong BA, Ward W (2021) Psychologists' contributions to medical education and interprofessional education in medical schools. J Clin Psychol Med Settings 28:666–678
Roediger HL III, McDermott KB (1995) Creating false memories: remembering words not presented in lists. J Exp Psychol Learn Mem Cogn 21:803–814
Servant-Miklos VFC, Norman GR, Schmidt HG (2019) A short intellectual history of problem-based learning. In: Moallem M, Hung W, Dabbagh N (eds) The Wiley handbook of problem-based learning, vol 1. Wiley. ISBN:9781119173243, pp 3–24. https://doi.org/10.1002/9781119173243
Stewart M (2012) Understanding learning: theories and critique. In: Hunt L, Chalmers D (eds) University teaching in focus: a learning-centred approach, vol 1. Routledge, London, pp 3–20
Tomiyama AJ, Carr D, Granberg EM, Major B, Robinson E, Sutin AR, Brewis A (2018) How and why weight stigma drives the obesity 'epidemic' and harms health. BMC Med 16(123)
Walker I, Hulme C (1999) Concrete words are easier to recall than abstract words: evidence for a semantic contribution to short-term serial recall. J Exp Psychol Learn Mem Cogn 25(5):1256–1271
Willey JM, Lim YS, Kwiatkowski T (2018) Modeling integration: co-teaching basic and clinical sciences medicine in the classroom. Adv Med Educ Pract 9:739–751
Witt EE, Onorato SE, Schwartzstein RM (2022) Medical students and the drive for a single right answer: teaching complexity and uncertainty. ATS Scholar 3(1):27–37

Part II

Inclusive Practice

The Evolution, Implementation, and Impact of Representative Research Practices in Anatomy

5

Siobhan M. Moyes, Katie M. Snow, Sara Sulaiman, Lily A. Evans, and Scott Abbott-Paterson

Abstract

While anatomists, globally, are working to improve inclusive practice, the unique challenges faced by this discipline can form a barrier to sustained progress. This chapter explores the critical importance of representative research practices in anatomy, discussing its principles, challenges, and implementation strategies. Examining the historical context of exclusion and bias in anatomical studies highlights how past practices have shaped current inequalities, contextualising the current move towards representative research.

Key topics include the evolution of inclusive practices in anatomy curricula, barriers to implementing representative research (such as institutional constraints and implicit biases), and practical guidance on conducting representative research, including participant recruitment and data analysis. A case study on demographic data collection emphasises the complexities of identity categorisation.

The chapter presents frameworks for evaluating the inclusivity of research practices and offers concrete suggestions for incorporating representative approaches into various teaching modalities. It explores the integration of representative research principles into anatomy education through the research-teaching nexus, mapping this onto established educational models like Bloom's Taxonomy and Miller's Pyramid.

Representative research is not only an ethical imperative but also crucial for advancing anatomical knowledge and improving healthcare outcomes for diverse populations. By applying principles of representative research, anatomists can contribute to a more inclusive and equitable understanding of human anatomy. The authors hope that by providing both

S. M. Moyes (✉) · L. A. Evans
Peninsula Medical School, University of Plymouth, Plymouth, UK
e-mail: Siobhan.moyes@plymouth.ac.uk; lily.evans@plymouth.ac.uk

K. M. Snow
Peninsula Medical School, University of Plymouth, Plymouth, UK

School of English, Drama and Film, University College Dublin, Dublin, Ireland
e-mail: katie.snow@ucd.ie

S. Sulaiman
Royal College of Surgeons in Ireland Medical University of Bahrain, School of Medicine, Busaiteen, Kingdom of Bahrain

School of Anatomy, University of Bristol, Bristol, UK
e-mail: ssulaiman@rcsi.com

S. Abbott-Paterson
School of Anatomy, University of Bristol, Bristol, UK
e-mail: scott.abbottpaterson@bristol.ac.uk

R. Door (ed.), *Accessibility, Inclusivity and Diversity in Education and Beyond*, Biomedical Visualization 8, https://doi.org/10.1007/978-3-031-81795-3_5

theoretical context and practical advice, this chapter serves as a practical guide for anatomists seeking to adopt more inclusive and equitable practices in their work.

Keywords

Representative research · Inclusive anatomy · Demographic data · Research-teaching nexus · Intersectionality · Equity in education

5.1 Reflexivity Statement

Reflecting on our positionalities, we acknowledge the profound influence our diverse backgrounds have on our research approach. As a team of researchers with varied identities—including queer/gay, women, Arab, and white individuals from different cultural and educational contexts—we recognise how our unique perspectives shape our work. Our collective experiences span multiple disciplines, countries, and educational systems, enriching our understanding while also presenting potential biases. We are acutely aware of our privileges, including able-bodiedness, neurotypicality, and access to higher education, and how these impact our interactions with the world and our scholarly pursuits.

We commit to ongoing critical self-reflection, acknowledging the limitations of our individual experiences and the potential influence of Western and colonial paradigms on our research. By explicitly stating our positionalities, we invite readers to engage critically with our findings, recognising that this work is part of a broader dialogue aimed at improving inclusivity in research and education. We strive for methodological rigour while remaining open to diverse voices and alternative interpretations, aiming to conduct research that is not only academically sound but also socially responsible and culturally sensitive.

5.2 Representation in Anatomy Curricula and Research: From Then to Now

5.2.1 Why Is Inclusive Anatomy Important? A Historical Narrative

From its genesis in the ancient worlds of Egypt and Greece through to its modern inception in the dissection halls of the eighteenth century, anatomy has been intrinsically shaped by society and culture. As an object of both scientific and social study, the body throws tensions between empirical, experiential, and ideological ways of being into sharp relief. Collective engagement with global histories and legacies of imperialism has increased awareness of Western bias in medicine, and the perhaps more subtle processes of colonisation and subjugation that persist today, for example through the exporting of knowledge developed in western contexts as universal or 'best practice'. Different branches of medical science, including anatomy, have their own specific histories as mechanisms of colonialism and oppression, or what Finn, Danquah, and Matthan call the 'messy, yet unchangeable, past' of anatomy (2022). However, these messy pasts scaffold modern anatomical education and practice, structuring how we work today.

At the time of writing in 2024, the language of inclusion and anti-exclusion has become a familiar—and at times superficial—phenomenon in education, research, and clinical practice. Attempts to engage with equity and diversity and to 'decolonise' medical curricula and research, itself a contested terminology, have taken place alongside calls to queer and crip—that is, to work against ableist frameworks and recognise more accessible and just alternatives—lessons and reading lists, to actively counteract the systemic marginalisation of women's contributions to science, and to read, cite, and teach scholars from the global south (Morley 2022).

While these calls to confront how medicine has been inflected by—and has also helped to sustain—sexist, ableist, heteronormative, and colonist cultures and practices are not new, their urgency has been renewed in the wake of the police murder of George Floyd in 2020 (Wong et al. 2021). The resulting #BlackLivesMatter protests increased attention to the violent inequalities at play across society, with individuals, communities, organisations, and institutions at least more nominally alert to their responsibility to engage with problems of social justice. This movement coincided with incontrovertible evidence that racialised people were dying of Covid-19 in disproportionate numbers, demonstrating how structural racism had created a burden of poorer health which made them more susceptible to the virus (Cooper et al. 2023; Thomas 2023). Having historically prioritised research and resources for problems more likely to affect white bodies, medical systems were directly implicated in this society-wide ethical failure. At no point in the history of anatomy, then, has anti-exclusionary education and research been more important; a reckoning with that history is a good place to start.

5.2.2 Histories of Exclusion in Anatomy Research and Education

Anatomical research and education have a troubling history of underrepresentation, exclusion, and violence against minority populations. To a certain extent, anatomy has always favoured some bodies over others. Early anatomists mobilised biological theories relating to race, gender, sexuality, class, and disability to regulate and pathologise bodies deemed 'other' to social and medical norms. In Europe, pre-existing social divisions were intensified by practices such as phrenology, which saw anatomical attributes linked to negative character traits, inferior mental faculties, and imagined aspects of class, race, and gender. More broadly, the biological sciences have played—and continue to play—a significant role in processes of social and medical exclusion. Classification systems based upon difference—and ideas about inferiority and superiority—have amplified issues of inequality, engendering colonialist and ableist structures. Research has established that the history of anatomy is entangled with histories of crime, eugenics, slavery, race, class, sexuality, gender, and disability (Comer 2022). Embedding these histories into anatomy can allow anatomists and the institutions they work within to acknowledge their complicity in processes of exclusion, as well as encouraging reflection on the political contexts of medicine more widely.

Whilst modern medicine might seem dislocated from practices such as eugenics, these histories continue to shape conversations around controversial subjects such as genetic screening and engineering. Ongoing debates about rapidly evolving scientific phenomena including the so-called designer babies raise important questions about the ethics of medical progress predicated on systems of worth. Inequity based injustices in clinical research also persist; whilst there are increasing efforts to represent diversity and mitigate biases in the development of medical treatments and pharmaceuticals, some demographics, for example women of colour, are still inadequately included or misrepresented in trials (Bierer et al. 2022). The messy histories and ongoing implications of such failings have been well documented. Women's exclusion from clinical research, for example, has been framed by discussions about the rationales behind the persistent lower priority given to women's and trans' health needs, the medical industry's cautious approach to including women of childbearing potential in early-stage drug trials, and questions about how to account for gender-based hormonal variabilities (Ravindran et al. 2020). Highlighting the contentious clinical failings underpinned by racism, Kouri et al. (2021) show how early research into pulmonary capacity was damaged by social constructs around racial difference in the 1930s, furthering racial bias and ultimately lending 'credibility to scientific racism'. Of the streams of similar studies that followed across the century, Kouri et al. note, the vast majority neglected to account for factors such as socioeconomic status and environmental influences and failed to even set out their understandings of race and ethnicity.

Historically, anatomy education has been more accessible to a particular body of students, teachers, and researchers, supporting the progress of the privileged whilst marginalising those with protected characteristics. In anatomy education today, the legacies of these exclusionary histories are painfully visible. Despite progress made globally, textbooks still teem with images of white, athletic, young, able-bodied males, with healthy variation underrepresented, anatomical models are overwhelmingly white, body donors do not fully represent the range of diversity within the community, eponyms are resistant to change, anatomical collections remain divorced from their colonial contexts, teaching materials do not always support neurodiverse learners and educators, and many anatomy labs remain inaccessible for those with physical disabilities.

Whilst the above injustices can never be fully repaired, or their impacts fully mitigated, growing awareness about the importance of diversity in research and education is motivating meaningful changes. Recognising this, researchers, reviewers, funders, sponsors, ethics committees, and recruitment teams behind clinical trials are developing strategies to improve demographic representation and subsequently make findings and scientific developments more applicable to diverse populations. The NHS-Galleri trial, for example, which aims to transform early cancer detection via blood testing, has been praised for its equitable recruitment of participants and its ongoing monitoring of the representativeness of its enrolled participants, a process which includes 'the dynamic adjustment' of invitation lists and the tactics used to target potential groups for inclusion (Swanton et al. 2022).

Of course, unethical, unrepresentative, and historically violent medical practices have painful and lasting legacies, and current anatomists must work imaginatively and flexibly to build trust amongst minoritised communities, and to overcome the complex cultural and socioeconomic barriers which can hinder diverse participation in clinical trials and research. If we are to address complex global health needs in the present, then we must learn from past mistakes; particularly concerning the systemic issues that underlie problems of exclusion and mistrust in medical institutions. Looking to history not only helps us to wrangle with the complexities of teaching and researching anatomy in the present but opens new horizons for a more inclusive future.

5.3 Increasing Representation Through Research

5.3.1 What Is Inclusive or Representative Research?

As the previous section has highlighted, not all 'high quality' research is representative or inclusive of the range of diversity present. For example, while observations during clinical trials and focus groups are robust research methods, data cannot necessarily be extrapolated to unrepresented populations. Then, even if a research project recruits a diverse group of participants, if the researchers themselves are not diverse, will their research questions be relevant to the participants and the findings meaningful or applicable to these groups of people?

With the best will in the world, a researcher's lived experience affects their priorities of what should be studied and how. These decisions can then impact the validity of the findings in a myriad of unexpected ways. While this could sound insurmountable, collaboration is the key to ensuring that research represents diversity.

At this point it can be helpful to define terminology. While 'inclusive research' is a catchy title, it is already applied to a specific research methodology. The term 'inclusive research' was coined by Walmsley (2001) when describing their approach to research with, and not on, people with learning disabilities, to make it more accessible. This approach has since been expanded to research with people with other disabilities, and beyond the limit of disability. However, in order to minimise confusion in a field that already contains complex, overlapping, and extensive terminologies, the term 'representative research' has been used here to describe a variety of research approaches that can increase representation of diversity.

A representative research approach aims to involve diverse participants, perspectives, experiences, and voices, particularly those of marginalised or underrepresented groups, at all stages of the research process. Representative research acknowledges and works to address existing power dynamics, biases, and privileges that may influence the research process, ultimately striving to produce more relevant, ethical, and impactful outcomes.

There is a plethora of established research paradigms (frameworks) that have inclusivity at their core, and more still that require a more proactive approach to representation. I (Siobhan Moyes) confess to being ignorant of many of these paradigms when starting my own research, but I have since found many of them to be of great inspiration. Defining your research paradigm also helps others understand your ontology, epistemology, methodology, and methods. If these terms are foreign to you, like they were to me, this section will provide an overview of some of the key paradigms that can be applied to anatomical research. This is by no means exhaustive, so for further detail, please see the following resources (Clark 2021; Hesse-Biber 2010).

5.3.2 Defining Research Paradigms and Theories

A research paradigm (or framework) provides the theoretical structure that will influence every stage of your rigorous research. It will influence how you ask your questions, the people you involve, and the methods you use. Therefore, before delving into individual paradigms, it is essential to understand their underlying theories of ontology, epistemology, and methodology (as illustrated by Fig. 5.1).

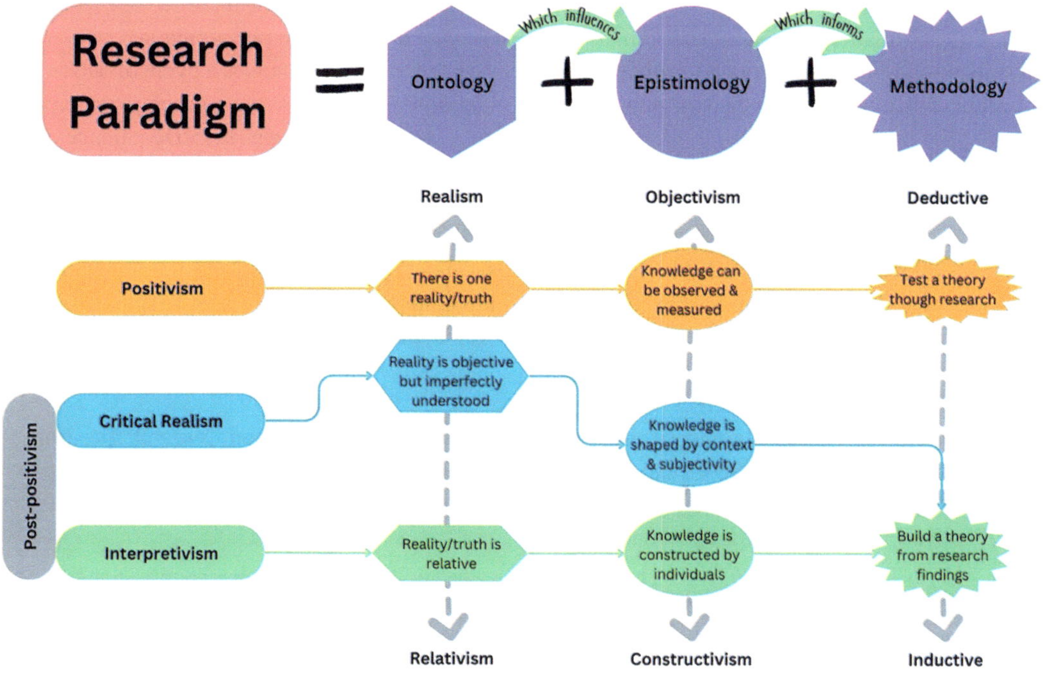

Fig. 5.1 An overview of the interrelationships between ontology, epistemology, and methodology, and how they inform research paradigms. Three examples of research paradigms have been used to highlight their different underlying theories

5.3.2.1 Ontology

Ontology (the study of being) is your belief about the nature of what you are studying, your 'reality', and there is no right or wrong answer. Do you feel that your research findings will be generalisable to the masses (one true reality = *realism*)? Or do you feel that your findings are contextual and may differ between groups (reality is relative = *relativism*)?

You are not tied to one ontology throughout your research career, as different ontologies can suit different research questions. Your ontology will also influence the epistemology you use.

5.3.2.2 Epistemology

Epistemology (the study of knowledge) is your opinion of how 'observable' the knowledge is. Again, there is no right or wrong approach, they all have their benefits and difficulties. Do you feel that, in your research, you are able to take objective measurements without influencing them (*objectivism*)? Do you feel that by observing and measuring, you are influencing the findings (*constructivism*)? Or are you somewhere in between, insofar as you would like to be objective, but pragmatically understand this is not always possible and will work to mitigate these influences? The epistemology you choose will inform your methodology.

5.3.2.3 Methodology

Methodology is not the methods used, but your approach to generating data and theories which, in turn, defines the methods used. Firstly, do you aim to start with a hypothesis, collect large amounts of data, and analyse it statistically, to decide if your hypothesis is correct (*deductive*)? Or do you start with the data and let them lead you to a theory (*inductive*). Will the data you collect be numerical and analysed with statistics (*quantitative*), will it be the thoughts and experiences of your participants (*qualitative*), or will it be a combination of both (*mixed methods*).

You typically choose your methodology based on your research questions, ontology and epistemology, and the nature of the data being investigated. Once you have decided whether you are following a deductive or inductive approach, and the type of data you are interested in collecting, this is where methodology differs to ontology and epistemology. Ontologies and epistemologies exist, more or less, on a spectrum (as illustrated by Fig. 5.1), while methodologies are far more varied. While this can seem daunting, methodologies can help direct you towards research approaches, and methods, that you may not have considered, and so, can be very helpful.

Below are a few examples to highlight the diversity of methodologies available.

Quantitative research is a methodology which involves the collection of large quantities of numerical data that are statistically analysed to identify patterns and make generalisations.

Experimental research aims to establish cause and effect by manipulating variables and then measuring the outcomes.

Qualitative research focuses on understanding human behaviour, experiences, and social phenomena through non-numerical data.

Descriptive research, as the name suggests, provides an in-depth description of an un(der) researched phenomenon, through systematic observation and documentation. The aim is not to answer questions or test theories, but to understand the nature of the phenomenon.

Action research involves a cycle of action and reflection aimed at solving a problem or improving a situation through iterative changes and learning from the outcomes.

Mixed methods research (confusingly named despite being a methodology!) combines qualitative and quantitative research approaches to provide a more comprehensive understanding of the phenomena.

> **Activity 5.1:** With your research question in mind, what ontology, epistemology, and methodology do you most identify with?

5.3.3 Research Paradigms and How They Can Be Applied Through an Inclusive Lens

A research paradigm is the sum of the ontology, epistemology, and methodology, as outlined in Fig. 5.1. I would encourage any researcher to invest time in determining where they sit along these spectra, as this can help you structure your research more effectively and will allow others to understand the theoretical underpinning that have influenced your findings. Journals ethics committees, and funding bodies will ask for this information too, and will, quite rightly, challenge approaches that do not fit within your stated paradigm. For example, carrying out statistical analysis on qualitative research, or making assumptions about why people ticked a box (without asking the question) is not the sign of sound research.

Below is a selection of research paradigms that can be readily used in anatomy (Clark 2021; Hesse-Biber 2010).

Positivism is grounded in the belief that reality is objective and can be observed and described from an external standpoint, independent of the researcher's influence. It emphasises the use of quantitative, deductive methods, such as experiments and quantifiable surveys, to test hypotheses and generate generalisable results. In anatomy research, this paradigm allows for precise measurements and statistically robust conclusions about anatomical structures and functions. For example, determining whether there is a correlation between pelvis shape and uterine position based on measurements on body donors or via cross-sectional imaging.

Post-positivism is often considered an umbrella term that encompasses a range of research paradigms that evolved from positivism. While it shares some core principles of positivism, such as the importance of empirical evidence and the use of scientific methods, it recognises the complexity of reality and the influence of context, including the biases and subjectivity of researchers. Therefore, it allows for the use of both quantitative and qualitative methods, and it supports a more nuanced understanding of research findings.

Post-positivist research paradigms acknowledge the fallibility of human knowledge and the need for a critical, reflective approach to scientific inquiry. In anatomy, post-positivist research might combine anatomical measurements of the uterus and pelvis with interviews to understand the impact of any anatomical variations on patient experience of pregnancy and birth.

Interpretivism and Social Construction research paradigm is the polar opposite end of the spectrum to positivism and asserts that reality is entirely socially constructed and subjective, emphasising the understanding of human experiences and meanings. The interpretivist epistemology is founded in the belief that knowledge is constructed through social interactions and the understanding of human experiences. It emphasises the importance of context, meaning, and the perspectives of participants in the research process. Researchers adopting this paradigm aim to understand how people make sense of their world and how their experiences influence their perceptions and behaviours. Interpretivist research in anatomy might involve exploring how individuals from different cultures experience the birthing process, using methods like ethnography or narrative analysis to capture the richness of these perspectives.

Critical Realism sits between positivism and interpretivism in the research paradigm spectrum. It argues the existence of an objective reality independent of human perception but acknowledges that our understanding of this reality is inevitably influenced by social, cultural, and historical contexts. It highlights the importance of identifying the underlying structures and mechanisms that generate observable phenomena, while also understanding how the researchers' own lived experience will influence the findings and actively address these biases. Critical realism allows for a comprehensive exploration of both the biological aspects of human anatomy and the sociocultural factors influencing medical practices and health outcomes. For example, to what extent are anatomical differences responsible for the increased mortality in postpartum Black women in the UK (Vousden et al. 2024), and how do sociocultural factors influence these outcomes? This approach facilitates a deeper understanding of complex biopsychosocial interactions by combining quantitative measurements with qualitative insights, ensuring a robust and multifaceted analysis.

5.3.3.1 Research Approaches

While research paradigms provide a theoretical framework of how you consider the world and what you will do to study it, your research cycle can vary depending on your research ethos. For example, a critical realist study could be conducted using any of the following research approaches.

Pragmatism focuses on the practical application of research findings and values both quantitative and qualitative methods. It is flexible and problem-centred, aiming to find solutions to real-world issues. In anatomy research, a pragmatic approach might integrate clinical trials with patient interviews to develop and refine surgical techniques or educational tools. Be warned however, that true pragmatism is rigorous research and is not a case of choosing a method because it is easier.

Transformative Research is dedicated to addressing social injustices and empowering marginalised groups through research. It combines qualitative and quantitative methods to challenge existing power structures and promote equity. Transformative anatomy research could involve studying the accessibility of medical education for underrepresented communities and advocating for inclusive practices.

Community-Based Participatory Research (CBPR) is a collaborative approach that involves community members in all phases of the research process. This paradigm values the expertise of the community and aims to produce research that directly benefits them. A CBPR study might include partnerships with local communities to address health disparities and develop culturally sensitive health interventions.

Emancipatory Research seeks to liberate and empower participants by addressing power imbalances in the research process. It involves participants as co-researchers and focuses on generating knowledge that leads to social change. This might involve working with disabled individuals to improve medical training and healthcare services that better meet their needs.

Participatory Research involves stakeholders, particularly those who are the subject of the research, actively in the research process, ensuring their perspectives and needs shape the outcomes. This approach values co-creation of knowledge and mutual learning. Participatory research might see neurodiverse educators and students co-developing new teaching approaches that enhance teaching, learning, and engagement for neurodiverse students.

Inclusive Research is a type of participatory research that prioritises making the research process accessible and empowering for all participants or stakeholders, especially those who are typically excluded from the research process. Inclusive research initially focused on research with people with learning differences but has since expanded to disabilities in general. It aims to produce more relevant and applicable findings. Inclusive research could involve collaborating with patients to design studies that address their concerns and priorities in medical treatments.

Critical Theory Research is concerned with critiquing and changing society by addressing power dynamics and social inequalities. It uses both qualitative and quantitative methods to uncover and challenge oppressive structures. It might be used to examine how societal norms and biases influence anatomy and healthcare education and practice, advocating for more equitable and just healthcare systems.

> **Top Tip** Please do not worry if you don't have all the answers as to the 'best' approach to conduct representative research. The fact that you want your research to be representative is the key first step.

5.3.4 Representative Research: A Practical Guide

Anatomy research includes multitudes of papers cataloguing the anatomy seen in cadavers or in living people. These are often considered to be objective measurements of the body with generalisable findings. But are they? If the population measured is not diverse, how valid or meaningful are the data to the un(der)represented groups?

Poor-quality lung capacity data, extrapolated beyond its validity, is still used today to inform spirometry calculations (Kouri et al. 2021) and anatomical research is no different, with flawed data on pelvic shape, collected in the 1930s, still informing many curricula today (Kuliukas et al. 2015; Caldwell et al. 1940).

As anatomy is a research-informed discipline, if the research we rely on is not representative of the range of diversity seen in the population, this can impact on the application of this knowledge, particularly in healthcare. Are we misinforming future healthcare professionals and therefore negatively impacting on the quality of care they can offer to underrepresented groups? Our research must reflect the rich diversity of the populations we study. Ensuring representation across all types of diversity is not only ethical but enhances the validity and applicability of our findings.

5.3.4.1 Step 1: Check Your Values, Privilege, and Biases

All research starts with a question. What would you like to understand better? What is the purpose, or aim of your research? But have you stopped to ask yourself why you are asking this particular question? Or why are you asking it in that specific way? Our own lived experience will influence the questions we ask and the way in which we conduct our research. This can be as true for positivist research paradigms as it is for interpretivist. However, being aware of the impact of privilege or bias on our research question and strategy is a solid step towards making your research more representative.

Our values play a crucial role in shaping the focus and scope of our studies. These values influence the selection of research topics, the framing of research questions, and the interpretation of findings. For example, a researcher who values social justice may prioritise studies on marginalised communities, aiming to highlight issues of inequality and advocate for change. Prioritising certain values over others can produce research agendas that either promote inclusivity and representation or overlook critical issues affecting diverse populations. This underlines the importance of reflection and reflexivity to maintain a balanced and inclusive approach.

Privilege can also create significant blind spots in research. Privilege can refer to advantages, benefits, or rights granted to individuals based on specific aspects of their identity or from societal structures and norms. Researchers from privileged backgrounds may unconsciously design studies that reflect their own experiences, inadvertently excluding perspectives of less privileged groups. For example, lacking awareness of the barriers faced by marginalised communities in accessing healthcare, leading to research that fails to address these critical issues (Delgado et al. 2017). Privilege can also relate to access to resources, funding, and networks, further skewing research priorities and outcomes in their favour. If you are fortunate enough to be in a position of privilege, for example, working within an institution where equality, diversity, and inclusion (EDI) research is supported and encouraged, can you use this fortunate position to include marginalised voices in the research process, ensuring that studies are informed by a wide range of experiences and are more representative of the broader population?

Biases, whether conscious or unconscious, can influence every stage of the research process, from the formulation of hypotheses to the interpretation of data. For example, confirmation bias may lead researchers to favour data that supports their preconceived notions while dismissing contradictory evidence, resulting in skewed findings (Nickerson 1998). Cultural biases can also lead to misinterpretations of data from diverse populations, as researchers might impose their own cultural norms and values on their subjects. Addressing these biases requires a commitment to reflexivity and the use of strategies such as peer debriefing, triangulation, and member checking to validate findings and ensure they are not unduly influenced by the researcher's biases. Additionally, incorporating diverse research teams and engaging with community stakeholders can help mitigate the impact of individual biases and enhance the inclusivity and representativeness of research outcomes.

In qualitative research, *positionality* and *reflexivity* are important for enhancing transparency, credibility, and ethical conduct, as they encourage researchers to be mindful of their influence on the research process and outcomes.

Positionality refers to your social position, identity, and background in relation to the research context and participants. This includes factors such as gender, race, class, cultural background, educational level, and other social identities. Positionality acknowledges that your perspective is shaped by your social location and experiences, which can affect how you interact with participants, interpret data, and draw conclusions.

Reflexivity refers to the process of critically examining your own assumptions, biases, and role in the research process. It involves being aware of how your background, experiences, and worldview might influence the research, from data collection to analysis and interpretation. Researchers practicing reflexivity actively reflect on their own position and how it shapes their understanding and approach to the subject matter.

Reflecting on your own values, positionality and biases can help develop a more transparent and ethical research environment and reduce any unintended consequences on the validity of the research. Section A in activity 5.2 provides some suggested questions that can allow you to better understand the decisions you have made and develop strategies to ensure your research question is representative.

Top Tip: Write a Reflexivity Statement
Understanding your positionality can help you identify areas where insight and contribution from others may strengthen your research. Exposing our biases, assumptions, and values can be both uncomfortable and enlightening. You may not feel comfortable sharing this information widely, for fear of being judged. But remember, everyone has biases and values, and acknowledging what they are is a step towards understanding why we have them and learning to recognise their impact. Olmos-Vega et al. (2023) have published an excellent guide on how to approach reflexivity. See ours in Sect. 5.1.

Activity 5.2: Reflecting Towards Representative Research
We are not always aware of how our own lived experience influences the decisions we make. As such, it can be helpful to ask ourselves certain questions to prompt reflection on the decisions we make, to try and identify any unintended consequences that could undermine the validity of the findings through lack of representation.

The questions below are not exhaustive, and many will not be relevant to your research paradigm. Instead, they are intended as a starting point for researchers to be able to explain the decisions made. This information can be useful when writing up your research.

A. Research design and methods

1. What are my own biases?
    (a) Have I reflected on how my own background and assumptions might influence my research questions, analysis, and interpretation of the data?
    (b) Am I seeking diverse perspectives, particularly from underrepresented groups, to challenge and refine my interpretations?
2. Is my research design representative?
    (a) Have I considered the needs and perspectives of diverse groups in my research design?

(continued)

   (b) Are my data collection tools and methods accessible to all potential participants (e.g. those who have different levels of physical or cognitive ability or speak different languages?)
3. How do I ensure I am culturally sensitive?
   (a) Are my research questions and materials culturally appropriate and respectful of all characteristics?
   (b) Have I sought input from representatives of the communities I am studying to ensure relevance and sensitivity?

B. Ethical considerations

4. How will I protect participant rights and autonomy?
   (a) How will I obtain informed consent in a manner that is understandable and accessible to all participants?
   (b) How can I ensure confidentiality and privacy for all participants?
5. Am I addressing power dynamics?
   (a) Have I considered the power dynamics between researchers and participants, especially when working with marginalised communities?
   (b) Am I working collaboratively with participants to ensure their voices are heard and valued in the research process?

C. Participant recruitment

6. Who is included in my study?
   (a) Have I considered the demographic characteristics of my target population?
   (b) What efforts have I made to include participants from diverse backgrounds?
7. How are participants being recruited?
   (a) Am I using recruitment methods that reach a diverse audience?
   (b) What barriers might prevent certain groups from participating (e.g. language used, access to technology, physical accessibility, and mistrust of research within certain communities)?

D. Data analysis and interpretation

8. How am I analysing my data?
   (a) Is it appropriate to disaggregate data and examine outcomes for different demographic groups (e.g. age, ethnicity, gender, disability, sexual orientation, and religion)?
   (b) Am I considering how intersecting identities might impact the results?
9. How can I fully understand the data?
   (a) Should data analysis be carried out by a diverse group (e.g. including members of the target population)?

E. Sharing my findings

10. Who is my audience?
    (a) Am I communicating my findings in ways that are accessible and relevant to diverse audiences?
    (b) Have I considered multiple formats for disseminating my research (e.g. community presentations, accessible reports, and multilingual materials)?
11. How will my findings impact different groups?
    (a) Have I considered the potential implications of my research for various demographic groups?
    (b) How can I highlight findings that are particularly relevant to underrepresented or marginalised groups?

5.3.4.2 Step 2: Ethical Considerations

Power dynamics between the researcher and participant can significantly impact on the inclusivity and representativeness of research. Researchers can often hold more power than the participants, especially in contexts involving marginalised or vulnerable groups. This imbalance can lead to situations where the voices and experiences of participants are overshadowed or ignored. For instance, researchers might impose their own interpretations and priorities, which may not align with the needs and perspectives of the communities being studied (Gaventa and Cornwall 2008). Conversely, if the participants do not identify with the researchers, they may be less willing to share their experiences. This can result in research that reinforces existing hierarchies and fails to genuinely represent the experiences of the participants. To counteract these dynamics, researchers need to actively seek to share power, by involving participants in decision-making processes and ensuring that their contributions are valued and reflected in the research outcomes.

In some studies, the researchers may have a different identity to the group being researched, for example, a male researcher studying the impact of menopause on women in the workplace. The challenge of accurately 'representing the other'—individuals or groups different from the researcher in terms of culture, ethnicity, socioeconomic status, or other factors—was explored by Wilkinson and Kitzinger (1996). Researchers may inadvertently project their own biases and stereotypes onto the subjects they study, leading to misrepresentation and a lack of nuance in understanding diverse experiences (Said 1978). For instance, cultural misunderstandings can result in the oversimplification of complex social phenomena or the exoticisation of certain groups. There is also an ethical tension between speaking for others or the research not being conducted due to the underrepresentation of certain demographics in the research community.

Co-production and co-design involve the active participation of community members in the research process, from the formulation of research questions to the dissemination of findings. This approach can significantly enhance research by ensuring that it is grounded in the lived experiences of those it aims to serve (Hickey and Mohan 2004). When community members are engaged as equal partners, their knowledge and perspectives can guide the research in ways that are more relevant and impactful. Co-production and co-design help to democratise the research process, making it more responsive to the needs and priorities of diverse populations. However, this approach requires a genuine commitment to collaboration and the sharing of power, as well as the resources and time to support meaningful engagement.

The traditional competitive research paradigm does not support a representative research approach. In a time where funding and time are increasingly limited, researching alone can be an impossible task. Collaborative research encourages the pooling of resources, diverse perspectives and expertise, which can lead to more comprehensive, sustainable, and inclusive studies. However, successful collaboration requires clear communication, mutual respect, and the negotiation of differing priorities and goals. By fostering a collaborative environment, researchers can produce work that is more reflective of the complexities of real-world issues and more attuned to the needs of diverse communities.

Sections B, C, and E in activity 5.2 provide some suggested questions that can allow you to better understand the decisions you have made and develop strategies to ensure your research question is representative.

5.3.5 Key Points

- 'Representative research' describes approaches that enhance diversity representation, addressing power dynamics, biases, and privileges to produce ethical and impactful outcomes.
- Research paradigms influence every stage of research, from question formulation to data analysis. Therefore, understanding your ontology, epistemology, and methodology is crucial.
- Researchers' lived experiences shape their priorities, influencing the study focus and outcomes, which can affect the validity and

generalisability of the research. Reflecting on your values, privileges, and biases is essential in shaping representative research.
- Collaborative and participatory research approaches democratise the research process and ensure it is grounded in the lived experiences of diverse populations. Collaborative research also allows for pooling of resources and expertise, leading to more comprehensive and sustainable studies.

5.4 Challenges to Representative Research Practice

There are many barriers and challenges to representative research practice. Some are related to individuals engaging in research or systems and processes within an organisation, while others are related to potential participants and the community. Overcoming these barriers will ensure that the study is accessible, equitable, and representative of diverse perspectives. Common barriers to representative research practices are discussed below.

5.4.1 Creation and Planning Process of Research

Planning representative research requires careful consideration of the project team members' characteristics, roles, relationships as well as project outcomes. Having a diverse team ensures that diverse perspectives are considered from the start of the design phase rather than as an afterthought (Garratt et al. 2022). However, it is not uncommon for research projects to be initiated by an academically trained researcher who then recruits people with lived experience as collaborators or co-researchers. While it seems more aligned with the needs of research institutions, it often results in limited co-researcher ownership and greater power imbalances. Commencing research with co-researchers with a lived experience ensures that the research question is more relevant and the resulting output actionable (Walmsley et al. 2018; Frankena et al. 2019). There are many ways to ensure this collaboration and engagement. Members of self-advocacy organisations can be hired as representatives on advisory groups and engaged from the beginning of research projects. Conversely, academic researchers can serve as advisors to self-advocacy organisations and provide guidance on funding methods or offer practical support (Garratt et al. 2022).

> **Top Tip** Think about who is in your research team. Do they come from a diverse range of backgrounds? Are they representative of the people you are researching? If not, consider collaborating with other researchers or stakeholders who can complement your team with a different positionality.

Historically, higher education institutions have been inaccessible and exclusionary, resulting in limited staff diversity. As these are often the institutions where research is discussed and planned, the people whom the research is carried out on, are not always represented. Representative research often requires researchers to collaborate with people outside of higher education institutions. However, institutions' 'collaboration agreement' processes often limit how researchers approach and collaborate with people with lived experiences outside their institutions (Armstrong et al. 2022).

5.4.2 Resources and Support

Lack of formal training on representative research practice, as well as inclusive engagement strategies can mean that researchers do not know where to start or how to apply inclusive principles to anatomical research (Durant et al. 2014; Niranjan et al. 2020; Passmore et al. 2022). Although general research training is offered by many institutions, customised training that can equip research staff with knowledge, tools, and skills to better understand and address exclusionary practices is key.

Support and mentorship have been associated with increased research engagement and productivity. Mentors support mentees, provide resources, and enhance mentees' self-confidence, which results in higher success rates in obtaining grants and greater preparedness for research (Sambunjak et al. 2006; van Schalkwyk et al. 2017). A lack of mentorship programmes supporting researchers' development and exploration of representative research practices can hinder their personal development and the impact of their research findings.

Researchers engaging with representative research practices have developed skills by doing and experiencing. However, limited opportunities are available for novice researchers to develop their skills and knowledge further (Garratt et al. 2022). Communities of practice focusing on representative research practices can help in creating spaces where this experience is communicated and shared (O'Brien et al. 2014). This aligns with collaborative and dialogical learning models where a group of people co-construct knowledge by actively participating in discussions (García Iriarte et al. 2023). These models challenge the traditional one-way delivery of information and encourage an exchange of knowledge. Participants get to ask questions, share perspectives, debate and reflect on different viewpoints, and collaborate to build understanding.

Top Tip Check if your institution offers a mentorship scheme or reach out to the wider community through social media. There may be researchers using inclusive methods that are willing to give you some pointers.

As inclusive practice requires researchers to engage with diverse groups as co-researchers, training opportunities should not only target university personnel or the principal investigators but all co-researchers involved in the study (Hughes and Schwartz 2024; O'Brien et al. 2022). The training should be tailored to their individual needs and wishes as well as created to fit the design and aims of the study (Walmsley et al. 2018). In their study, Vega-Córdova et al. (2020) investigated barriers that their co-researchers who had intellectual disabilities encountered while performing research. The study reported that the lack of customised training to enable the co-researchers to perform certain tasks was one of the barriers. Some researchers with intellectual and/or developmental disabilities highlighted that receiving training from academically trained researchers helped them feel more skilled and prepared to conduct research (Herer and Schwartz 2022; St. John et al. 2018). Training should also include strategies to recognise and address inherent power imbalances resulting from the different skill set and experience of the research team. Recommendations from a toolkit discussing inclusive strategies to support and guide co-researchers include non-physical support, such as providing positive reinforcement and additional time to finish a task, and physical support such as providing information in an accessible way (Pfeiffer et al. 2024). Other studies recommended having a specific contact person for co-researchers which not only facilitated meaningful participation and engagement of co-researchers but also had an added benefit of increased advocacy around accessibility and equity of the research procedure (Bigby et al. 2014; Kim et al. 2022; Hughes and Schwartz 2024).

Moreover, a lack of knowledge can lead to fear. Researchers can be fearful about potentially hurting others by using the wrong language or not having enough knowledge to approach a situation correctly. Also, as representative research practices can be resource-intensive, fear can also extend to include concerns about not having enough resources, such as time, money, or people, to allow researchers to engage with representative research practices fully and appropriately.

5.4.3 Costs and Funding

Representative research practice can have an impact on costs. Costs not only associated with money, but time, effort or even anxiety and discomfort for research staff (Passmore et al. 2022).

It also requires more people than research undertaken by trained university researchers alone (Walmsley et al. 2018). While often unintentional, exclusionary practices often prioritise researcher convenience and can exacerbate power imbalances, while representative research practices can be both time-consuming and resource intense (Gilmore-Bykovskyi et al. 2022; Walmsley et al. 2018; O'Brien et al. 2022). Costs can also be related to additional steps required in representative research such as translation into different languages, creating accessible interview tools and/or research materials, payment for supporters, and elongated timelines. However, this increase in effort is often accompanied with outcomes that have a greater positive impact.

Lack of funding resources to support representative research is rarely discussed in the literature (O'Brien et al. 2022), but is a sad reality that needs to be addressed. This includes persuading funding bodies that the additional cost associated with representative research practice are justified (Walmsley et al. 2018). Representative research teams must carefully consider fair compensation for their co-researchers with transparency about how much and which tasks are compensated. This ensures the collaboration is equitable and financially viable for the co-researcher. This can require careful consideration if co-researchers rely on governmental benefits. In their study, Kim et al. (2022) reported working closely with co-researchers with developmental disabilities. During the initial phases of the project, the principal investigator discussed their roles and participation in the project, payment method, and the number of hours required in the project (Kim et al. 2022). As research projects often have varying level of effort across the duration of the project life, it can be challenging to determine the appropriate number of hours that the project requires without jeopardising the co-researchers benefits (Kim et al. 2022; Hughes and Schwartz 2024).

It is also important to carefully consider the compensatory value to participants as a measure to improve engagement from un(der)represented groups. While financial compensation is not always necessary, it can be a barrier to participation. Participants may have to take time off work or arrange for transportation and child/family-care in order to be able to take part in a research project that would often require multiple visits and follow-ups (Passmore et al. 2022). Careful thought to how participants will be compensated for their time so that their lives are not disrupted is key in inclusive recruitment practices.

> **Top Tip** Be transparent and communicate with your co-researchers, stakeholders, and participants from marginalised groups. Representative research is underfunded so you may not have enough money for payment or resources, so let them make an informed decision on whether they have capacity to contribute to the research.

5.4.4 Academic Culture

Systemic barriers within higher education institutions still exist and can challenge meaningful and authentic engagement of co-researchers in the research process. These include institutional policies and procedures. For example, for work to be published and shared with the larger learning community, ethical approval must be obtained. However, the ethical approval process for research can be complex to navigate and is often obtained from a university, which generally excludes people with disabilities. Also, people with disability are considered 'vulnerable' with reference to the ethical approval processes. This can make it challenging for academics to include them as co-researchers or even a principal investigator. Funding can also affect how the roles within the research team are established and how much time the team can work together. Inflexibility of granting bodies with regard to funding timelines and limited bridge funding between projects can present as a barrier to in-depth collaboration for some representative research teams together (Hughes and Schwartz 2024).

Academic culture is competitive. Researchers are under immense pressure to increase their research productivity, obtain grants and publish, which can affect how they adopt and prioritise inclusive practice. Time is an important factor that can affect proper engagement with representative research practice. For example, co-writing can be time-consuming and may occur after the funding had ended. Often co-researchers would not be compensated for the laborious task of writing up the outcomes of the project, thus subsequently would decline the opportunity. Also, the benefits gained from publishing the work from an academic perspective might not motivate co-researchers to engage with the activity. Representative research practices require considerable time, dedication, and effort that are not often recognised in the traditional assessment of research output. Walmsley et al. (2018) reported from a personal communication with a researcher engaged in representative research practice that they estimated that 2 days of their 5-day work week was dedicated to supporting the co-researcher they were working alongside. Interestingly there are global efforts to change how research is being recognised in higher education institutions. For example, the League of European Research Universities is calling to recognise the multidimensional aspects of a researcher's work looking into not only the output, but also how research is generated (Overlaet 2022). Researchers would be encouraged and more likely to engage in inclusive practices if their institutions recognised them for their innovation, leadership, and collaboration with diverse communities.

5.4.5 Communication and Dissemination of Research

Sharing research findings is an important, but often overlooked, stage of any project, particularly with the current focus on research impact. Research findings are typically presented at conferences, or published in journals behind a paywall, making it inaccessible to the communities that the findings relate to. Even if accessible, research is usually communicated in an inaccessible language filled with jargon, addressed to academics and professionals (O'Brien et al. 2022). Outcomes must be communicated in an accessible way to engage the diverse audiences and communities they affect. The use of radio, podcasts, social media, art, and photographs are all important and successful means of communication (García Iriarte et al. 2023).

Top Tip Consider other ways of presenting and distributing your data that are accessible to your community and participants. This could be other forms of media alongside your research paper, including posters, podcasts, radio, social media and photographs, blogs, etc.

Attending conferences is a vital component of professional and academic growth for researchers. It offers many benefits beyond immediate experience. One of the main advantages of attending conferences is networking opportunities and knowledge exchange. Conferences bring together experts and professionals from various regions and offer interdisciplinary insight into a subject. Engaging in discussions, exchanging ideas, and building relationships with like-minded individuals can lead to future collaborations, mentorship, and personal growth. These benefits are lost to co-researchers if funding is not available which is often the case for co-researchers (O'Brien et al. 2022).

The role of team members and how the research outcomes are distributed should be established at the initiation of any representative research project (Hughes and Schwartz 2024). It takes a firm belief in representative research and a solid self-identity to be able to fully adopt and promote representative research practices, including willingly handing over power to another researcher and negotiating who is best placed to present the work, be the first author or the principal investigator (Armstrong et al. 2022).

Top Tip Decide at the start of the research project the roles of your team members. Who could benefit from more career development opportunities like presenting at conferences? What is the order of authorship?

5.4.6 Assumptions and Implicit Bias

The literature highlights a gap between the awareness of the importance of diversity and representative research practices among researchers and the actual implementation of these practices (Passmore et al. 2022). Academics can hold strong beliefs about how representative research should be done, and while they may be unaware of their own biases or fears of failure, these can affect how they engage with the research process.

An important factor that presents a challenge to inclusive efforts is the belief that increasing diversity in research is an unattainable goal (Passmore et al. 2022). This belief can stem from the assumption that underrepresented groups are unwilling to participate, ineligible, limited in numbers or thinking that diversity in data is 'not worth the effort'. It can also be a result of viewing representative research strategies as optional rather than central to advancement in science. As a result, researchers do not pursue inclusive practices or consider inclusive strategies while conducting their research. In their study, Passmore et al. (2022) reported that even though participants expressed strong beliefs about the importance of inclusive practice within research, they engaged with no efforts to promote or practice it because they saw the barriers as inherently 'insurmountable'. The perception of 'there's not much we can do about it' is an important barrier that could be broken down initially by recognising it.

Additionally, some researchers may feel anxious and uncomfortable working with underrepresented groups. This can be a result of unconscious biases or lack of confidence in discussing topics related to diversity and inclusion, even if not directly related to their projects.

> **Top Tip** Leverage reflexivity to help you explore how your experiences may affect the research you are conducting. Keeping a research diary will allow continuation of reflexivity throughout the duration of your project—it is an ongoing process! See activity 5.2 for a guided reflexivity activity.

5.5 Demography as a 'Disruptive' Case Study

Most surveys start by requesting our demographic identity through a series of tick boxes. But how does this data enhance the research findings and, given the range of global diversity, are the boxes even meaningful? As we see demographic questions so frequently, we can become blind to their efficacy. This case study highlights the factors that should be considered when asking for demographic data and challenges the idea of the tick-box approach commonly seen.

While comprehensive demographic data can be valuable, it is not always used responsibly. As such, researchers must justify the collection of each piece of demographic information by identifying the relevance to the study's objectives and whether it contributes meaningfully to the findings. If demographic information is, indeed, beneficial to the data analysis, then accurate and inclusive demographic identifiers are crucial. When participants cannot see themselves represented in questionnaire options, they may feel marginalised, potentially leading to reduced participation from un(der)represented and oppressed groups (Hughes et al. 2022). This is particularly concerning given that past research, especially in healthcare, already shows bias due to the historical exclusion and exploitation of marginalised communities (see Sect. 5.2). Perpetuating such oversights would only reinforce existing imbalances in the field.

Often these data are used to contextualise the subsequent answers and assess whether a diverse group of participants have been represented. If certain groups are un(der)represented, researchers can adjust their data collection strategies accordingly, as demonstrated in the Galleri study (Neal et al. 2022).

> **Top Tip** Consider the reason why you are collecting demographic data and how this will link to your research aims and questions. What demographic data is essential to collect to meet these aims? What demographic information is not essential? How do you hypothesise these demographics will influence your data?

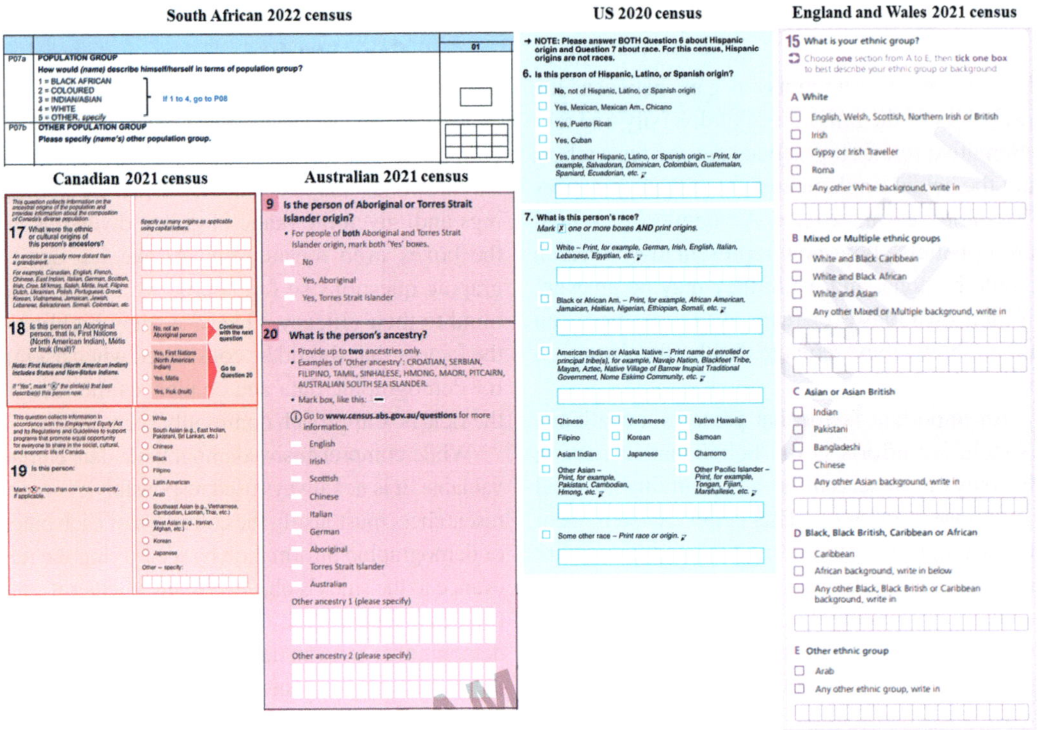

Fig. 5.2 Questions focusing on ethnicity taken from the most recent census in a selection of countries where English is the official language. Examples are taken from government or bureau websites (Australian Bureau of Statistics 2021; Statistics South Africa 2022; Office for National Statistics 2021; Statistics Canada 2021; United States Census Bureau 2020)

5.5.1 Tick-Box Demography

Currently, many demographic elements of surveys use binary tick boxes to collect data, which forces the participant to categorise themselves into a box that was created by the researchers. More recently, surveys have included an 'other' box to allow open-text answers for those who do not fit in the designated categories. This includes the largest survey delivered nationally, the census, in which the government collects information about the population once per decade. As an example, the question pertaining to ethnicity from Australia, Canada, South Africa, USA, and England and Wales are illustrated in Fig. 5.2.

While the ethnicity categories in many census forms have expanded slightly since their introduction, these countries are becoming significantly more diverse over time, calling into question whether this demographic data collection method reflects that shift.

While the addition of open-text boxes has broadened the scope of data collection, certain categories still lack nuance. For instance, the 'White and Asian' option conflates half of an individual's ethnic identity with the world's largest and most diverse continent. This raises important questions: Does this categorisation truly capture a person's identity? Are we overlooking crucial subtleties? How might indigenous peoples situate themselves within such a framework?

Despite these limitations, the tick-box approach can offer practical advantages when dealing with large datasets. Given that millions participate in the census, pragmatically this method may indeed be the most feasible and appropriate, particularly when complemented by an open-text option for more detailed responses.

It is important to acknowledge that the census examples discussed here are primarily drawn from English-speaking countries and may not represent the diversity of census questionnaires worldwide. Western and colonial perspectives have likely influenced how governments have constructed these categories. Therefore, we must critically examine the processes and rationales behind these demographic groupings.

This complexity in ethnic categorisation highlights the need for ongoing refinement and sensitivity in demographic data collection, especially in field of anatomy research where accurate representation is crucial for inclusive practice.

Other organisations, such as the National Institutes of Health (NIH) in the United States, employ standardised demographic forms for all their funded projects. These forms often rely solely on forced-choice tick boxes for ethnicity data collection (National Institute of Health 2019). Woolverton and Marks (2023) conducted a study comparing how adolescents self-describe their ethnicity with and without these confined options. Their findings revealed that the tick-box format more accurately captured the ethnic identity of White participants compared to non-White participants. Interestingly, the study also highlighted that cultural identity often differed from ethnic identity, emphasising the complexity of demographic categorisation.

This issue extends beyond ethnicity to other demographic groups, particularly affecting non-binary, gender non-conforming, and transgender individuals. Cameron and Stinson (2019) found that only 15 out of 106 journal articles from Psychological Science between 2016 and 2018 provided inclusive gender/sex measures in their demographic data collection. Moreover, the authors discovered not only a lack of inclusive options for non-cisgender people but also a failure to describe gender measures in the articles or supplementary materials. This lack of transparency necessitated direct communication with the authors to determine the measures used.

Top Tip Think about how you plan to present your demographic data and describe the measures for each demographic category. What is obvious to you may not be obvious to your readers!

If, after considering these points, you decide that a tick-box approach is necessary for your demographic data collection, there are accessible online resources to help create more inclusive questionnaires. The Diversity and Inclusion Survey (DAISY 2022) question guidance offers practical applications of more inclusive demographic categories. Similarly, action groups and charities, such as Stonewall, provide advice on representative research practices for specific demographics.

These findings underscore the importance of thoughtful, inclusive demographic data collection in anatomy research and education. Researchers must balance the need for standardisation with the imperative to accurately represent diverse populations, ensuring that all participants feel seen and valued in the research process.

5.5.2 Othering

In any surveys that offer tick boxes, if the participant does not fit those prescribed boxes, then they must categorise themselves as ‘Other’. Forcing unrepresented people to identify as ‘other’ can exclude and stigmatise these participants (Alpers 2018) at both an individual level and institutional level. Although Othering may have had good intentions like offering support to underrepresented groups, often it reinforces an ‘Us and Them’ mentality and social hierarchy (Akbulut and Razum 2022). It can misrepresent those participants as they may have completely different perspectives and demographics but are both classed as ‘Other’. In Hughes et al.’s (2016, 2022) inclusive survey framework, there is still an 'other' tick box which should perhaps be reconsidered. Woolverton and Marks even named their 2023 paper 'I just check "other"' which highlights the expectation of Othering that underrepresented people face when participating in surveys.

> **Top Tip** Consider how you can use open-text demographic data collection in your own research. How many participants do you expect to have? Is it feasible to make the demographic data completely open text with no tick boxes? If not, can you add an option for open text alongside tick boxes, so participants have the opportunity to describe themselves in their own words?

5.5.3 Ambiguity and Interpretation

A significant challenge when using open-text demographic questions is the potential for ambiguity, sarcasm, and misinterpretation—a common complication with qualitative data. For instance, in a previous study (Evans et al. 2025), multiple participants misinterpreted the question 'How would you describe your ethnicity?' by stating their nationality or providing a generic definition of ethnicity rather than stating their own. This resulted in some data being unusable for analysis, potentially leading to missed nuances during interpretation. To mitigate such issues, it is valuable to pilot your questionnaire with a diverse group of individuals from varying backgrounds. These beta testers could include people from outside your discipline, different institutions, members of the public, or those with demographic backgrounds different from your own.

> **Top Tip** Before opening your questionnaire, send it to as many people as possible from a variety of backgrounds to test it first. This may pick up some potential ambiguities that you can troubleshoot before collecting data. If people outside of the research team can understand the questions, then chances are your target audience will too!

When responses are communicated via text box, they can be interpreted differently by data analysts depending on their background. The interpretivist paradigm dictates that our individual realities are social constructions shaped by our lived experiences and socio-geo-political contexts. It acknowledges that researchers' subjectivity inevitably influences the inquiry process, including the formulation of research questions and data analysis (Alharahsheh and Pius 2020).

Reflexive thematic analysis can be employed to identify the positionality of the analysts and understand how they interpret the data. Braun and Clarke (2022) recommend that researchers engage in an exercise before initiating a research project, considering their own backgrounds, including aspects of privilege/marginality, life experiences, education, personal values, and political views (activity 5.2). They emphasise that reflexivity is an 'ongoing process of reflection', suggesting the maintenance of a reflexive diary throughout the project.

Employing multiple coders can enhance the analytical process, as data is interpreted by individuals with diverse positionalities who can identify more nuances and provide broader insights (Braun and Clarke 2022). Additionally, inviting participants to focus groups or interviews allows them to further articulate their ideas and experiences, adding depth and context to the data.

By implementing these strategies, researchers in anatomy education can improve the accuracy and inclusivity of their demographic data collection, leading to more representative and nuanced understanding of diverse populations in their field.

> **Top Tip** Although it is perfectly feasible to have a single person coding the data, you may want to have multiple coders from a diverse range of backgrounds. This will depend on capacity, funding, and the field of research, but consider how varying perspectives could make your results richer and more nuanced.

5.5.4 Intersectionality

A prevalent issue in demographic data collection is the tendency to categorise participants into discrete groups based on one or two aspects of their identity, often without considering social context. For instance, data from all Black participants might be analysed for common themes within this distinct group. While sometimes additional factors are considered, such as 'Black and female' or 'Black and queer', this approach still provides only a linear representation of complex and overlapping social categories, falling short of capturing true intersectionality (Rogers and Syed 2021).

The inclusion of free-text boxes can allow participants to comment on their identity within specific social contexts. For example, in the author's recent study (Evans et al., 2025), when asked about ethnicity, one participant responded:

> Ethnicity is an extremely loaded concept in my cultural background. I grew up in the diaspora in a very international environment, the ethnicity I identify with the most varies according to the context I find myself in.

While asking participants to determine which aspects of their identity are most relevant in a specific social context (e.g. the anatomy classroom) may be beneficial, Azmitia et al. (2008) found that directly questioning college students about their intersectionality did not yield sufficiently detailed responses. It was only during conversations about their everyday experiences that true intersectional experiences were revealed. Consequently, Rogers and Syed (2021) note that some scholars argue only in-depth interviewing can genuinely capture a participant's intersectionality.

This suggests that individuals may not always be consciously aware of their own intersectionality, further complicating efforts to measure it accurately. The challenge for researchers is to develop methods that can capture these nuanced, context-dependent aspects of identity while maintaining practicality in data collection and analysis. To address this complexity, researchers might consider employing mixed methods approaches, combining quantitative demographic data with qualitative interviews or focus groups. This could provide a more holistic understanding of participants' identities and experiences within the context of anatomy research and education. Additionally, longitudinal studies might reveal how individuals' perceptions of their own intersectionality evolve over time and across different educational contexts.

> **Top Tip** Questionnaires may not be enough to understand the true intersectionality of the participants. This will require deeper conversation in interviews or focus groups. If this is not feasible for your research, you can acknowledge as a limitation.

5.5.5 Anonymity

Before delving into the analysis of qualitative demographic data, it is crucial to address the issue of participant anonymity. Depending on your research questions, participants may share personal information that they wish to keep confidential. This concern should be addressed in your ethics application, as a combination of specific descriptive factors could potentially enable participant identification. This risk is particularly high if you are personally acquainted with your participants, such as colleagues or students. So, how can you ensure participant anonymity while still capturing valuable demographic insights?

One potential solution is to employ a data custodian who is unfamiliar with your participants. This custodian could be responsible for removing demographic data from the main dataset and storing it under a unique participant number. When researchers need to understand how demography might influence certain themes emerging from the data, they can provide the custodian with a list of participant numbers. The custodian can then identify any demographic trends among those participants without revealing individual identities.

This approach offers several advantages:

- It maintains a higher level of anonymity for participants, as their demographic information is separated from their responses.
- It reduces the risk of researcher bias, as the primary analysts are not directly privy to individual demographic details.
- It allows for demographic analysis when needed, without compromising participant confidentiality.
- It adheres to ethical standards by providing an additional layer of data protection.

However, this method also presents some challenges:

- It requires additional resources and personnel to manage the data custodian role.
- It may slow down the analysis process, as researchers need to coordinate with the custodian for demographic insights.
- It could potentially limit the depth of intersectional analysis, as researchers might not have immediate access to the full complexity of each participant's demographic profile.

To implement this approach effectively, clear protocols should be established for communication between researchers and the data custodian. Additionally, secure data storage and transfer methods must be employed to protect sensitive information.

Ultimately, the goal is to strike a balance between protecting participant anonymity and gathering rich, contextual demographic data that can inform inclusive practices in anatomy research and education. By carefully considering these aspects of data management, researchers can ensure both ethical integrity and analytical depth in their work.

Top Tip Consider how anonymous your data will be if you collect open-text demography. Do you need to include this in your application for ethical approval? Is there a way to minimise the risk of revealing the identities of your participants?

5.5.6 Analysis

While several papers have recommended using open-text demography as a representative research method (Fernandez et al. 2016; Hughes et al. 2022), there is a paucity of literature describing effective analysis techniques for this data. The challenge lies in avoiding the pitfall of collecting rich, nuanced data only to revert to categorisations influenced by our own positionality and biases during analysis. While it can reveal how particular ways of thinking might influence people's self-identification within demographic categories (Hughes et al. 2016), Hughes cautions against assuming that a person's demographic background automatically determines their actions or beliefs.

To address this, we propose an approach that prioritises the analysis of other questionnaire responses before examining demographic data. This method allows for the identification of themes independent of demographic considerations. For instance, if your reflexive thematic analysis reveals anxiety as a significant emotion among participants, you can then retrospectively examine the demographic data to identify any patterns or commonalities among those expressing anxiety.

This approach offers several advantages:

- It reduces the risk of demographic stereotyping during initial analysis.
- It allows for the emergence of unexpected patterns that might be overlooked in a demographic-first approach.
- It provides a more holistic view of participants' experiences before considering demographic factors.

However, it is crucial to maintain awareness of the complexities inherent in demographic analysis. A final, critical note for researchers is to guard against using demographic data to stereotype groups of participants. Not all members of a demographic group will share the same opinions or experiences, and we are all complex, multifaceted individuals existing within diverse social and political contexts.

To enhance this approach, consider:

- Using inter-rater reliability checks to ensure consistency in theme identification.
- Employing mixed methods approaches, combining quantitative analysis with qualitative insights.
- Engaging in regular reflexivity exercises to maintain awareness of personal biases.
- Involving participants in the interpretation of results, where appropriate and feasible.

By adopting these strategies, researchers in anatomy education can leverage the richness of open-text demographic data while maintaining an inclusive and nuanced approach to analysis. This method aligns with the goal of representing the majority through inclusive practices in anatomy research and education, acknowledging the diversity within demographic groups and the unique experiences of individual participants.

> **Top Tip** Grouping participants by demographic can limit your data interpretation. Instead, you could identify themes in your subsequent questionnaire questions first and then look at the demographics of those participants to find commonalities.

5.6 The Research-Teaching Nexus: Models for Representative Research-Informed Teaching

This subsection aims to explore how representative research practice can be integrated into education. This is a complex and multifaceted challenge, comprising inclusive education, representative research, inclusive teaching about research, and representative research-informed teaching and the interconnections between these. An important part of inclusive practice is to examine and change what and how we teach (Grue 2021).

A full review of the literature relating to each of these rich areas is beyond the scope of this chapter. Instead, we aim to draw out principles relating to how research can be broadly integrated into educational practice, and how this can be amended to consider the principles of EDI.

Universities and higher education institutes (HEIs), as common settings for both knowledge advance and discovery (research) and knowledge sharing (education), facilitate a dynamic and interactive relationship between these two activities (Griffiths 2004). This symbiotic relationship is the very core of higher education (Brew 2006; Robertson 2007) and is known as the Research-Teaching Nexus, a term first coined by Neumann (1994).

While the research-teaching nexus has been subject to much effort in terms of conceptualisation, optimisation, and refinement (as well as examining it from specific disciplinary and philosophical perspectives), summarised by McLinden et al. (2015), these models provide a useful framework for considering representative research practice and its integration into teaching.

Acknowledging the rich opportunities for research-teaching links, Healey (2005) conceptualised a model offering an opportunity for standard terminology. Healey integrated the research-teaching nexus with curriculum based on the design of the teaching in terms of student participation and the focus. The resultant model (adapted in Fig. 5.3) consists of two axes: one relating to a continuum between research content/output and research processes, and another relating to a continuum of student participation from active to passive audience. Research content was characterised by a focus on the output of research, the subject knowledge, and data derived from studies. At the opposite end of this continuum is a focus on research process and problems, including exploration of methods and methodologies. The continuum between student and teacher focused education can also be conceptualised as one that considers students as participants, active in the research process (student focused) and student as observers and audience members, with students being taught more didactically. Each quadrant (research-tutored, research-based, research-led, and research-oriented education) offers an opportunity to embed repre-

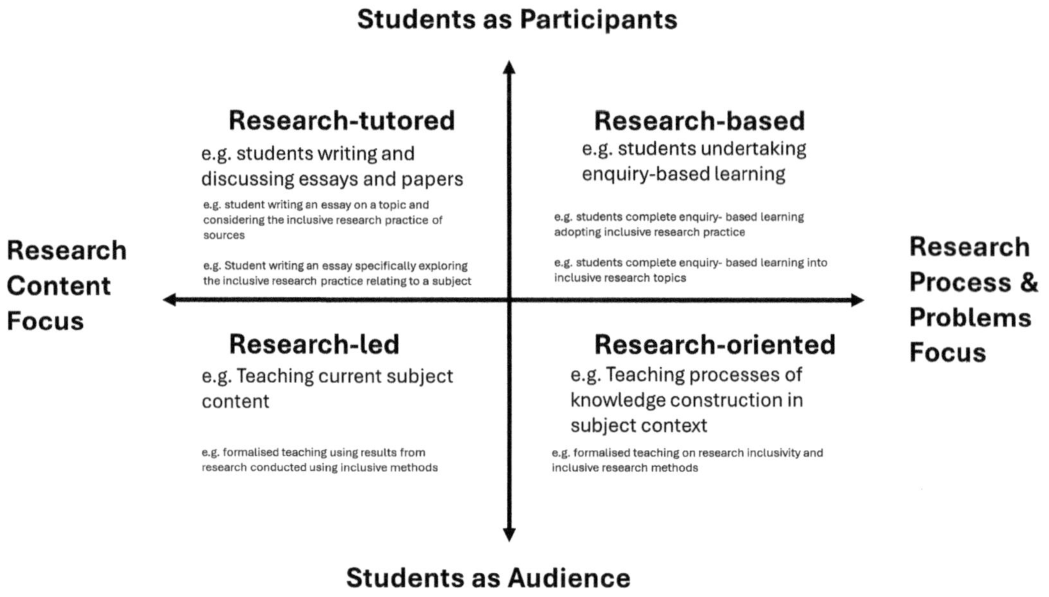

Fig. 5.3 The research-teaching nexus. Adapted from Healey (2005)

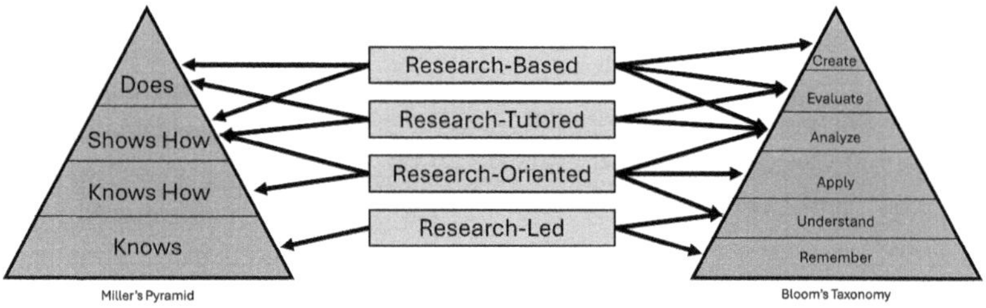

Fig. 5.4 Healey's conceptualised modes of research-informed teaching (centre boxes), mapped to Miller's Pyramid (left) and Bloom's Taxonomy (right)

sentative research practice and knowledge in the curriculum.

In addition to Healey's, other models for curriculum design are notable in biomedical education. Bloom's Taxonomy (Bloom 1956) and Miller's Pyramid (Miller 1990) are fundamental frameworks in science, anatomy, and clinical education, guiding curriculum design and assessment strategies. These models provide a hierarchical structure for developing and evaluating learning outcomes, ensuring a progression from basic knowledge acquisition to complex skill application and clinical competence.

While the validity and nuance of the model outlined by Healey (2005) may be debated, the aim of this subsection is to demonstrate ways that the research-teaching nexus can be adapted to embed representative research principles.

Mapping Healey's conceptualisation of research-rich teaching to Bloom's Taxonomy and Miller's Pyramid offers a valuable framework for representative research teaching in anatomy and applied anatomical education (Fig. 5.4). This integration allows educators to design curricula that accommodate diverse learning needs and abilities while progressively developing research skills. By aligning research activities with cognitive levels and competency stages, instructors can create a scaffolded approach that engages all students, from novices to advanced learners. This structured method ensures that research experiences in anatomy are accessible and meaningful to students with varying backgrounds, learning styles, and career goals. The following mapping demonstrates how Healey's model can be integrated with Bloom's and Miller's frameworks to create a comprehensive and inclusive approach to research-rich anatomical education.

Healey's conceptualisation of research-rich teaching can be effectively mapped onto both Bloom's Taxonomy and Miller's Pyramid, highlighting the progressive nature of research integration in education. Research-led teaching aligns with the lower levels of Bloom's Taxonomy (remembering and understanding) and the 'Knows' level of Miller's Pyramid, as students are introduced to current research findings. Research-oriented teaching corresponds to the middle levels of Bloom's Taxonomy (applying and analyzing) and the 'Knows How' level of Miller's Pyramid, focusing on developing research skills and techniques. Research-tutored teaching relates to the higher levels of Bloom's Taxonomy (analyzing and evaluating) and the 'Shows How' level of Miller's Pyramid, as students engage in critical discussions and analysis of research. Finally, research-based teaching represents the pinnacle of both models, aligning with the highest level of Bloom's Taxonomy (creating) and the 'Does' level of Miller's Pyramid, where students actively conduct research projects. This alignment demonstrates how Healey's framework can be used to scaffold students' progression from passive consumers of research to active researchers, mirroring the cognitive development outlined in Bloom's Taxonomy and the competency acquisition described in Miller's Pyramid.

5.6.1 Representative Research-Led Teaching

Based on Healey's model, research-led teaching occupies the lower left quadrant, representing an approach that has an emphasis on research content and output, and that is led by the educator, with students acting primarily as an audience.

From the original model, this can be seen in a curriculum that centres current subject knowledge. Activities typical of research-led teaching might be seminars or workshops where students are learning about specific academic developments and current knowledge in their discipline. In this way, lecturers may be presenting and sharing their own research experience and expertise with students, sharing their contributions with the cohort.

This approach is likely to look different depending on the cohort being taught. While early year students may be focusing on discipline knowledge as presented in textbooks, more advanced years are likely to be exploring primary literature and considering newly formed knowledge against their developed narrative of the discipline. For example, in anatomy, a group may be examining clinical, electromyographic, and functional primary literature to expand understanding of supraspinatus muscle function.

To embed inclusive practice into this format of teaching, you may wish to include studies that have been carried out on a specific but historically overlooked sample population. This can be integrated into already existing sessions exploring research outputs and discipline knowledge, or as a basis for novel sessions exploring discipline knowledge from an inclusivity perspective.

For example, in the University of Bristol, students on the Applied Anatomy and Functional and Clinical Anatomy programmes have timetabled seminars exploring prescribed reading in their learning. In these sessions, in addition to exploring the topic at hand (e.g. bone remodelling and growth) they also explore how generalisable the output data is, by identifying assumptions and omissions.

> **Top Tip** Acknowledging that as a discipline, inclusive studies are not always available for discussion, consider using this as an opportunity to discuss the lack of diversity in extant literature and possible solutions. This can help highlight assumptions and gaps in current discipline knowledge.

5.6.2 Representative Research-Tutored Teaching

In representative research-tutored teaching, the focus of the delivery remains on the output of the research and the research content (as in Research-Led teaching, above) but the method of delivery shifts to considering the students as participants in research.

In order to engage students as participants in this quadrant of the model (Fig. 5.3), students are often engaged in active tasks (e.g. tutorial discussions) or coursework (e.g. essays or presentations), where they are required to demonstrate familiarity with the literature and its content on a specific topic.

Research-tutored teaching can adopt representative research practices in a number of ways. Discussions and tutorials exploring research paper findings can incorporate discussing the findings from an inclusive lens. Additionally, prescribed reading could be selected based on the inclusive applications of the reading.

Top Tip When discussing the findings of results, consider the use of questions to introduce the exploration of inclusivity. For example:

- How might the study's conclusions differ if applied to groups not well-represented in the sample?
- Did the researchers consider intersectionality in their analysis? If not, how might this impact the interpretation of results?
- Are there any cultural or socioeconomic factors that could influence the generalisability of the findings?
- What assumptions about gender, race, ethnicity, or ability are implicit in the study's conclusions?
- How do the study's findings reinforce or challenge existing power structures or social inequalities?
- How might the dissemination and application of these findings affect different social groups?
- What further research questions arise when considering the study's implications for promoting equity and inclusion?
- How could the study's conclusions be applied to create more inclusive policies or practices in relevant fields?
- Are there any potential unintended consequences of implementing the study's recommendations that could disproportionately affect certain groups?

An example of a representative Research-Tutored approach is the seminar 'The Anatomy of Sex and Gender' delivered by Scott Abbott Paterson at the University of Bristol, Hull York Medical School, The University of Edinburgh (and other institutes) where the academic literature considering sex and gender and how it relates to minoritised populations is explored through a combination of discussion and independent and small group reflection and research.

Academics may encourage students to include an examination of inclusive applications or considerations of studies in responses to essay questions. This can be used to evidence criticality, analysis, and argumentation. Additionally, essay questions themselves might be expanded to include inclusive applications. For example, an essay title used at the University of Bristol on the BSc Functional and Clinical Anatomy programme has been 'A 35-year-old transfeminine patient presents for elective genital ('bottom') gender affirmation surgery following 12 months of continuous feminising hormone therapy and referral from pre-operative counselling. Discuss the anatomical basis for the procedures available to this patient'. This title centres the inclusive discussion by asking students to consider a historically marginalised group at the centre of a clinical and anatomical essay, necessitating students to consider minoritised groups and literature relating to them in their essay construction.

5.6.3 Representative Research-Oriented Teaching

In research-oriented teaching, the focus of the delivery shifts from the output of the research and the research content to the methods, process, and problems. However, aligning with research-led teaching, this is more passive in terms of student participation, with a focus on didactic delivery.

Formalised teaching here may be found in skills units/modules, commonplace as a supplement to research-based teaching (Sect. 5.6.4). Sessions may explicitly consider research methods common in disciplinary research, for example tutorials examining the value of dissection as a research method, or the design of imaging-based data collection.

Inclusivity may be embedded into this delivery by either (1) embedding the consideration of inclusivity into the presentation of standard methods, or (2) specifically including sessions on representative research practices. In option 1, academics may discuss their research methods and apply a critical lens exploring the inclusivity of these. They may highlight challenges faced in terms of diverse and inclusive participant recruitment, for example. Alternatively, a session may be designed to teach participant recruitment, incorporating the advice offered previously.

> **Top Tip** Occasionally, audiences may be resistant to representative research practices or discussing these in educational settings. Pairing inclusive practice with research quality, i.e. linking representation and data accuracy with validity and quality of findings, and generalisability of conclusions can be a useful way of demonstrating the importance and inviting people into the conversation.

5.6.4 Representative Research-Based Teaching

In the Healey (2005) model of research-informed teaching, research-based teaching occupies the upper right quadrant, representing an approach that emphasises research processes and problems while engaging students as active participants. This quadrant provides rich opportunities for embedding inclusive practices into the research process itself.

In research-based teaching, students are typically involved in conducting their own research projects, often in the form of dissertations or capstone projects. To make this approach more inclusive, educators can guide students towards adopting (1) representative research practices and (2) exploring inclusive subjects.

1. Research projects adopting inclusive practices:

Students can be encouraged to design research projects that incorporate inclusive methodologies from the outset. This might involve:

- Developing diverse and representative sampling strategies
- Creating accessible data collection tools and methods
- Considering intersectionality in data analysis
- Engaging with community stakeholders in participatory research designs

For example, a student could be encouraged to recruit a diverse participant pool that includes individuals from various ethnic backgrounds, age groups, and ability levels. They could also be guided to design data collection methods that are accessible to participants with various physical or cognitive abilities.

2. Research on inclusive subjects:

In addition to adopting inclusive practices, students can be encouraged to focus their research on topics that directly address issues of inclusivity, equity, and diversity. Examples might include:

- Dissertations examining the process of decolonising curricula in anatomy education
- The anatomical basis of clinical interventions in diverse patient groups
- Anatomical conceptualisations that perpetuate stereotypes in marginalised populations

For instance, at the University of Bristol, dissertation topics offered to the BSc Applied Anatomy and iBSc Functional and Clinical Anatomy programmes include titles such as: 'Sex segregation in sport: an anatomical examination' or 'Anatomical Complications of Gender Affirming Surgery'. These projects not only adopt representative research practices but also focus on an inclusive subject matter.

To support students in these endeavours, educators can:

- Provide guidance on representative research methodologies during research methods courses
- Offer a list of potential research topics that address inclusivity in the discipline
- Connect students with mentors who have experience in representative research
- Encourage critical reflection on the inclusivity of research processes throughout the project

By engaging in research-based teaching with an inclusive focus, students not only learn about their discipline but also develop skills in conducting ethical, representative, and socially responsible research. This approach can help prepare students to be more inclusive and equitable researchers and practitioners in their future careers.

Top Tip When guiding students in research-based teaching, encourage them to consider the potential impact of their research on different communities. Prompt them to reflect on questions such as: How might your research findings benefit or potentially harm various groups? What steps can you take to ensure your research process and outcomes are equitable and inclusive?

Top Tip While the model presented has suggested that these approaches are mutually exclusive, significant overlap can be observed between them. For example, discussions of data findings commonly involve an examination of the methods. Depending on the level of the course or unit/module, and curriculum, it is likely that single sessions will adopt practices that align with all quadrants of the model. Look for opportunities to highlight representative research practice across the quadrants, even in the same session. If inclusive considerations are made whenever research practices are mentioned, students will also realise that inclusive and good research practice are inter-related.

5.6.5 Research-Teaching Nexus and Spectrum of EDI Responsiveness

While examining equity and inclusion in research funding, Gladstone et al. (2022) created an EDI responsiveness spectrum, proposing this framework to help support accountability in terms of EDI responses. This framework conceptualises EDI responsiveness on a spectrum from 'EDI Unequal' (where activity perpetuates inequality) to 'EDI transformative' where actions consider EDI issues and eliminate the impact of structural inequality.

The integration of representative research practices into the Healey (2005) research-teaching nexus model offers a compelling framework for enhancing science education. Mapping the research-teaching nexus relative to the Gladstone et al. framework and inclusivity allows us to suggest how this practice might evolve (Fig. 5.5). By considering EDI principles across research-tutored, research-based, research-led, and research-oriented approaches, educators can create more inclusive and representative learning experiences. Relative to the EDI Responsiveness Spectrum, truly representative research education should aim for the 'EDI Transformative' level. At this level, educators not only address causes of inequality but also actively work to transform harmful norms and eliminate structural barriers within the discipline. We can recognise that representative research-led practices, where the current subject content is cognisant of inequalities can be a useful mechanism to move from an 'EDI unequal' or 'EDI unaware' to 'EDI

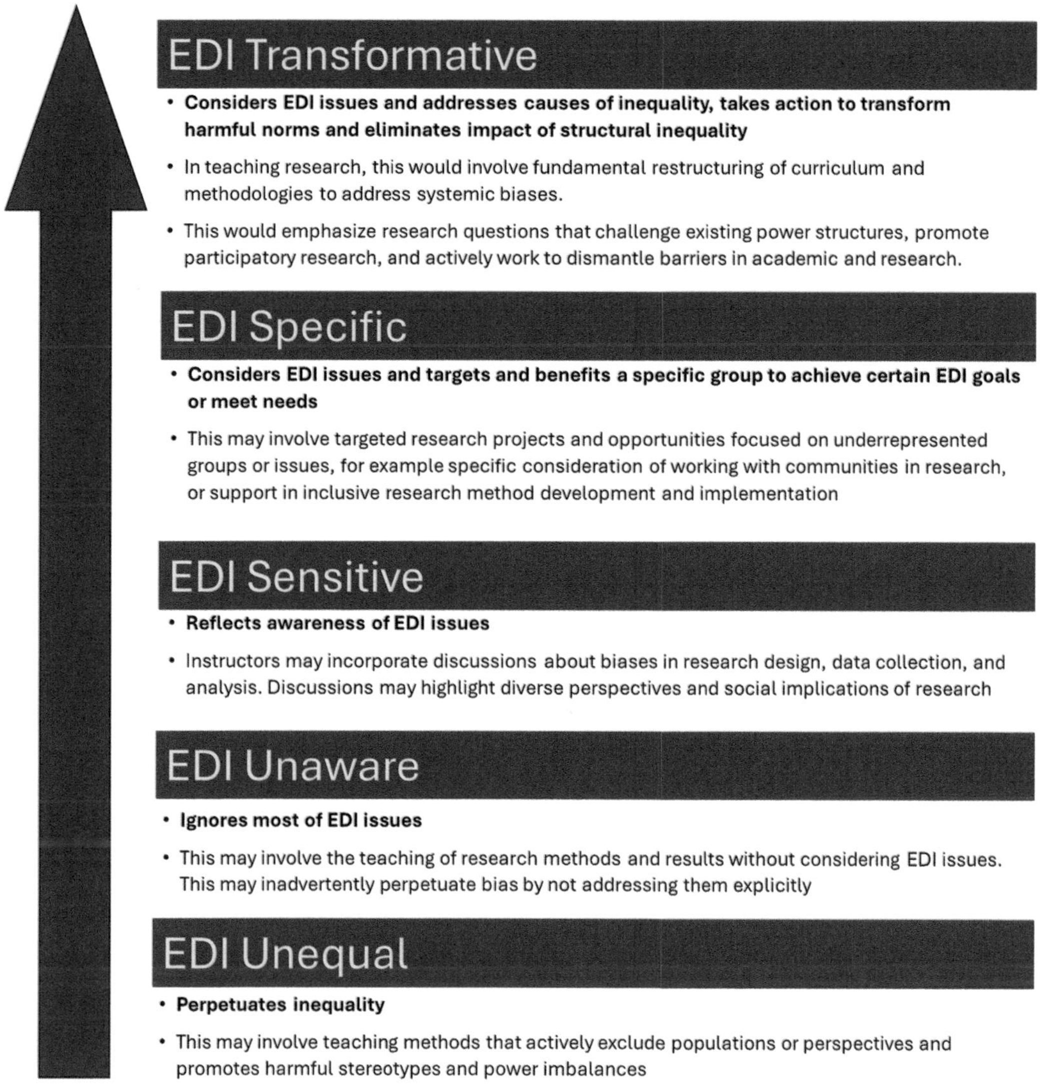

Fig. 5.5 The EDI responsiveness spectrum as applied to the representative research-teaching nexus (spectrum adapted from Gladstone et al. (2022))

sensitive' position. Similarly, research-based practices where research opportunities specifically relate to equality and to overcome systemic inequity (for example, the representative research dissertations discussed previously) are opportunities for 'EDI specific' and 'EDI transformative' practice. These may be especially impactful when supported by training in representative research methods.

There remain significant opportunities to critically examine and refine what we teach and how we teach it, particularly in terms of representing diverse populations in anatomical and scientific research and education. Future work should focus on developing concrete strategies to move anatomy departments along this spectrum, ensuring that representative research practices become intrinsic to both the content and methodology of anatomical education and research.

In their review of the integration of teaching and research, Brew (2006) offers that academics consider that their research contributes to their understanding of themselves and the world, additionally they note that as students explore research, they develop their personal epistemologies and emotional dimensions—this observation highlights the important personal and social dimensions and benefits for students and academics. These align so well with goals of inclusivity and equality and speak to the continued need for a representative research-teaching nexus.

5.7 A Call to Action for Representative Research in Anatomy

As a discipline, anatomy is easily dismissed as being 'dead', as having few new discoveries left and little new knowledge to gain. However, science's historical and non-remitting omission of minoritised populations and poor consideration of sample diversity means that as a discipline, our current baselines may not represent the range of diversity seen in the population. By embracing representative research practices, we not only fulfil our moral obligation but also enhance the quality and universality of our scientific understanding.

Let us take actionable steps towards this goal:

- Embrace reflexivity: Understand your positionality and biases. Use tools like reflexivity statements and research diaries to maintain awareness throughout your projects.
- Diversify your team: Seek collaborators from varied backgrounds to enrich your research perspective.
- Engage with your community: Be transparent with stakeholders and participants, especially those from marginalised groups.
- Innovate in data presentation: Explore accessible formats like podcasts, social media, and visual media to reach wider audiences.
- Plan for equitable authorship and recognition: Determine roles and authorship order early in the project.
- Refine your demographic data collection: Critically assess what information is essential and how it relates to your research aims.
- Implement open-text options: Allow participants to describe themselves in their own words when possible.
- Pilot your questionnaires: Gather feedback from diverse individuals to identify and address potential ambiguities.
- Consider multiple coders: Incorporate varying perspectives in data analysis when feasible.
- Acknowledge limitations: Recognise and discuss the constraints of your methods, particularly regarding intersectionality.
- Prioritise ethical considerations: Ensure participant anonymity, especially with open-text demographic data.
- Adopt flexible data interpretation: Consider themes before demographics to avoid limiting your analysis.
- Address gaps in literature: Use the lack of diverse studies as an opportunity to discuss and propose solutions.
- Link representation to research quality: Demonstrate how inclusive practices enhance validity and generalisability.
- Encourage student reflection: Prompt students to consider the broader impact of their research on different communities.
- Integrate representative practices across teaching: Highlight inclusive research methods throughout various aspects of anatomy education.

The journey towards truly representative research is ongoing. Each step we take, no matter how small, contributes to a more inclusive and equitable scientific community. By committing to these practices, we not only advance our discipline but also pave the way for a more comprehensive and nuanced understanding of human anatomy.

Let us embrace this challenge and work together to reshape the future of anatomical research and education.

Acknowledgments The authors would like to thank Dr Jaysan Charlesford, School of Psychology, University of Plymouth, for the constuctive feedback he provided for Sect. 5.3. His knowledge and insight helped clarify the points raised in this section.

References

Akbulut N, Razum O (2022) Why othering should be considered in research on health inequalities: theoretical perspectives and research needs. SSM Popul Health 20:101286–101286

Alharahsheh HH, Pius A (2020) A review of key paradigms: positivism VS interpretivism. Glob Acad J Humanit Soc Sci 2:39–43

Alpers L-M (2018) Distrust and patients in intercultural healthcare: a qualitative interview study. Nurs Ethics 25:313–323

Armstrong A, Collis A, Walmsley J (2022) The messy realities of inclusive research: personal. Reflections 2022:311–328

Australian Bureau of Statistics (2021) Census of population and housing: census dictionary [online]. Available at: https://www.abs.gov.au/census/guide-census-data/census-dictionary/2021/downloads [Accessed: 25 July 2024]

Azmitia M, Syed M, Radmacher K (2008) On the intersection of personal and social identities: introduction and evidence from a longitudinal study of emerging adults. New Dir Child Adolesc Dev 2008:1–16

Bierer BE, Meloney LG, Ahmed HR, White SA (2022) Advancing the inclusion of underrepresented women in clinical research. Cell Rep Med 3(4):100553

Bigby C, Frawley P, Ramcharan P (2014) Conceptualizing inclusive research with people with intellectual disability. J Appl Res Intellect Disabil 27:3–12

Bloom BS (1956) Taxonomy of educational objectives: the classification of educational goals

Braun V, Clarke V (2022) Thematic analysis: A practical guide. London: SAGE Publications

Brew A (2006) Research and teaching: beyond the divide

Caldwell WE, Moloy HC, D'Esopo DA (1940) The more recent conceptions of the pelvic architecture. Am J Obstet Gynecol 40:558–565

Cameron JJ, Stinson DA (2019) Gender (mis)measurement: guidelines for respecting gender diversity in psychological research. Soc Personal Psychol Compass 13:e12506

Clark T (2021) Bryman's social research methods. Oxford University Press

Comer AR (2022) The evolving ethics of anatomy: dissecting an unethical past in order to prepare for a future of ethical anatomical practice. Anat Rec (Hoboken) 305(4):818–826

Cooper F, Dolezal L, Rose A (2023) COVID-19 and shame: political emotions and public health in the UK. Bloomsbury, pp 59–76

DAISY (2022) Diversity and inclusion survey (DAISY) question guidance - working draft (V2) [Online]. Diversity and Inclusion Survey (DAISY) Question Guidance (Version 2) - EDIS (edisgroup.org). Accessed 10 July 2024

Delgado R, Stefancic J, Harris A (2017) Critical race theory (third edition): an introduction. NYU Press

Durant RW, Wenzel JA, Scarinci IC, Paterniti DA, Fouad MN, Hurd TC, Martin MY (2014) Perspectives on barriers and facilitators to minority recruitment for clinical trials among cancer center leaders, investigators, research staff and referring clinicians. Cancer 120:1097–1105

Evans L, Snow K, Jayanth A, Moyes SM (2025) From attention to action: advancing inclusivity in the landscape of anatomy education. Method Innov 0(0). https://doi.org/10.1177/20597991251313878

Fernandez T, Godwin A, Doyle J, Verdín D, Boone H, Kirn A, Benson L, Potvin G (2016) More comprehensive and inclusive approaches to demographic data collection

Finn GM, Danquah A, Matthan J (2022) Colonization, cadavers, and color: considering decolonization of anatomy curricula. Anat Rec 305:938–951

Frankena TK, Naaldenberg J, Cardol M, Garcia Iriarte E, Buchner T, Brooker K, Embregts P, Joosa E, Crowther F, Fudge Schormans A, Schippers A, Walmsley J, O'Brien P, Linehan C, Northway R, Schrojenstein Lantman-De Valk H, Leusink G (2019) A consensus statement on how to conduct inclusive health research. J Intellect Disabil Res 63:1–11

García Iriarte E, Díaz Garolera G, Salmon N, Donohoe B, Singleton G, Murray L, Dillon M, Burke C, Leddin N, Sullivan M, Spelman M (2023) How we work: reflecting on ten years of inclusive research. Disabil Soc 38:205–227

Garratt D, Johnson K, Millear A, Picken S, Slattery J, Walmsley J (2022) Celebrating thirty years of inclusive research. Soc Sci (Basel) 11:385

Gaventa J, Cornwall A (2008 Power and knowledge. In: P Reason, H Bradbury (eds) The SAGE handbook of action Research 2nd edn SAGE Publications Ltd. pp 172–189. Available at: https://doi.org/10.4135/9781848607934

Gilmore-Bykovskyi A, Croff R, Glover CM, Jackson JD, Resendez J, Perez A, Zuelsdorff M, Green-Harris G, Manly JJ (2022) Traversing the aging research and health equity divide: toward intersectional frameworks of research justice and participation. Gerontologist 62:711–720

Gladstone J, Schipper L, Hara-Msulira T, Casci T (2022) Equity and inclusivity in research funding: barriers and delivering change

Griffiths R (2004) Knowledge production and the research-teaching nexus: the case of the built environment disciplines. Stud High Educ 29:709–726

Grue M (2021) Diversify or decolonise? What you can do right now and how to get started. In: Doing equity and diversity for success in higher education: redressing structural inequalities in the academy. Springer

Healey M (2005) Linking research and teaching: disciplinary spaces. In: Reshaping the university: new relationships between research, scholarship and teaching, vol 30. Open University Press, p 42

Herer A, Schwartz AE (2022) How being a researcher impacted my life. Soc Sci 11:127

Hesse-Biber SN (2010) Mixed methods research merging theory with practice
Hickey S, Mohan G (2004) Participation: from tyranny to transformation: exploring new approaches to participation in development. Zed Books, p x
Hughes OE, Schwartz AE (2024) Planning inclusive research: a comparison of two inclusive dissertation research projects. Inclusion (Washington, DC) 12:40–54
Hughes JL, Camden AA, Yangchen T (2016) Rethinking and updating demographic questions: guidance to improve descriptions of research samples. Psi Chi J Psychol Res 21:138–151
Hughes JL, Camden AA, Yangchen T, Smith GPA, Domenech Rodríguez MM, Rouse SV, McDonald CP, Lopez S (2022) INVITED EDITORIAL: guidance for researchers when using inclusive demographic questions for surveys: improved and updated questions. Psi Chi J Psychol Res 27:232–255
Kim J, Florian L, Pantić N (2022) The development of inclusive practice under a policy of integration. Int J Incl Educ 26:1068–1083
Kouri A, Dandurand RJ, Usmani OS, Chow C-W (2021) Exploring the 175-year history of spirometry and the vital lessons it can teach us today. Eur Respir Rev 30:210081
Kuliukas A, Kuliukas L, Franklin D, Flavel A (2015) Female pelvic shape: distinct types or nebulous cloud? Br J Midwifery 23:490–496
McLinden M, Edwards C, Garfield J, Moron-Garcia S (2015) Strengthening the links between research and teaching: cultivating student expectations of research-informed teaching approaches. Educ Pract 2
Miller GE (1990) The assessment of clinical skills competence performance. Acad Med 65:S63–S67
Morley L (2022) Queering higher education: troubling norms in the global knowledge economy
National Institute of Health (2019) NCCIH clinical research toolbox [online]. NCCIH (nih.gov)
Neal RD, Johnson P, Clarke CA, Hamilton SA, Zhang N, Kumar H, Swanton C, Sasieni P (2022) Cell-free DNA-based multi-cancer early detection test in an asymptomatic screening population (NHS-Galleri): design of a pragmatic, prospective randomised controlled trial. Cancers (Basel) 14:4818
Neumann R (1994) The teaching-research nexus: applying a framework to university students' learning experiences. Eur J Educ 29:323–338
Nickerson RS (1998) Confirmation bias: a ubiquitous phenomenon in many guises. Rev Gen Psychol 2:175–220
Niranjan SJ, Martin MY, Fouad MN, Vickers SM, Wenzel JA, Cook ED, Konety BR, Durant RW (2020) Bias and stereotyping among research and clinical professionals: perspectives on minority recruitment for oncology clinical trials. Cancer 126:1958–1968
O'Brien P, McConkey R, García-Iriarte E (2014) Co-researching with people who have intellectual disabilities: insights from a National Survey. J Appl Res Intellect Disabil 27:65–75
O'Brien P, Garcia Iriarte E, McConkey R, Butler S, O'Brien B (2022) Inclusive research and intellectual disabilities: moving forward on a road less well-travelled. Soc Sci 11:483
Office for National Statistics (2021) Census 2021 paper questionnaires [online]. Available at: https://www.ons.gov.uk/census/censustransformationprogramme/questiondevelopment/census2021paperquestionnaires (Accessed: 24 July 2024)
Olmos-Vega FM, Stalmeijer RE, Varpio L, Kahlke R (2023) A practical guide to reflexivity in qualitative research: AMEE guide no. 149. Med Teach 45:241–251
Overlaet B (2022) A pathway towards multidimensional academic careers—a LERU framework for the assessment of researchers. League of European Research Universities
Passmore SR, Kisicki A, Gilmore-Bykovskyi A, Green-Harris G, Edwards DF (2022) "There's not much we can do..." researcher-level barriers to the inclusion of underrepresented participants in translational research. J Clin Transl Sci 6:e4
Pfeiffer B, Hallock T, Tomczuk L, Kramer J (2024) Peer support provided by people with intellectual and developmental disabilities: a rapid scoping review to develop a toolkit for inclusive research. Social Sci (Basel) 13:47
Ravindran TS, Teerawattananon Y, Tannenbaum C, Vijayasingham L (2020) Making pharmaceutical research and regulation work for women. BMJ 371:m3808
Robertson J (2007) Beyond the 'research/teaching nexus': exploring the complexity of academic experience. Stud High Educ 32:541–556
Rogers LO, Syed M (2021) Conceptualizing the multiple levels of identity and intersectionality
Said EW (1978) Orientalism. Pantheon Books, New York, pp xi–368
Sambunjak D, Straus SE, Marušić A (2006) Mentoring in academic medicine: A systematic review. JAMA 296:1103–1115
St. John B, Mihaila I, Dorrance K, Dawalt LS, Ausderau KK (2018) Reflections from co-researchers with intellectual disability: benefits to inclusion in a research study team. Intellect Dev Disabil 56:251–262
Statistics Canada (2021) Census of population: questionnaire(s) and reporting guide(s) for 2021 [online]. Available at: https://www23.statcan.gc.ca/imdb/p2SV.pl?Function=getSurvInstrumentList&Id=1283315 [Accessed: 24 July 2024]
Statistics South Africa (2022) Census 2022. [online] Available at: https://census.statssa.gov.za/#/ (Accessed: 24 July 2024).
Swanton C, Neal RD, Johnson PWM, Dur CC, Hamilton SA, Zhang N, Kumar H, Sasieni P (2022) NHS-Galleri trial design: equitable study recruitment tactics for targeted population-level screening with a multi-cancer early detection (MCED) test. J Clin Oncol 40:TPS6606
Thomas M (ed) (2023) Race, ethnicity, and the COVID-19 pandemic. University of Cincinnati Press

United States Census Bureau (2021) Improvements to 2020 census race and hispanic origin question designs [online]. Available at: https://www.census.gov/newsroom/blogs/random-samplings/2021/08/improvements-to-2020-census-race-hispanic-origin-question-designs.html [Accessed: 24 July 2024]

Van Schalkwyk GI, Marin CE, Ortiz M, Rolison M, Qayyum Z, McPartland JC, Lebowitz ER, Volkmar FR, Silverman WK (2017) Social media use, friendship quality, and the moderating role of anxiety in adolescents with autism Spectrum disorder. J Autism Dev Disord 47:2805–2813

Vega-Córdova V, Álvarez-Aguado I, Jenaro C, Spencer-González H, Díaz Araya M (2020) Analyzing roles, barriers, and supports of co-researchers in inclusive research. J Policy Pract Intell Disab 17:318–331

Vousden N, Bunch K, Kenyon S, Kurinczuk JJ, Knight M (2024) Impact of maternal risk factors on ethnic disparities in maternal mortality: a national population-based cohort study. Lancet Reg Health Eur 40:100893

Walmsley J (2001) Normalisation, emancipatory research and inclusive research in learning disability. Disab Soc 16:187–205

Walmsley J, Strnadová I, Johnson K (2018) The added value of inclusive research. J Appl Res Intellect Disabil 31:751–759

Wilkinson S, Kitzinger C (1996) Representing the other: a feminism & psychology reader

Wong SHM, Gishen F, Lokugamage AU (2021) 'Decolonising the medical curriculum': humanising medicine through epistemic pluralism, cultural safety and critical consciousness. Lond Rev Educ 19

Woolverton GA, Marks AK (2023) "I just check 'Other'": evidence to support expanding the measurement inclusivity and equity of ethnicity/race and cultural identifications of U.S. Adolescents. Cultur Divers Ethnic Minor Psychol 29:64–73

6 Incorporating Inclusivity and Accessibility in Anatomy Education: A Toolkit for Early Career Anatomy and Medical Educators

Munesh Pal Khamuani and Elena Patera

Abstract

Inclusivity and accessibility are important concepts in higher education as all students are entitled to a learning experience that is free of barriers. The following chapter aims to act as a toolkit for early career anatomists and medical educators who aspire to make their teaching practice more inclusive and accessible. The chapter consists of simple tips that can be incorporated into an educator's daily teaching practice. The chapter also signposts the readers to some freely available resources that can be used by early career anatomy and medical educators to create inclusive and accessible educational content. In terms of inclusivity, the chapter will discuss the importance and use of inclusive images that represent a diverse range of bodies and skin tones. Additionally, the use of inclusive language across the anatomy curriculum, inclusive assessment strategies as well as how anatomy educators can become more familiar with their students and different cultures and faiths will be discussed. In terms of accessibility, the chapter will focus on how educators can make their teaching content and material accessible to students with or without visible or hidden disabilities to ensure equitable learning without any significant barriers or challenges.

Keywords

Inclusivity · Accessibility · Anatomy education · Medical education · Medical students

6.1 Introduction

According to Advance Higher Education (AdvanceHE), inclusive teaching and learning is when every student is entitled to a learning experience where diversity, learning needs, and preferences are respected, participation is enabled, and barriers that impede the students' learning process are removed (AdvanceHE 2020a). Inclusive teaching and learning is a result of a multi-step process. Educators need to thoroughly consider and plan how to design and deliver educational content that is inclusive for their students. Furthermore, educators must ensure that the assessment methods they use to assess students' knowledge allow students to meaningfully

M. P. Khamuani (✉)
School of Medicine, Faculty of Health Sciences and Wellbeing, University of Sunderland, Sunderland, UK
e-mail: Munesh.Khamuani@sunderland.ac.uk

E. Patera
Human Anatomy Resource Center, University of Liverpool, Liverpool, UK

Section of Anatomy, St George's University of London, London, UK
e-mail: E.Patera@liverpool.ac.uk

R. Door (ed.), *Accessibility, Inclusivity and Diversity in Education and Beyond*, Biomedical Visualization 8, https://doi.org/10.1007/978-3-031-81795-3_6

engage with their curriculum to help them achieve their full potential. Inclusive teaching implies that the educator values student diversity and attempts to represent this student diversity within their educational content (Verbree et al. 2023).

In the United Kingdom, the Equality Act (2010) applies across employment and education. The act outlaws direct and indirect discrimination on the grounds of the protected characteristics including, but not limited to, age, disability, ethnicity, gender reassignment, religion/belief, sex, and sexual orientation (Gov.uk 2010). Students in a higher education setting come with a wide variety of characteristics relating to their educational background, their personality, their current and previous circumstances, and their cultural experience. Hence, curricula that do not take sufficient account of student diversity may lead to unintentional exclusion. In addition, when curricula do not reflect student diversity in their content, some individuals may feel disadvantaged. Educators may not always remember that students come from various backgrounds and have various needs. As educators, the best we can do is to not make assumptions based on our unconscious biases but ask students if they have specific academic concerns they would like to communicate or adjustments that they would need.

In the context of anatomy, educators can ensure that they use illustrations in their educational material (lecture slides or laboratory practical worksheets) that represent the demographics of their student cohorts. The resources used in their teaching sessions, such as anatomical plastic models and mannequins must represent a wide range of skin tones and body sizes and shapes. Moreover, the use of inclusive language is key to students' inclusive learning experience. Students' protected characteristics should be respected, and different ethnic backgrounds and cultures should be valued. Educators' attitudes towards inclusivity can serve as an example of a positive learning experience for students. When students experience inclusive teaching and learning, they can potentially be a step closer to becoming inclusive professionals themselves.

It is evident from the literature that access to education may not always be easy for individuals with disabilities or additional learning needs (Drexel University 2024). This may be due to lack of inclusion or lack of awareness within institutions. It is important to educate university staff and raise awareness to promote inclusive education so there are no barriers to learning for all learners (Fernández-Batanero et al. 2022).

Golden and Petty (2022) highlight the under-representation of people with physical and sensory disabilities in the medical and healthcare profession. They reveal that misconceptions and unconscious biases are a major factor in limiting opportunities for such individuals to access education.

To overcome the above, the General Medical Council (GMC), UK, has stated the following:

> No long-term health condition or disability by virtue of its diagnosis automatically prohibits an individual from studying or practising medicine (GMC 2024a).

Based on the above, in this book chapter, we provide a few examples and tips from our learning as early career anatomists on making anatomy education inclusive and accessible to all learners. We also provide some recommendations on resources that are available online that can help educators ensure that their teaching is inclusive and accessible.

6.2 Using Images and Schematic Illustrations Representative of Diverse Human Bodies

A contemporary issue in the field of anatomy education is decolonization of anatomy education. According to Finn et al. (2022), decolonizing anatomy education primarily involves acknowledging the unethical practices anatomists and medical professionals delved into in the past to acquire bodies for dissection, acknowledging how specific groups of individuals were used without their consent to advance science as well as how minorities were never represented in the healthcare or medical education context, except as learning resources. Despite the presence of various articles published in the literature on how anatomy and medical curricula can be decolonized (Nazar et al. 2015; Wong et al. 2021; Mbaki et al. 2021; Finn et al. 2022), there is still a long way for anatomy curricula to become decolonized as decolonization is not a single act or event (Finn et al. 2022).

From our perspective as early career anatomy educators, while learning, studying, and teaching anatomy, we noticed that most images in anatomy and medical textbooks represent white-male or female able-bodied individuals. Evidence from the literature highlights that in medical textbooks female and male bodies are usually represented as toned and muscular, respectively (Parker et al. 2017). The use of such images that represent a specific population group is concerning, as future medical practitioners will be exposed to a wide range of patients with different protected characteristics. It is documented in the literature that individuals from specific ethnic groups are underrepresented in the medical curricula (Louie and Wilkes 2018; Finn et al. 2022). In one study, Martin and associates (2016) found that lecture slides predominantly represent white individuals which is not representative of the racial demographics of the United States (Martin et al. 2016). Finn et al. (2022) emphasize that a major flaw in instructional design is the absence of images and models that represent the range of skin tones that exist. While educators might not be responsible for the models available at their institution, they can individually ensure the accurate representation of a range of skin tones and body types in their teaching material.

In our teaching material, we aim to incorporate images that represent different body types, skin tones, ages, and anatomical variations. Instead of using the "traditional" body outlines present in most anatomical textbooks which often represent able-bodied white biological female or male individuals, educators can use a body outline that represents individuals of different skin colours, ethnic groups, or ages. We delivered a lecture on the subject of anatomy and anatomical terminology to first-year medical students. In our lecture, we used images from the SERVIER. medical art (Lyon, France) website to represent a range of individuals with different demographic characteristics. Examples of our lecture PowerPoint slides can be seen in Figs. 6.1, 6.2, 6.3, and 6.4.

Servier Medical Art is a freely accessible resource that educators can use and access using the following link: https://smart.servier.com. Servier Medical Art consists of 3000 free images that represent anatomical, cellular, and molecular structures. In terms of the anatomical illustrations present on this website, the content is divided into 10 body systems (auditory, cardiovascular, digestive, locomotor, lymphatic, nervous, reproductive, respiratory, urinary, and visual) (Servier).

The images on Servier Medical Art are licensed under Creative Commons License Y4.0 (https://creativecommons.org/licenses/by/4.0/) meaning the user is free to share and adapt the images with attribution to the original creator. The images can be downloaded in both PNG format and as a PowerPoint file, the latter allowing edits to be made. For instance, Fig. 6.5a, b was obtained from the Servier Medical Art upon searching the term "genitalia". Even though Fig. 6.5a, b represents the biological female and male genitalia in a white-skinned individual, respectively, educators can remove the colour and simply use the body outline that demonstrates the pelvis and the biological female and male genitourinary systems. An example of this can be seen in Fig. 6.6a, b.

"Introduction to Anatomy: Anatomical Movements" by the University of British Columbia (UBC) is a great example of a resource that represents individuals of different skin colours, body types, and ages. Link to access resource: https://www.clinicalanatomy.ca/labs/411movements.pdf

UBC also provides additional representative resources, including a 3D animation video showing the cardinal movements of labour with the foetus having black skin. The animated video can be accessed using the following link: https://clinicalanatomy.ca/embryology/CardinalMovements.html

In 2021, medical student Chidiebere Ibe became a Forbes featured medical illustrator and

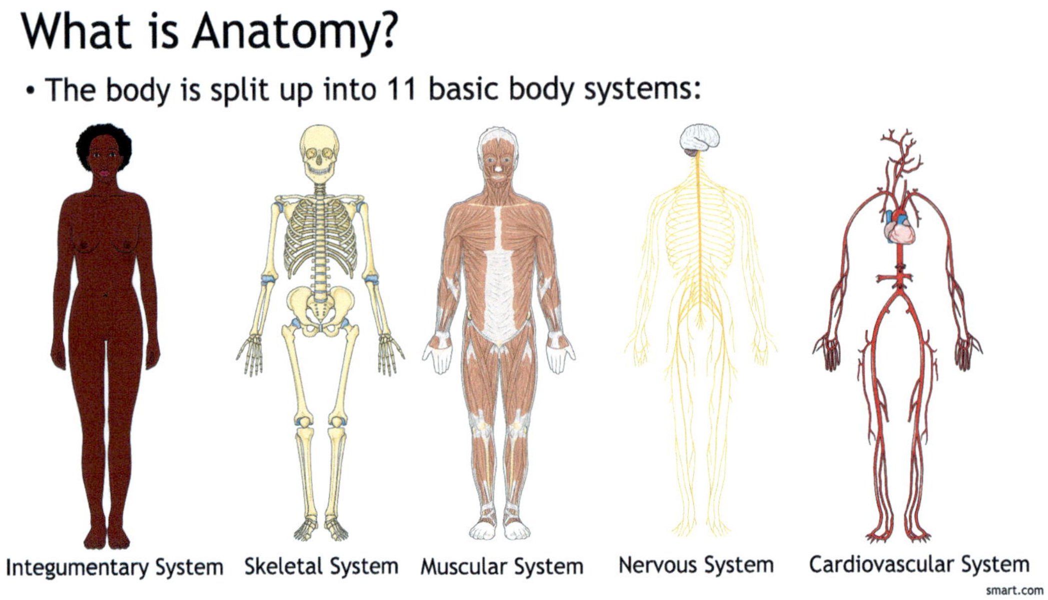

Fig. 6.1 A PowerPoint slide with a schematic illustration of the body of a middle-aged biological female of colour (Servier)

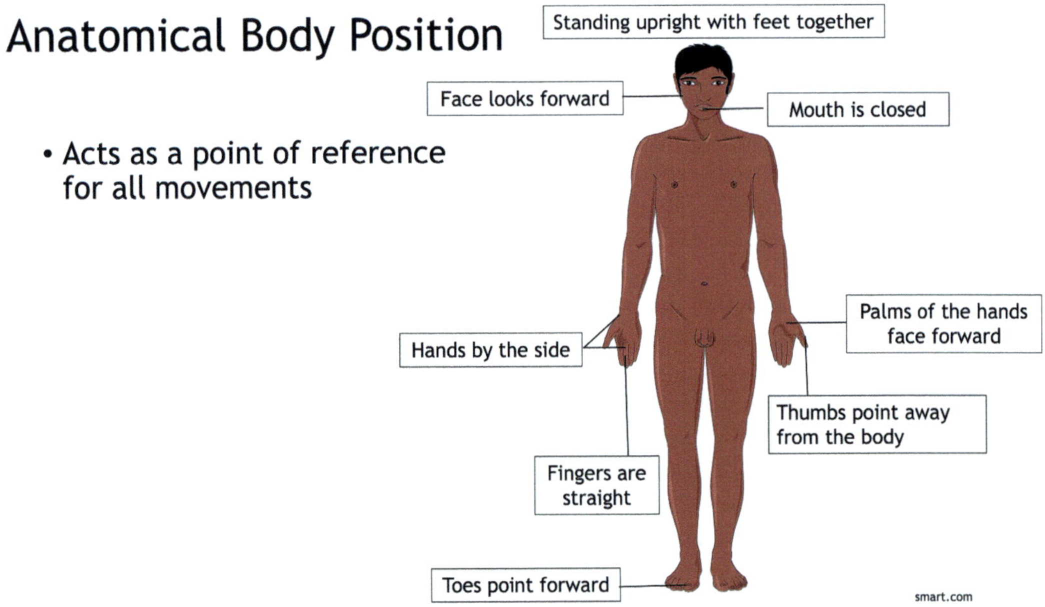

Fig. 6.2 A PowerPoint slide with a schematic illustration of the body of a young biological male with darker skin (Servier)

became known for being the illustrator of the black foetus (Ibe 2024). The medical student and illustrator was the first medical illustrator to have created an illustration of a black pregnant individual and their black foetus. The illustration can be accessed using the following link: https://www.chidiebereibe.com/the-black-fetus-illustration/?doing_wp_cron=1717430829.1570999622344970703125

Notably, traditional anatomical textbooks are starting to change. For instance, a key feature in the latest edition of Gray's Anatomy for Students

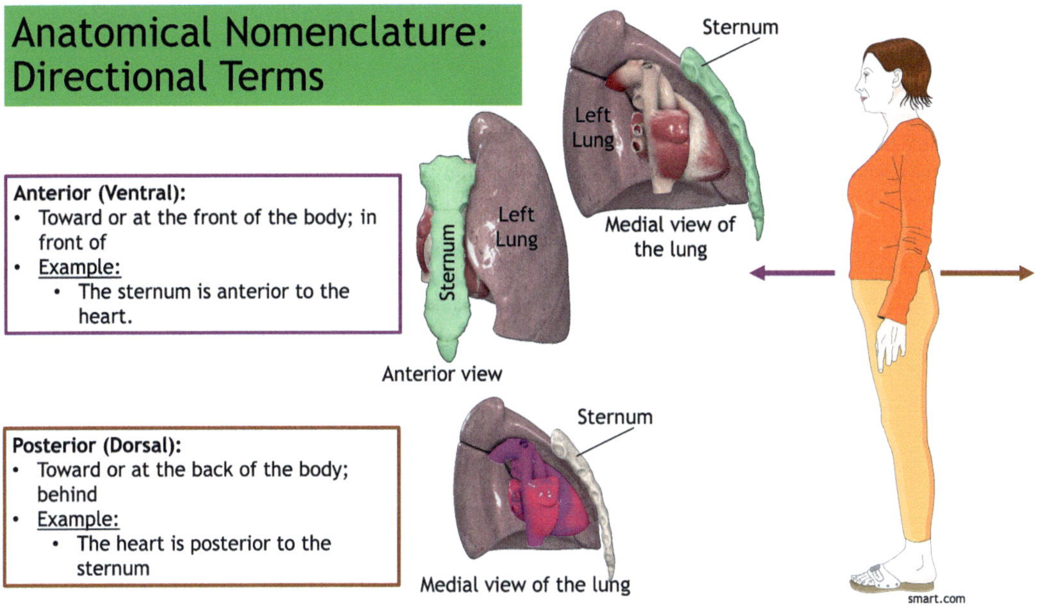

Fig. 6.3 A PowerPoint slide with a schematic illustration of an elderly biological female of white-coloured skin (Servier)

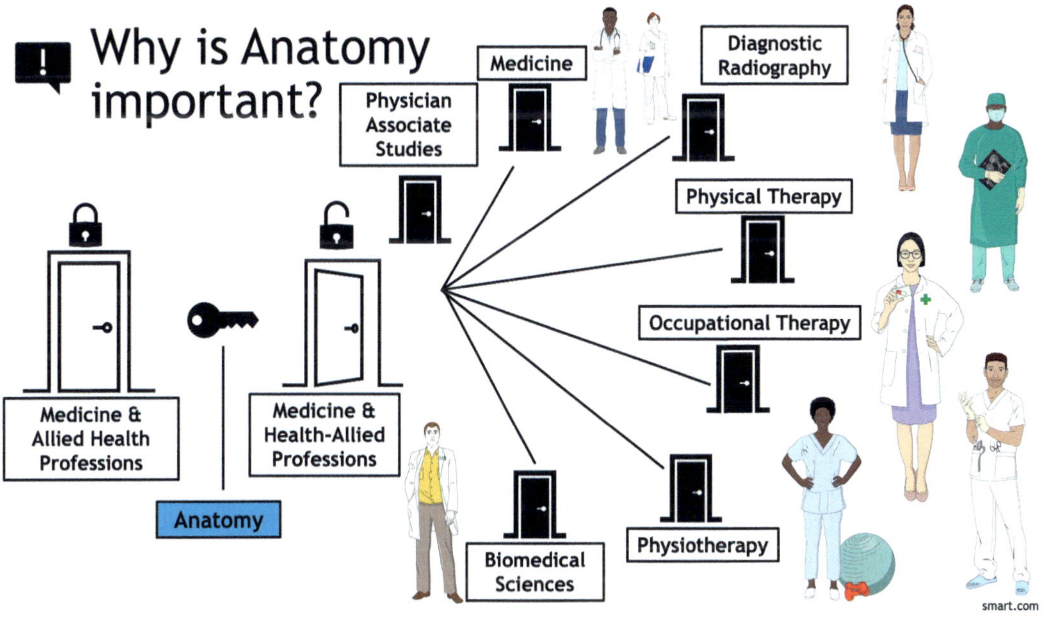

Fig. 6.4 A PowerPoint slide with multiple images representing individuals of different ethnicities in a range of healthcare professions (Servier)

(5th edition) textbook is the increased representation of diverse population groups. A great range of skin tones can be observed throughout the textbook. Other features of the textbook include clarifications over the inclusive use of anatomical terminology regarding trans and non-binary individuals. Lastly, clinical anatomy considerations in relation to intersex and transgender individuals

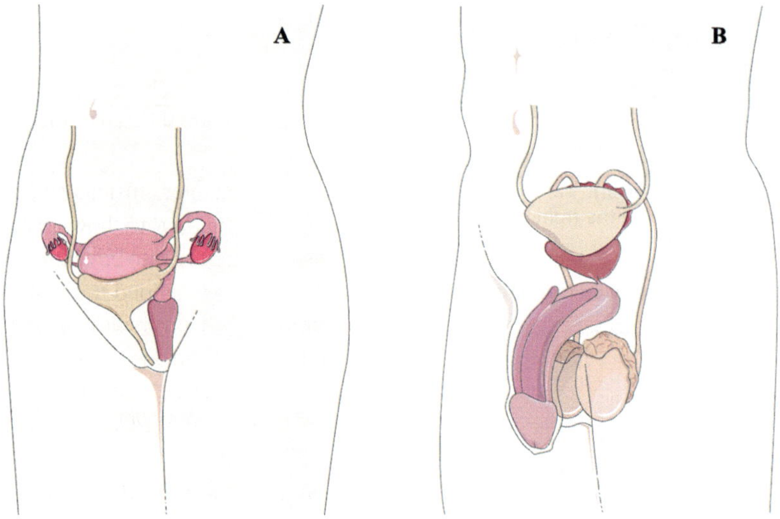

Fig. 6.5 (**a**) and (**b**) Medical illustrations obtained from Servier Medical Art representing the genitourinary system in a white biological female (**a**) and a white biological male (**b**) (Servier)

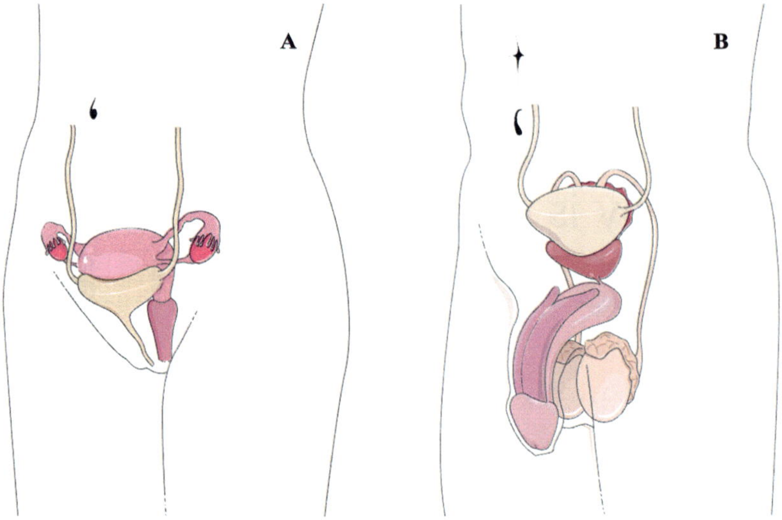

Fig. 6.6 (**a**) and (**b**) Original illustrations obtained from Servier have been modified in Microsoft PowerPoint by the educator to represent the genitourinary system in a biological female and male without any indication of the individual's skin colour (Servier)

are another important feature of the textbook (Drake et al. 2023; Elsevier 2024). Despite this, illustrations on the diversity of human bodies are not extensive; therefore, this can be a call for educational companies and institutions to actively work on creating resources that demonstrate such diversity.

A similar resource to Servier Medical Art is BioRender (Biorender, Toronto, Canada). BioRender is a Canadian software company that can be accessed using the following link: http://biorender.io.

On the BioRender website, there are various scientific templates and more than 50,000 icons from a range of science fields that scientists and educators can use to create scientific figures or illustrations for their content. There is a specific category titled "Anatomy and Clinical" where anatomy educators can find icons and templates relating to gross anatomy and microanatomy. In addition, individuals can use specific search strings to identify structures of interest.

For example, when the search strings "skin tones", "body", or "human body" are used, relevant illustrations depicting the human body in different skin tones, sizes, and shapes are shown. Individuals can sign up for free using the BioRender Basic Plan and will be eligible to create up to five illustrations.

Some medical institutions have a license for the students and staff to use the Complete Anatomy app (3D4Medical, ELSEVIER, Amsterdam, Netherlands). This app is a 3D virtual anatomy platform which now offers a feature where the user can adjust the model's skin based on their preferences. In our teaching practice, we ensure that we adjust the facial characteristics and skin colour of the model throughout a course module to ensure the adequate representation of facial anthropometric differences in individuals of different ethnicity, age, and biological sex (Fig. 6.7).

Until recently, white-skin-coloured biological male anatomy models were the standard resources used in most anatomy laboratories despite manufacturers offering models in different skin colours (Finn et al. 2022). With this ever-increasing accessibility of physical resources, it is imperative that medical institutions should invest in medical models (plastic anatomical models or mannequins) that are non-white (Finn et al. 2022). In case the anatomy laboratory at each institution consists only of plastic models that represent white skin, educators can propose to obtain plastic models that show skin colours other than white. If there are financial constraints regarding purchasing educational models that represent non-white individuals, medical institutions can potentially apply via capital bidding. Most universities have a designated Equality, Diversity and Inclusion (EDI) committee that is responsible for the promotion of various EDI aspects within a department. If an EDI committee is available, anatomy departments could potentially lobby for funding or apply for a grant from internal university resources such as the departmental EDI committee. If lobbying for funding is not possible via internal resources, anatomy departments can approach external funders. Another option is for universities to donate such educational models to other universities if they are no longer being used. Alternatively, universities can temporarily loan educational models that are not needed, to universities in need.

In recent years, a lot of attention has been paid on genital anatomy and its representation in anatomical and medical textbooks (Finn et al. 2022; Hayes and Temple-Smith 2022). Hayes and Temple-Smith (2022) investigated how the vulva

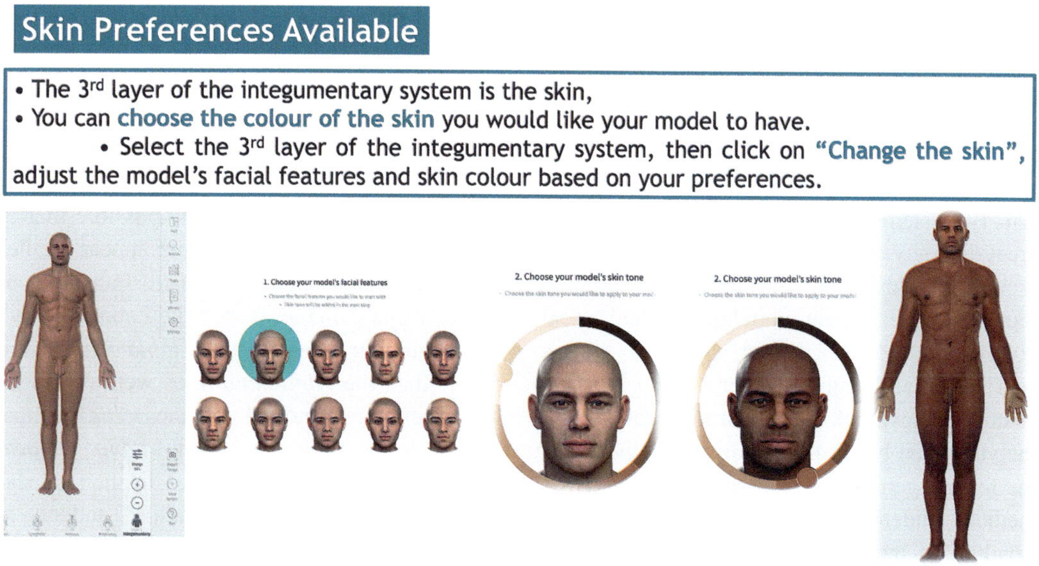

Fig. 6.7 Example of a PowerPoint slide with guidelines prepared by the educator (EP) for the students on how they can adjust the skin colour and facial characteristics of the Complete Anatomy app's model based on their preferences (3D4Medical from Elsevier 2024)

and breasts have been portrayed across old and newer editions of anatomical and medical textbooks. The authors analysed and compared the images and written information for each of these two anatomical structures. Notably, the authors reported common observations across most of the textbooks they investigated. Newer editions of anatomy and medical textbooks that are used in anatomy and medical education were found to not provide students with a realistic representation of how diverse vulvae can be in terms of their shape, size, and colour. In some textbooks, images of the vulva where labial protrusion was evident were used however, there was no textual description referring to it. The authors highlight that none of the textbooks provided any reference to vulvar diversity, range of vulvar measurements, or labial asymmetry. Accentuating the lack of representation of vulvar diversity is the representation of the vulva of non-white biological females. According to Beni et al. (2022), the representation of female external genitalia in a biological female of colour has only been evident in a single anatomical textbook titled "Gray's Surgical Anatomy". In our experience, in previous lectures we received as students on the female genitourinary system, the "typical" image showing the external genitalia of a white biological female is used for descriptive purposes. While this may simplify the concept, it may perpetuate negative associations that female external genitalia that appear different are abnormal. Hence, when we deliver a lecture on the female genitourinary system, we use images that show the great diversity of female genital anatomy, which promotes inclusive practice and more accurately represents the human population (Beni et al. 2022).

There are a limited number of images across anatomical textbooks demonstrating how diverse the vulva can be. In the lecture we delivered on the female genitourinary system to second-year medical students at St. George's University of London, an introductory lecture slide was used before discussing the anatomical structures that make up the external genitalia in a biological female. Examples of the images used in the lecture slides on the anatomy of the female external genitalia were taken from "The Labia Library" (Women's Health Victoria 2024). The use of such images aimed to make students aware that the appearance of the biological female external genitalia can be extremely diverse across different individuals. Initially, the images from "The Labia Library" demonstrated the vulva from an anterior and inferior view from 24 different biological females. Recently, the library has increased the number of images available on the gallery in an attempt to increase the representation of vulvar diversity. The gallery now consists of images of the vulva of 52 different individuals two of which are non-white. The gallery highlights that the individuals who have been photographed for the purposes of the "labia gallery" are individuals that represent a wide range of genders and ages as well. The gallery notes that when the individuals were photographed for the purposes of the "labia gallery" some of them either had experienced in the past, or were currently experiencing one of the following: peri-menopause, menopause, birth-related injury, episiotomy as a surgical intervention during child-birth, vaginal birth, pregnancy, taking testosterone for longer than six months or gender-affirming surgery. Nevertheless, when the images are accessed no information is provided on whether the individual has experienced any of the above. The images can be accessed online using the following link: https://www.labialibrary.org.au/labia_gallery/

The standard anatomical textbook definition of the Latin term "*labia majora*" translates to "bigger lips". Despite this, it is possible that in some individuals the *labia majora* appear smaller than the "*labia minora*" which translates to "smaller lips". Images from "The Labia Library" show individuals where their *labia majora* are bigger than the *labia minora*, as well as images where in some individuals the *labia minora* appear to be larger than the *labia majora*. Educators can use such images in their educational material, as they can allow students to appreciate how variable the size, shape, and appearance of these structures can be. The gallery is not exhaustive in terms of showcasing how the vulva appears across different individuals of different age, ethnicity, or specific conditions that may affect the appearance of the vulva (i.e. lichen sclerosus and pregnancy) (RCOG 2013; Dominoni et al. 2022). Therefore, there is still a need for representation of the vulva of biological females that are not white. Nevertheless, "The Labia Library" is a resource that can help both healthcare professionals and the lay audience become more familiar with the diversity of the vulva.

The lack of representation of anatomical structures across different ethnic groups is concerning as the schematic illustrations present in anatomical and medical textbooks are not always representative of the human population. This could eventually be a challenge for future healthcare professionals because they may not be familiar with how anatomy varies across patients or individuals of different ethnicities which they are likely to encounter in their practice. The images of the vulva we used in the lecture on the anatomy of the female external genitalia represented the diversity of the vulva across white individuals. Hence, to ensure that students get to see what the vulva looks like, in non-white individuals, an image of the vulva of a biological female of colour (fig. 74.1 in Chapter 74) available in the Gray's Surgical Anatomy textbook was used during the equivalent anatomy practical (Brennan et al. 2019).

Access to anatomically accurate medical illustrations that are representative of the human population can be challenging. In recent years, many accounts have been created across social media where individuals of different professions aim to share educational content regarding genital anatomy. There are specific Instagram (Meta) accounts that offer schematic medical illustrations that represent the biological female external genitalia and genitourinary system across different individuals. One example of an Instagram account of an artist who creates schematic illustrations of the vulva of different biological females is "The Vulva Gallery" which can be accessed using the following links: https://www.instagram.com/the.vulva.gallery/ and https://www.thevulvagallery.com.

"The Vulva Gallery" consists of illustrations that showcase how diverse the vulva can be across a range of biological females of different ages, ethnicities, and genders. In addition, some illustrations schematically demonstrate the vulva of individuals with conditions such as endometriosis, pregnancy, cyst of the greater vestibular glands, atopic dermatitis, or who underwent specific procedures (i.e. episiotomy). In case educators would like to use such illustrations to demonstrate to students the great diversity that exists across the female genitalia, they must ensure the anatomical accuracy of such illustrations since not all artists who create such illustrations have an anatomical or medical background. Despite the caveats surrounding the use of illustrations from social media, it is important to acknowledge that through those accounts, the use of illustrations that demonstrate specific anatomical variations that are not available on anatomical or medical textbooks or websites is feasible. Such illustrations can be used in self-directed learning (SDL) content that can be uploaded to the university's learning management system. For example, we used a schematic illustration that showed the vulva of a living individual who had an episiotomy scar as part of the SDL content on the anatomy of the female genitourinary system for second-year medical students. The schematic illustration showing the vulva of an individual who has an episiotomy scar can be accessed here: https://www.instagram.com/p/CoFeGLlokC4/?img_index=1

Another resource that aims to increase awareness of how female bodies should be more accurately represented in healthcare is "The Reframing Revolution Digital Gallery". This is a digital gallery that uses schematic medical illustrations that portray the diversity of biological female bodies under different circumstances including pregnancy, breastfeeding, and menopause. The aim behind this gallery's creation is to represent anatomical diversity across biological female bodies in healthcare since in the past the female body was often represented as white-coloured, hairless, slim, and able-bodied (O'Callaghan Tassia 2024).

The Reframing Revolution Digital Gallery can be accessed using the following link: https://www.peanut-app.io/blog/reframing-revolution.

Illustrations from the reframing revolution digital gallery can be downloaded using the following link: https://www.dropbox.com/sh/

t272zg9zvskuewa/AACDlGSSys2gtZpzz8WrW9SWa?e=1&dl=0

The gallery consists of illustrations that represent mostly the diversity of female bodies under different real-life experiences. The types of illustrations present in the gallery can be seen in Fig. 6.8.

Another example of inclusive anatomy education is "Anatome" which can be accessed using the following link: https://anatomeinclusivity.wixsite.com/anatome. Anatome is a website and a YouTube channel founded by Aisia Lea, a medical student studying at the University of Nottingham (Lea Aisia 2024). The YouTube channel provides access to anatomy videos that showcase what inclusive anatomical education looks like. For instance, a video on anatomical terminology represents individuals of different ethnic backgrounds as well as individuals with a disability (See YouTube video: Anatome Teaching Series: Anatomical Terminology, at 4:00 min). https://www.youtube.com/watch?v=waupFHQBeeM&t=165s).

The website also provides access to podcasts that aim to give a voice to individuals who discuss the changes that need to be made for healthcare to be inclusive for all individuals. Examples of topics discussed in the podcasts include diversification of dermatology resources, censorship in women's health advocacy, students promoting diversity within research, and diversity in medical academia. Anatome also has a blog that consists of posts that focus on anatomy, medicine, inclusivity, medical education, and diversity. Examples of blog posts include LGBTQIA+ bias in healthcare, gender bias in healthcare, and differing maternal outcomes based on ethnicity.

Other individuals who are underrepresented in anatomy and medical textbooks are transgender and intersex (Kelly and Rodriguez 2022). Schematic illustrations representing transgender individuals could be used in chapters of anatomy and medical textbooks other than the genitourinary system. For example, an illustration of a transgender individual who has not undergone gender-reaffirming surgery can be used when introducing an anatomical concept that is not related to pelvic anatomy (i.e. dermatomes, anatomical movements and muscles of the leg). The representation of such individuals is essential as future healthcare practitioners will be expected to treat such patients; therefore, the early exposure of students to a variety of patients they will treat could potentially help reduce the bias associated with these individuals as highlighted by Morris et al. (2019). The use of illustrations of such individuals can also help initiate discussions around the correct use of an individual's pronouns, as well as how assumptions should not be made based on an individual's external appearance or biological phenotype. Another way for medical institutions to ensure that individuals who identify as LGBTQIA+ are represented within the medical curriculum would be to include exam-

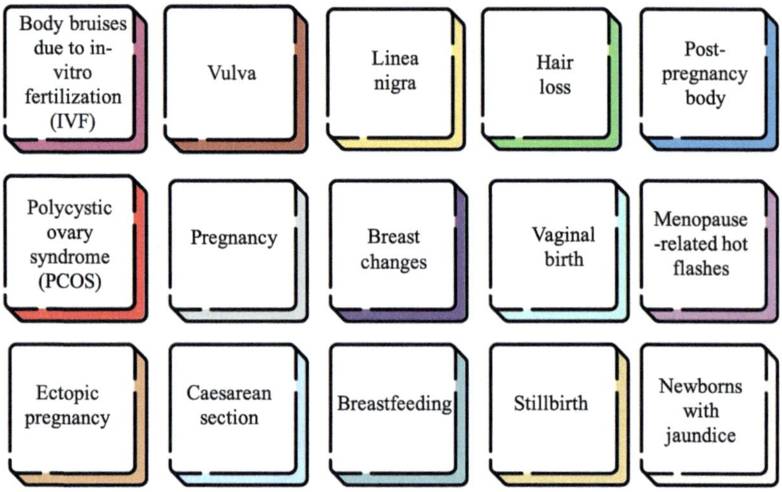

Fig. 6.8 Types of illustrations available on the Reframing Revolution Digital Gallery

ples or cases within case-based learning (CBL), problem-based learning (PBL), or team-based learning (TBL).

According to Endres et al. (2022) the resources on intersex individuals currently available to the medical community are scarce, hence, the authors highlight the need for educational material on intersex individuals to be created and used during medical education training. Additionally, the authors emphasize how the individuality of intersex people should be acknowledged as the medical treatment for some intersex individuals may be associated with specific challenges; hence, the need for stand-alone training resources dedicated to these individuals (Endres et al. 2022).

6.3 Using and Encouraging the Use of Inclusive Language

It is important to use language that respects the identity of all individuals as well as the diversity of the student cohort. Finn et al. (2022) highlight that one consideration for the anatomy curriculum to be decolonized is the use of inclusive language and appropriate terminology. Verbree et al. (2023) emphasize the need for educators, physicians, and students to receive training on the use of inclusive language. The authors also stress the importance for educational content and material to not be offensive to any individual (Verbree et al. 2023). The use of inclusive language is essential when discussing topics related to protected characteristics including gender identity, sexual orientation, any disabilities, and ethnicity. Using inclusive language can be challenging however, it may help individuals tackle their own implicit biases and eventually allow most if not all individuals to feel represented.

Easterling and Byram (2022) published an article titled "Shifting language for shifting anatomy: Using inclusive anatomical language to support transgender and nonbinary identities". In this article, the first author discusses her experience with anatomical language during her transition as a transgender woman. The author shares how positive her experience has been because her doctors have been using inclusive language by using terms that were purely focussing on the anatomical structures of interest instead of the biological sex of the individual in which such structures are found. Whilst this is only the experience of a single individual and might not be the case for all individuals who are transitioning, it showcases that healthcare providers can intentionally use anatomical language and other scientific terms in a way where individuals of all genders feel represented. Lastly, the authors advocate that resources dedicated to transgender and non-binary individuals and corresponding healthcare should be developed and should be freely accessible. The authors recommend that existing resources should undergo modifications to include context written in inclusive anatomical language, as well as adding crucial information regarding how hormonal replacement therapy and gender-reaffirming surgeries will modify the patient's anatomy (Easterling and Byram 2022).

In traditional anatomy textbooks, anatomical structures such as the gonads that pertain to a specific biological sex were often described as being present in "women" and "men". In Table 6.1, we list examples of such descriptions and how they could be rephrased to ensure the use of inclusive language. Such examples can be used by educators while discussing an anatomical topic. Educators can revisit educational material outside anatomy teaching, such as during CBL, PBL, and TBL cases if they are part of the curriculum to ensure the consistent use of inclusive anatomical language throughout. Newer versions of anatomy textbooks have replaced the terms "reproductive system" with the terms "urogenital or genitourinary system" or "pelvis and perineum". This is a great step towards being inclusive as not all individuals use their genitalia to reproduce.

Matthan and Finn (2020) stress how the hidden anatomy curriculum can be intentionally used by educators to address social issues and promote thoughtful discussions. A hidden message in the anatomy laboratory is how we give respect to donors. Hafferty and Finn (2015) outline that there are hidden messages in every-

Table 6.1 Examples of how anatomical information can be rephrased to ensure the use of inclusive language

Textbook/Source	Original sentence	Rephrased inclusive sentence
Human Anatomy, 8th ed. by Marieb, Wilhelm and Mallat (2016)	"In *women*, subcutaneous fat accumulates first in the thighs and breasts, whereas in *men* it first accumulates in the anterior abdomen as a "beer belly""	"In *biological females*, subcutaneous fat accumulates first in the thighs and breast, whereas in *biological males* it first accumulates in the anterior abdomen as a "beer belly""
	"Decreased folic acid levels in *pregnant women* increase the risk of neural tube defects in the developing fetus"	"Decreased folic acid levels in *pregnant individuals* increase the risk of neural tube defects in the developing fetus"
Gray's Anatomy: The Anatomical Basis for Clinical Practice, 41st ed. (2015)	"The anterior border of each thyroid lamina fuses with its partner at an angle of approximately 90° in *men* and approximately 120° in *women*"	The anterior border of each thyroid lamina fuses with its partner at an angle of approximately 90° in *biological males* and approximately 120° in *biological females*"
Gray's Anatomy for Students, 5th ed. (2023)	In *men*, the prostate gland is situated immediately anterior to the rectum, just above the pelvic floor	In *individuals* with a prostate, the prostate gland is situated immediately anterior to the rectum, just above the pelvic floor

thing that educators do or do not do. In the past, one of the authors worked at an institution where they noticed that an anatomy educator was placing surgical instruments all over the donor during a dissection-based practical. Subsequently, the students supervised by them repeated this model behaviour. It is of utmost importance that the anatomy teaching staff set an example of professional behaviour by actively showing students how to handle prosected human tissue specimens safely and respectfully. In addition, the use of inclusive language by instructors, such as using the term "donor" instead of "cadaver" is one way to help students reinforce the concepts of respect and compassion (Weeks et al. 1995). Another example of a hidden message in the anatomy laboratory in relation to giving respect to the donors is the way donors with increased adiposity are perceived. According to Rompolski and Pascoe (2024), the inclusion of anatomical variation and the representation of a wide range of bodies are essential examples of the hidden curriculum since they aim to increase the representation of different body types future healthcare practitioners are likely to encounter in their practice. In their study, Rompolski and Pascoe (2024) aimed to measure weight bias among Doctor of Physical Therapy students enrolled in a dissection or prosection-based anatomy course. In their study, the authors reported that 90% of the study participants identified larger-body donors as "difficult" to work with and that dissection was more time-consuming. Undoubtedly, human dissection can be a challenging skill to master especially for students who are novice learners, since it requires a great amount of adipose tissue to be removed especially when deeper anatomical structures are to be studied. In the experience of the second author, medical students enrolled in a dissection-based program reported similar experiences to the study participants in the study by Rompolski and Pascoe (2024). Students would express mild dissatisfaction if the donor was larger as that required them to remove a greater amount of fat which made dissection more laborious. As the authors emphasize, it is important that instructors in the anatomy laboratory do not address increased donor adiposity negatively as this can negatively influence students' perceptions of the usefulness of larger-body donors. On a similar note, it might be worth asking whether the model patient volunteers that are used during students' ultrasound or clinical skill sessions, are representative of the patient population students will encounter in their practice. The use of model patient volunteers with minimal levels of subcutaneous fat during these sessions may simplify the ultrasound demonstration or the clinical examination; nevertheless, this may perpetuate the notion that all patients will be slim. Hence, once students need to examine a patient that does not fit the "standard" patient they have been previously exposed to, they may struggle. Perhaps, the use of larger-body model patient volunteers within such sessions may help reduce the stigma around weight but also prepare future healthcare practitioners better in terms of examining this group of patients. Equally important is the use of inclusive language outside the anatomy laboratory. As aforementioned, some medical institutions employ CBL, PBL, or TBL and some of the cases cover sensitive topics such as eating disorders, suicide, or substance abuse. Due to the sensitivity of these topics, facilitators must prepare and plan accordingly on the terminology they will be using during these sessions as well as how they will communicate information to students. Facilitators must also be prepared that some students may experience emotional distress upon hearing specific information as it may remind them of a situation they themselves have been in or someone they know might have experienced. Hence, facilitators can inform students before having the session that the case will discuss a sensitive topic and ask if any students would need specific support. Facilitators can consider signposting students to wellbeing or other services available at their institution.

For some students, sarcasm and humour can be confusing during teaching; hence, it must be used with care to avoid alienation, and to avoid offending social and cultural sensitivities. In addition, students may become overwhelmed or confused when new or alternative terminology is introduced. Hence, new terms should be introduced carefully and thoroughly explained. The repetition and explanation of specialized terminology will help students understand and learn more effectively. In terms of anatomical terminology, educators can encourage students to use a glossary. Some anatomical textbooks like the Human Anatomy by Marieb, Wilhelm and Mallatt (2016) global edition have a glossary section at the end of the textbook that provides the reader with the pronunciation of a term and its definition. Educators can prompt students to create their own glossary where they can add new terminology they come across as they progress with their studies. The glossary can contain the definition of a term and its etymology as well. Most anatomical terms stem from Greek or Latin; therefore, familiarity of students with the etymology of a term may help them understand anatomical concepts better (Patera 2021; Mehta et al. 1996). Examples of some online resources that educators can use themselves or signpost students to learn more about the etymology of anatomical term can be seen in Table 6.2.

In the experience of the second author (EP), students were more likely to remember an anatomical or pathological term if they knew where the prefix or suffix of an anatomical term stems from and what it means in the language it stems from. Because of that, the second author always uses a purple box when introducing a new anatomical term in a lecture slide, along with the etymology and definition of the term, as well as how the term is pronounced in singular versus plural. Such examples can be seen in Fig. 6.9a, b.

Most universities in the UK require newly recruited staff to undertake specific training as part of their induction. There are specific induction courses such as EDI Training that inform individuals on the protected characteristics, highlighting the importance of using inclusive language and avoiding discrimination at the workplace so everyone feels welcomed and safe. While these induction courses are available to individuals working at such institutions, various universities make their resources available and accessible for free to everybody.

In Table 6.3, there are resources where individuals can learn more about the use of inclusive language in anatomical, medical, and healthcare education.

Another aspect of inclusive language and its broader use within education is for educators to avoid using colloquial terms as there might be non-native students who might not be familiar with them. For example, instead of saying *let's have a chinwag*, say: let's have a chat. Or, instead of saying *it's all gone pear-shaped*, say something has gone wrong. Educators must avoid using English idioms as some non-native students might not be familiar with them, but also because some students whether native or not may find specific idioms offensive. For example, the idiom "killing two birds with one stone" can be rephrased to "releasing two birds with one key".

Table 6.2 Online resources that can help educators and students become familiar with the etymology of specific anatomical terms

Resource	Link	QR Code
KENHUB: Anatomical and Medical Prefixes and Suffixes	https://www.kenhub.com/en/library/anatomy/commonly-used-prefixes-in-anatomy	
University Libraries: Medical Terminology for Healthcare Professions: Prefixes and Suffixes	https://pressbooks.uwf.edu/medicalterminology/chapter/prefixes-and-suffixes/	
Tallahassee Community College: Latin and Greek Word-Part List (prefixes, suffixes, roots)	https://www.tsc.fl.edu/media/divisions/learning-commons/resources-by-subject/science/anatomy-and-physiology/bsc2085/Terminology.pdf	
UCL: Medical Terms-Dissected, Defined and Explained	https://www.ucl.ac.uk/lapt/medterms.htm	
Douglas College Human Anatomy & Physiology II (4th ed.): Appendix I: Anatomical Prefixes and Suffixes	https://pressbooks.bccampus.ca/dcbiol120312094thed/back-matter/appendix-i-anatomical-prefixes-and-suffixes/	
Bodyviz: Anatomical Terminology	https://www.bodyviz.com/learn-more/blog/anatomical-terminology	

It might be common for educators to use specific terms to address a group of students. While one of these terms is "guys", according to the University of Newcastle, the term "guys" is not one that is perceived to be gender-neutral by all individuals living in the United Kingdom (Newcastle University 2024). Hence, educators can attempt to use terms that are perceived to be gender-neutral and inclusive, such as "everyone", "everybody", or "folks".

Furthermore, using an individual's preferred pronouns is a way of acknowledging and respecting their gender identity (UNISON 2024). One approach educators could employ to ensure they are inclusive is to get to know their students' names or preferred names as well as their pronouns should they wish to share them. Some may argue that this is not feasible, as a medical student cohort may consist of more than 300 students. While this might be true, there might be cases where students attend a small group teaching session or a large group teaching session in the anatomy laboratory, where they could wear a name tag. The name tag can include the student's preferred name, their surname, how their name is pronounced and/or their pronouns. This can allow educators to address students individually when they ask a question and can potentially allow students to feel acknowledged by their educator as well. An example of an inclusive name tag for students and educators can be seen in Fig. 6.10. Students can be encouraged to wear pronoun pins that they can attach to their lab coats or lanyards.

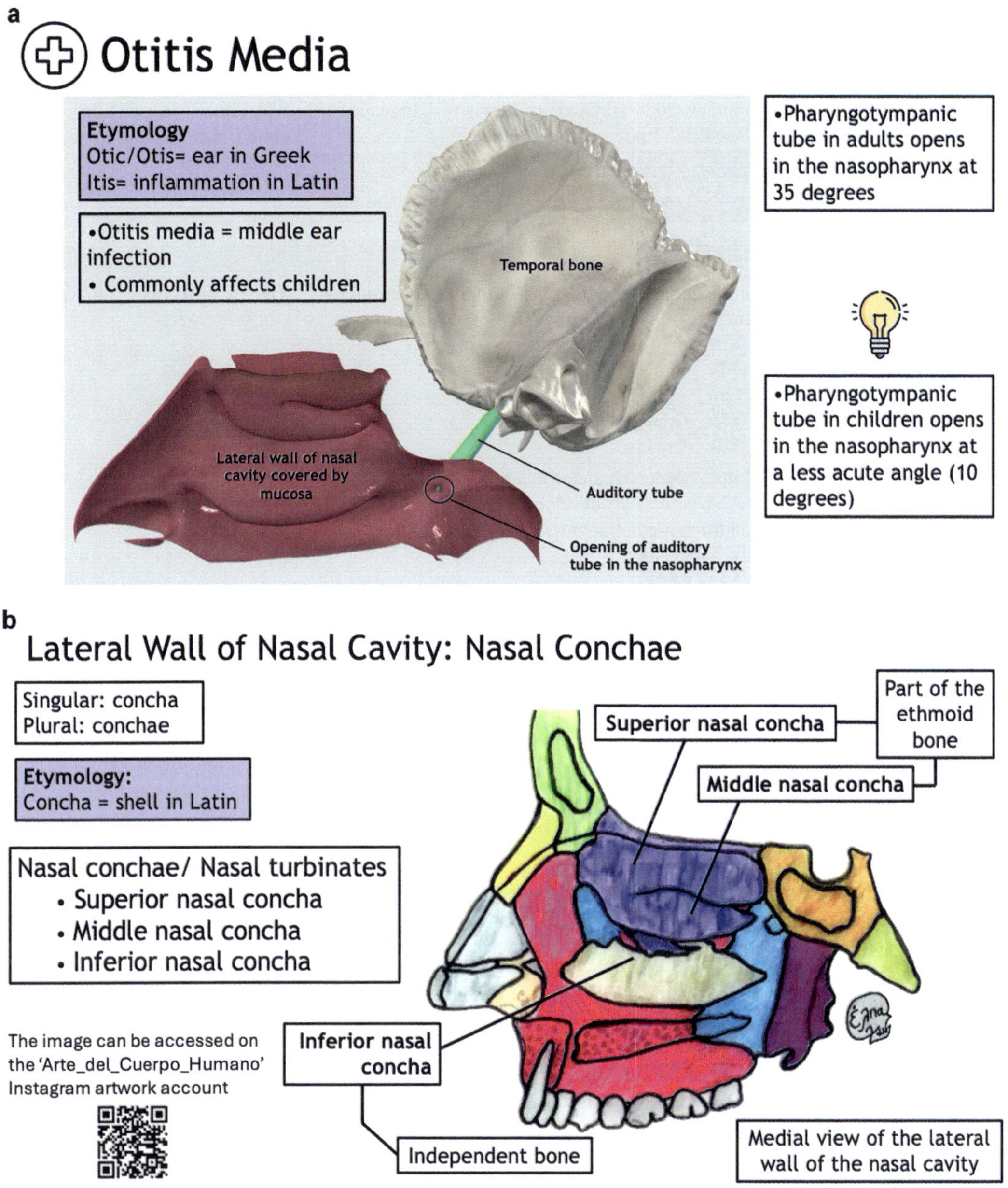

Fig. 6.9 (**a**) A PowerPoint slide where the etymology of a pathological term is explained in the purple box. (**b**) A PowerPoint slide where the etymology of an anatomical term is explained in the purple box. The singular and plural versions of the anatomical term are provided as well. The image showing the lateral wall of the nasal cavity was created by the second author (EP) and can be accessed on their Instagram artwork account "Arte_del_Cuerpo_Humano"

In case medical institutions use a physical register to monitor student attendance, an inclusive attendance register could include a student's name and surname and their preferred name and pronouns. This is important as within a class there might be some students who identify as a trans-man or a trans-woman but may not have had their name officially changed in legal documents, such as ID or passport; therefore, their name before their transition might still be shown on the class register.

Pronouns can also be part of an electronic signature that is used while individuals communi-

Table 6.3 Available online resources regarding the use of inclusive language in higher education

Resource	Link and QR Code	Pertains to:
Educator's Guide to Inclusive Language by Saskatoon Sexual Health	https://saskatoonsexualhealth.ca/wp-content/uploads/2020/10/Guide-to-Inclusive-Language-in-Sex-Ed-2-1.pdf	The use of inclusive language when discussing anatomy, contraceptives, puberty, and relationships
Inclusive Language Guidance by the University of Leeds	https://equality.leeds.ac.uk/support-and-resources/inclusive-language-guidance/	The use of inclusive language
Additional EDI Resources by the University of Leeds	https://leadershipandprofessionalpractice.leeds.ac.uk/home/staff-development/equality-inclusion/additional-edi-resources/	Race, microaggression, disability, gender and LGBTQIA+
EDI resources by the University of Oxford	https://www.dpag.ox.ac.uk/work-with-us/equality-diversity-inclusion/edi-resources	EDI, race, colonialism, gender, male centric view of medicine, gender identity and sexual orientation, disability
Inclusive Language Resources by University of British Columbia Equity and Inclusion Office	https://equity.ubc.ca/resources/inclusive-language-resources/	The use of inclusive language
Using Inclusive Language in Education by University College London	https://www.ucl.ac.uk/teaching-learning/publications/2023/jul/using-inclusive-language-education	Inclusive etiquette and language
Improving lives together	https://improvinglivesnw.org.uk/our-work/equality-diversity-and-inclusion-resource-hub/	EDI resources, anti-bullying and harassment, cultural awareness, de-biasing toolkit, health and wellbeing

(continued)

Table 6.3 (continued)

Resource	Link and QR Code	Pertains to:
Race Disparity Unit (2022) "Why we no longer use the term 'BAME' in government"	https://equalities.blog.gov.uk/2022/04/07/why-we-no-longer-use-the-term-bame-in-government/	The use of terms alternative to the term Black, Asian, and minority ethnic "BAME"
Mypronouns	https://pronouns.org	Inclusive Use of Personal Pronouns
GeekyMedics	https://geekymedics.com/introduction-to-trans-definitions-and-faqs/	Introduction to Trans Healthcare-Definitions and FAQs
Fenway Health	https://fenwayhealth.org/documents/medical/transgender-resources/Handout_7-C_Glossary_of_Gender_and_Transgender_Terms__fi.pdf	Glossary of Gender and Transgender Terms
Intersex Justice Project	https://www.intersexjusticeproject.org/intersex-for-medical-students.html	Intersex individuals
University of British Columbia	https://guides.library.ubc.ca/c.php?g=714749&p=5256746	Reading list on Intersex-2SLGBTQI+ individuals

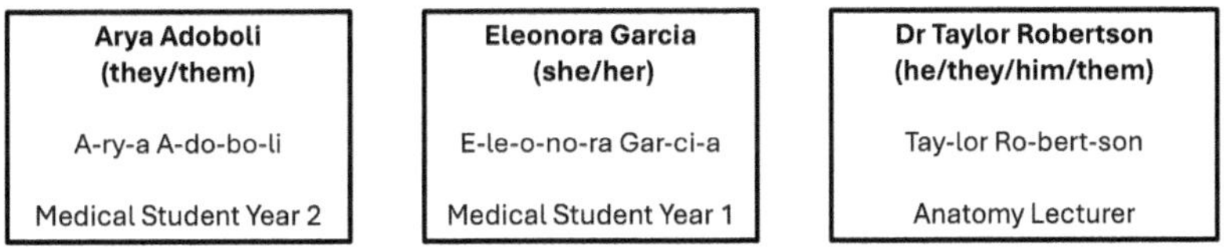

Fig. 6.10 Examples of inclusive student and staff name tags

cate via emails. According to Philippopoulos (2023), when staff include their pronouns in an electronic signature during email communication, it signifies that their organization has taken action towards normalizing discussions regarding the health of LGBTQIA+ individuals in an attempt to create a safe space.

6.4 Inclusive Assessment and Feedback

Student assessment is a fundamental element of education across all disciplines. Extensive research has been conducted on student assessment to investigate how educators assess students' learning as well as identify which type of student assessments promote student learning. According to the Equality Challenge Unit ECU (2014) inclusive practice is when educators acknowledge student diversity, and make educational content accessible to all students, allowing them to fully immerse themselves into the available learning activities, eventually allowing them to demonstrate their learning by performing well in assessment. While student assessment is multifaceted and can be challenging to modify for a number of reasons, including resources available and student cohort numbers, educators could take action towards ensuring that their assessment methods are as inclusive as possible.

First, students need to be aware of the structure of their formative and summative anatomy exams. Anatomy educators can create an anatomy syllabus that will include information about the anatomy component of the medical curriculum and how it is taught at their institution. Some universities teach anatomy regionally while others teach it systemically. Some universities teach anatomy using cadaveric tissue which may or not be accompanied by non-cadaveric resources while others solely rely on non-cadaveric resources, such as plastic models and/or digital anatomy models. The anatomy syllabus could contain information about all the anatomy lectures students will receive each term and the learning objectives (LOBs) for each lecture. Subsequently, the educators could explain the structure of the formative and summative student assessment. In case the summative student assessment consists of multiple elements, educators should provide the percentage that each element is worth. Equally important is for students to be aware of the dates of their exams from the beginning of the year so that they can work around their timetable and responsibilities and start preparing in advance.

In our experience as early career anatomy educators, we noticed that many medical students learn anatomy in a compartmentalized manner where they view and comprehend the body systems or anatomical structures in isolation instead of viewing them as intertwined structures.

Below in Fig. 6.11, we provide an example from our experience while facilitating students that showcases how some students learn anatomy in a compartmentalized manner (Fig. 6.12):

As it can be noted, in the questions above, the potential answer options are mentioned in alphabetical order which can make it easier for the readers to read. In our experience, we have received training on how to write single-best answer questions (SBAs) prior to becoming involved in exam writing and we recommend that educators receive training to ensure no unintentional errors or mistakes are made that may indirectly influence student scores.

During the lecture, many students answered correctly that the nerve indicated by the arrow was the internal laryngeal nerve which is a sensory branch of the superior laryngeal nerve that supplies sensory innervation to the mucosa of the larynx up until the level of the vocal cords.

In the corresponding anatomy laboratory practical session, students were shown a cadaveric specimen of the larynx where all laryngeal cartilages were present but some of these cartilages, such as the epiglottis and the paired laryngeal cartilages were covered by mucosa. There was a pin on the epiglottis which was covered by mucosa and students were verbally asked the following question:

"What is the sensory innervation to the pinned structure?"

Students were able to identify that the pinned structure was the epiglottis covered by mucosa; however, they were unable to answer the ques-

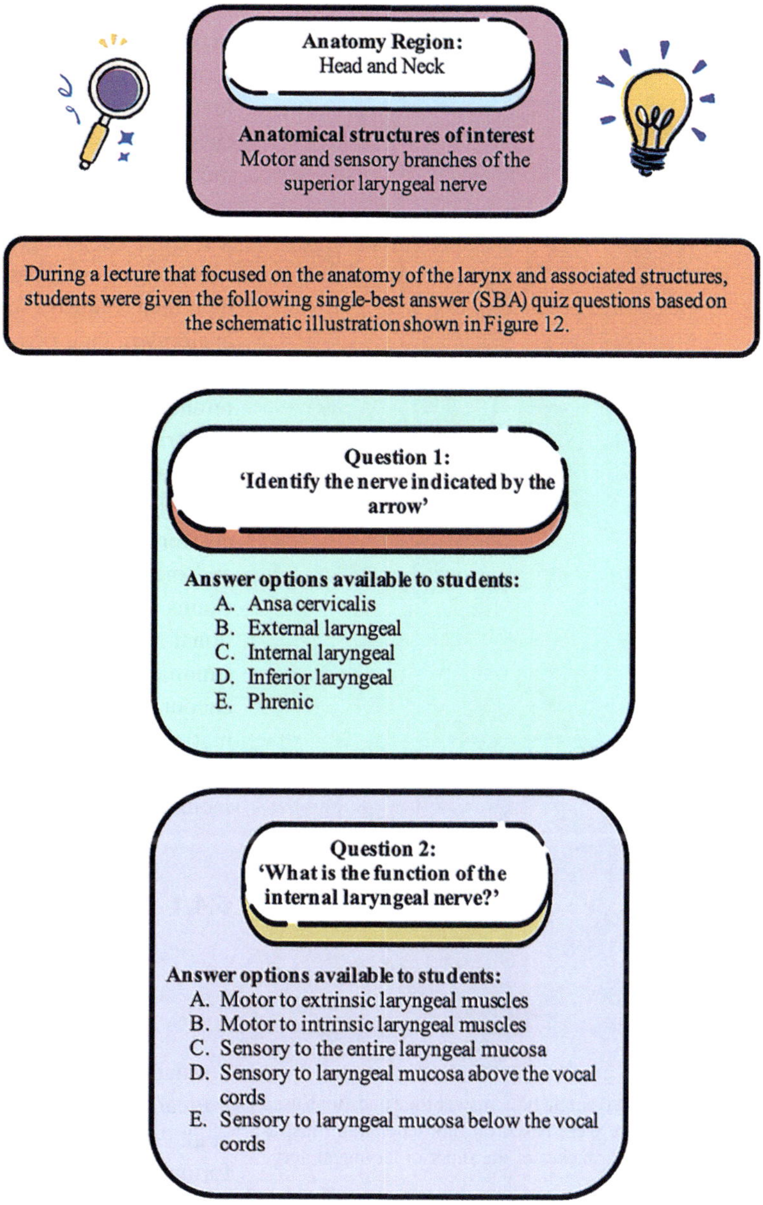

Fig. 6.11 Two single best answer quiz questions used during the lecture that focused on the anatomy of the larynx

tion. Hence, the students were verbally asked to consider the following questions shown in Fig. 6.13 to answer the initial question:

The example above demonstrates that while students might be able to recall structural and functional anatomical information, they may not always be able to combine their structural and functional anatomical knowledge to answer a question that does not directly ask "What is structure 'X'? Or What is the function of structure 'X'?"

Informal verbal and formal written student feedback we received throughout our practice has highlighted that many students find it challenging to link the different types of anatomy knowledge and subsequently link them to other non-anatomy

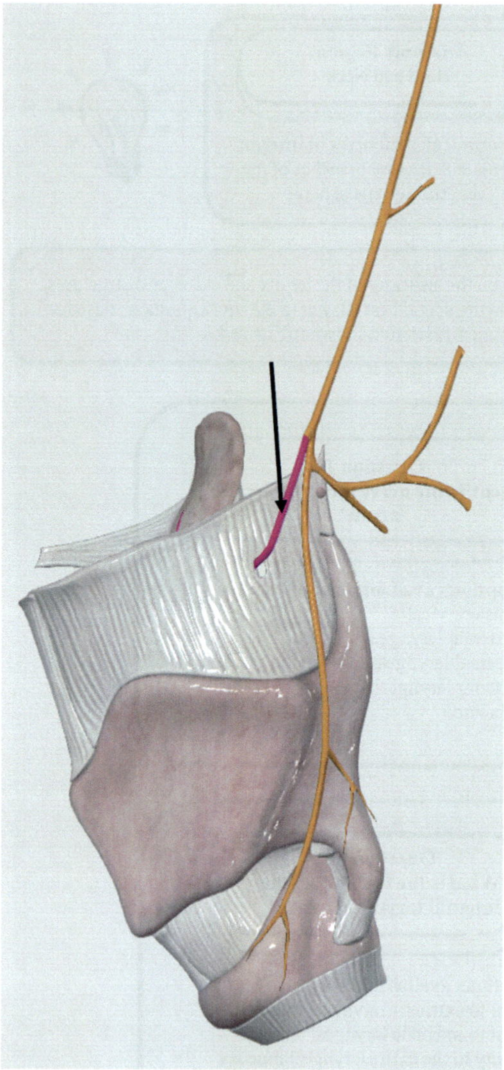

Fig. 6.12 The two single best answer quiz questions used during the lecture were based on this schematic illustration showing the branches of the superior laryngeal nerve

disciplines. Some students may struggle to apply anatomy in a contextualized manner; therefore, anatomy educators need to ensure that the anatomy assessment methods used reflect the way students need to learn anatomy which allows them to be able to progress from lower-order skills to higher-order skills. Also, the content that students are assessed on should constructively align with the content that students have been taught (AdvanceHE 2003).

In the following paragraphs, we provide a hypothetical example of a traditional anatomy formative and summative student assessment for first-year medical students. Subsequently, we discuss how these two types of assessments could be modified for assessment to be more inclusive, providing students with more opportunities to test their knowledge and identify gaps in their knowledge as well as receive feedback that may help them improve the way they study and learn anatomy in their own time. We acknowledge that the way students are assessed differs across different institutions within a country and across countries. In some universities, students might be assessed on anatomy via face-to-face or online multiple-choice questions (MCQs), single-best answer questions (SBAs), essay-based questions, and oral exams known as viva voce, or spotter exams. The provision of the hypothetical traditional versus inclusive anatomy formative and summative anatomy assessment below aims to encourage anatomy educators to rethink the format of the anatomy exam or feedback at their institution, and how they can make it more inclusive.

6.4.1 Hypothetical Example of a Traditional Anatomy Formative and Summative Assessment

Formative assessment belongs to the "assessment *for* learning" category and is low-stakes (Sambell et al. 2012; Elkington 2020). One of the aims of formative assessment is to provide students with feedback that they can act on (feed-forward) (Elkington 2020).

Graduate first-year medical students have a formative anatomy exam at the end of semester 1. This formative anatomy assessment aims to familiarize the students with the structure of the summative exam students will have at the end of semester 2 and allow them to check their knowledge of the learning outcomes covered during semester 1. The formative exam is a MCQ exam where students are assessed not only in anatomy

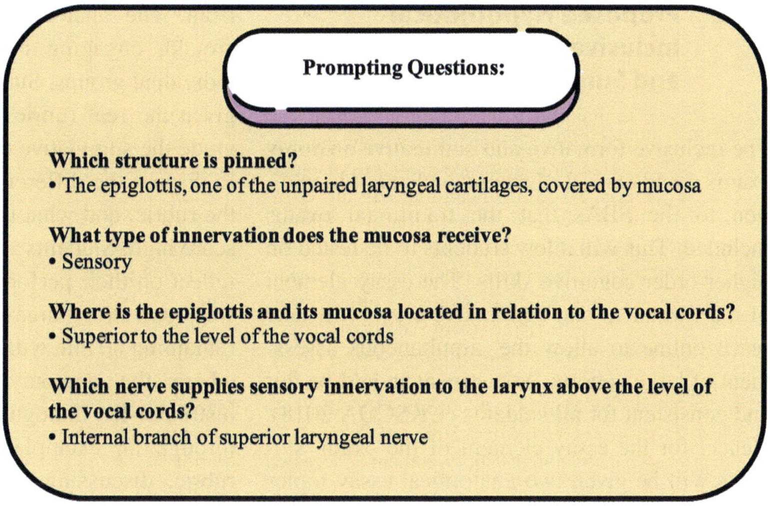

Fig. 6.13 Prompting questions that students were asked by the educator to help them guide them towards finding the correct answer

but other disciplines as well. There are a specific number of anatomy SBA questions, and each question is based on a schematic illustration or a cadaveric image. Students do not receive feedback on their exam performance other than the overall grade they have scored for the whole exam.

In SBA-style exams like the above where students receive an overall grade, students do not often see how much they scored in each exam or subject domain. They also do not always get to know which individual questions they answered incorrectly. If students could see which questions they answered incorrectly, this could potentially allow them to make an inference on whether they have specific gaps in a discipline (i.e. anatomy) or within the discipline (i.e. arteries). Additional concerns regarding the structure of the aforementioned anatomy exam include the fact that the anatomy SBAs can perhaps only test students on lower-order cognitive skills, such as recalling the name, location, and function of an anatomical structure. Nevertheless, students need to be able to perform higher-order cognitive skills by which they can demonstrate an in-depth understanding of anatomical structures and relationships as well as be able to link anatomy to other disciplines, such as physiology and pathology.

Inclusive assessment means that the educators provide students with different types of assessments and more specifically summative assessments that will contribute to a student's final degree classification. Ideally, an inclusive assessment in the context of anatomy would be an assessment that allows students to apply their anatomical knowledge to a "real-life" scenario. An inclusive approach towards designing student assessment is to take into consideration when the assessment will take place and under what circumstances. It will be considerate of educators to try and schedule exams at a time when students do not have multiple assessments due. The format of the exam is also something educators need to take into account. For instance, having the exam on-site might mean that the educator needs to secure a big exam hall or book multiple rooms at the same time to accommodate the number of students. In addition, some students may have to travel long-distance to arrive at the exam hall. On the other hand, if the exam is online this might be difficult for students who may not have access to a personal computer or reliable Internet connection. There could be financial implications, health limitations, transport strikes, and other issues that some students may face, which is why it is important to consider all such aspects to ensure the assessment is inclusive.

Below is a proposed anatomy formative and summative assessment that is more inclusive than the traditional one.

6.4.2 Proposed Hypothetical Inclusive Anatomy Formative and Summative Assessment

The inclusive formative and summative anatomy exams could consist of an essay element in addition to the SBAs that the traditional exams included. This will allow students to be tested on higher-order cognitive skills. The essay element of the formative and summative exam will take place online to allow the simultaneous assessment of large cohorts. Assessment should be fair and consistent for all students (UKSCQA 2018). Hence, for the essay element of the exam, students will be given two anatomical essay topics both of which will be based on the same level of higher-order LOBs. The students will have around 40 min to complete their chosen anatomy essay topic unless they have reasonable adjustments in place that allow them extra time.

This is also emphasized by the GMC, "*Medical schools are responsible for arranging reasonable adjustments for medical students*" (GMC 2024b).

Below is an example of an essay topic that will require students to demonstrate knowledge over structural, functional, and clinical anatomy as well as anatomical relationships.

Essay topic *Describe the topographical anatomy of the pancreas in relation to the duodenum, its function, and consequences of pancreatic head tumour.*

The blue, purple, and green colours correspond to structural, functional, and clinical anatomy knowledge, respectively.

In our opinion, the ideal feedback strategy for the essay element will be a 1-h and 45-min "feedback workshop" 2 weeks after the exam, delivered by each anatomy staff member. During the workshop, students will be divided into two groups and each anatomy educator will be part of the workshop. Students will have 35 min to compose their own rubric where they will be expected to write down and discuss with their pair the relevant anatomical information that was required for each of the two topics they could choose from. The anatomy educator will be floating around, engaging in a conversation with the individual groups. Subsequently, students will be given the real rubric anatomy educators use to grade the summative essay and will have 15 min to discuss the differences, become familiar with the rubric, and what they need to do to pass and score high. Students will then be given 15 min to reflect on their performance by identifying their weak and strong areas of their written essay. The remaining 40 min will be divided into two halves, where the anatomy educator will lead an instructor-led dialogue where all groups will go through an exemplar essay sample using the rubric, discussing key anatomical information that should have been mentioned in each of the two available topics, revising key concepts and allowing students to ask questions for any clarification. The feedback workshop could take place either online or face-to-face. The small number of students and strong teacher–student interactions during the workshop are factors that could enhance its overall quality (Carless 2015). Despite this, practical implications that could affect the quality of the workshop include staff's insufficient training, lack of experience, or a standardized form for student facilitation and leading the discussion.

Formative assessment could occur throughout a learning process or at the end; however, students will find the feedback process most useful when they receive it at a time that allows them to act upon it (QAA 2018). Untimely feedback is a main issue reported in open national student surveys (Williams et al. 2008). Hence, educators should aim to provide feedback soon after the assessment has taken place (Carless 2015). The feedback workshop described for the inclusive formative and summative anatomy assessments is based on the "assessment *as* learning" category which aims to make students aware of the criteria that instructors use to distinguish differential student achievement (QAA 2018). Students' early introduction to the rubric and associated criteria that will be used is an effective method for providing transparency in assessment (Jönsson and Prins 2019; Nicol 2021). This fit-to-purpose approach enables students to understand the lower- and higher-order cognitive skills we aim to evaluate via the essay assessment (Elkington 2020).

The formative essay assessment and its corresponding feedback workshop were designed upon the learning-oriented assessment model proposed by David Carless (2015). The model suggests that an effective learning-oriented assessment is based on three principles:

- Learning-oriented assessment tasks
- Student engagement with feedback
- Developing evaluative expertise

The workshop aims to bring greater transparency into the assessment's criteria and quality standards, as well as give students the ability to develop a productive student learning process which is the aim of learning-oriented assessment (Carless 2015). The structure of the workshop intends to assist students in developing an understanding of how when they are studying and learning anatomy in their own time, they need to be able to perform lower-order cognitive tasks for their factual anatomy knowledge (i.e. recalling a structure's name or location) eventually moving to higher-order cognitive tasks for their conceptual anatomy knowledge (i.e. describing a structure's function and spatial relationships) ultimately allowing students to apply their factual and conceptual knowledge in a clinical context. The structure of the essay along with the information that it requires students to provide on the different types of anatomy knowledge could be considered a form of "ways of thinking" (WTP) as WTP helps students focus their learning on various forms of knowledge that are necessary for disciplinary-based learning (Barradell et al. 2018). To support the development of WTP, real-world clinical problems should be integrated into an assessment (Carless 2015). The addition of real-world clinical problems in the formative essay might not be possible due to the introductory level of knowledge students acquire by the end of semester 1. Some medical schools employ a spiral curriculum where key concepts are revisited across different disciplines throughout the curriculum but with the layers of complexity deepening each time a concept is revisited (Bruner 1960). Therefore, this could be done for the summative essay as by that time students would have substantial anatomy knowledge.

The provision of assessment quality criteria, feedback, feed-forward, and engagement in a dialogue are characteristics of assessment *for* learning (Nicol 2021). The feedback workshop after the formative essay assessment aims to bring awareness to students on how they should approach their anatomy learning as well as provide some sort of structure to their learning and potentially help them prepare and perform better in the summative essay at the end of semester 2. In the feedback workshop for the formative essay, students will receive feed-forward feedback as they will identify the weak and strong areas of their essay, perhaps bringing their focus during their own self-study time to the areas that they need to work on.

Overall, the workshop is crucial as the whole procedure intends to make students develop evaluative expertise by developing self-evaluative capabilities (i.e. students building their own rubric), engaging with the formal assessment criteria (i.e. students seeing the actual rubric that will be used to assess and evaluate the formative and summative essays, respectively, helping them appreciate the grading approach from the educators' perspective), and lastly making informed decisions regarding their work and their peers (i.e. students receiving feed-forward feedback on their essay, identifying their essay's weak and strong areas, and planning on where to emphasize more while studying and learning anatomy in the future). The peer-to-peer discussions during the workshop are a great opportunity for students to develop skills such as critical analysis that are considered essential for medical practice (Mukhopadhyay and Choudhari 2024). Furthermore, the peer-to-peer discussions can allow students to learn how to provide effective peer-to-peer feedback and how they can utilize peer feedback to identify the areas they are strongest or weakest in (Lerchenfeldt et al. 2019). Lerchenfeldt et al. (2019) conducted a systematic review to examine the utilization, effectiveness, and quality of peer feedback during collaborative learning in undergraduate medical education.

Fig. 6.14 Examples of formative assessment that educators can use throughout the semester and the type of feedback students are provided with

Examples of formative assessment	Feedback provided
Quizzes on the university's learning management system (i.e. Canvas)	• Instant feedback • Students can re-take the quiz
Asynchronous and synchronous Mentimeter quizzes	• Instant feedback • Students can re-take the quiz if it is asynchronous
Quizzes with iClicker	• Instant feedback • Verbal feedback from the educator
Kahoot quizzes	• Instant feedback • Verbal feedback from the educator
Student group discussions or peer-to-peer discussions	• Peer feedback and feedback from the educator

Out of the 31 studies included in this systematic review, 11 studies reported that students were instructed on how to provide high-quality peer feedback as lack of guidance and training can result in students feeling unsure on what they should do or how to assess their peers (Lerchenfeldt et al. 2019).

The feedback workshop will not only be beneficial to students but to educators as well, as students can raise any concerns they have in relation to the structure of the assessment. Educators need to ensure that the written instructions on assessment are clear as English might not be the first language of all students, and unclear instructions can result in both native and non-native students losing points because they may have not understood what they were required to do.

In terms of formative feedback, educators can attempt to provide students with different types of formative assessment and feedback throughout the semester. Figure 6.14 provides some examples of formative assessment that educators can use.

Educators need to take into consideration that some of their students might have a visible or hidden disability. Hence, specific adjustments must be made for these students to enable them to take their exams. Further in this chapter, we provide more recommendations on how to make the learning environment and material more accessible for students who may have visible or hidden disabilities.

6.5 Supporting Mature, Graduate and International Students

Educators need to take into consideration that some of their students might be mature students and may need additional support. In specific countries, such as the United States, higher education academics have designated office hours. Office hours are defined as non-structured instructional time where students can visit their instructors outside class time to receive assistance with their studies (Hsu et al. 2022). These office hours can be particularly important for students whether they have disabilities or not. Additionally, office hours can be crucial for mature students or any student who might have missed a lecture and/or a practical due to other responsibilities and may require additional support. According to Habbal (2009), the teaching load of anatomy educators is higher compared to the teaching load of educators who teach other basic sciences. If having office hours is not feasible, alternative options that may assist students could be the following:

An Anonymous Discussion Board on the Learning Managing System For example, on Canvas (Instructure, Salt Lake City, Utah, USA), educators can create an anonymous discussion board that gives students the opportunity to post questions anonymously in relation to content they have received. Furthermore, there is an option that allows students to give a thumbs up; therefore, they could use this feature for a question they have that has been posted already. This can allow the educator to identify if a question is popular among students. Educators can post the answers to those questions or address them during a consolidation/revision session. Consolidation sessions can enable instructors to establish a teacher–student interaction pattern that fosters student engagement and enhances motivation (Ramsden 2003). A limitation surrounding discussion boards is that not all students may feel the need to engage with them; nevertheless, it would be good for educators to provide students with various options.

6.5.1 Padlet Board

An alternative to a learning management system discussion board could be the use of "Padlet". The free version of Padlet allows educators to create up to three padlets. Students can post anonymously any questions they may have or any topics they feel confused about. Subsequently, educators can address those during a consolidation session. A limitation of using this is that students may use inappropriate terms to describe their situation; hence, it is important to remind them to be professional.

Regarding international students, they may not always be familiar with the UK higher education assessment processes and standards; hence, it is important to provide clear instructions on assessment from the beginning. Verbree et al. (2023) have reported that educators should get the assessment checked by non-native or bilingual staff to ensure the assessment instruction is understandable for non-native students. It may not be possible to have non-native staff as part of the team and as such Verbree et al. (2023) emphasize that all educators should receive training in cultural diversity to be able to support students from diverse cultures and backgrounds due to the stereotypes and biases associated with certain cultures and backgrounds which may impact student learning and wellbeing.

6.5.2 Supporting Students with Neurodivergence

Neurodiversity is a term that is often used interchangeably with the term neurodivergence (Duong and Vogel 2022). The term is used to refer to a range of neurodevelopmental conditions including but not limited to attention-deficit-hyperactivity-disorder (ADHD) and autism (Duong and Vogel 2022). Duong and Vogel (2022) highlight that the rationale behind the neurodiversity movement was to view these neurodevelopmental conditions as differences instead of as deficits. These neurodevelopmental conditions may cause learning difficulties for neurodivergent individuals. Sienna Castellon, the founder of the "Neurodiversity Celebration Week" (NCW) emphasizes that while there are challenges associated with these conditions, the narrative should change to focusing on the strengths and talents of neurodivergent individuals instead (Castellon 2018). The "Neurodiversity Celebration Week" is a website which can be accessed using the following link: https://www.neurodiversityweek.com/

On this website, educators can find information about neurodiversity and the differences between different neurodevelopmental conditions such as autism, ADHD, developmental language disorder (DLD), dyslexia, dyspraxia, dyscalculia, and Tourette's syndrome. Understanding how each of these are different may help educators to understand how best they can support students with these conditions.

Sienna Castellon in association with a psychologist consultancy known as Lexxic runs the annual "Neurodiversity Celebration Week" (Castellon 2018) usually in March. During this week, various speakers, professionals, and advocates come together to run online events that aim to inform healthcare professionals, educators, and the lay audience about neurodiversity. Below in Fig. 6.15 is a list with the events of Neurodiversity Celebration Week 2024. These events were free, and anyone could register to attend them online via Zoom. Access links to the recordings became available after the events took place and can be accessed using the following link: https://www.neurodiversityweek.com/2024-event-recordings

Table 6.4 lists online resources educators can use to learn more about neurodiversity and neurodiversity-inclusive teaching practice.

In Table 6.5, we signpost anatomy and medical educators to peer-reviewed articles which they can use to improve their understanding of neurodiversity and the adjustments that can be introduced to support neurodivergent students.

6.6 Making the Anatomy Learning Material and Resources Accessible

Anatomy in the current era is being taught with a wide range of resources, from traditional methods, such as donated bodies to more modern, such as 3D technological software.

Every learner has different learning needs, and as educators, we must support learners and make learning materials and resources accessible to them. For example, when delivering a lecture, when possible, we should consider recording the lecture so that students who missed the lecture due to illness, caring responsibilities, or other reasons are not disadvantaged. This recorded material can then be reviewed by students later through the learning management system. Equally, some students might not be able to keep up with the pace of the educator delivering a lecture and may experience cognitive overload. Recorded lectures would make it easier for them to go over the content again at their own pace. In addition, if educators deliver pre-recorded lecturers, they can consider recording lectures in small parts or chunks, so that students do not feel overwhelmed or exceed their cognitive capacity (Van Merriënboer and Sweller 2010).

In the case of universities that use cadaveric human tissue for their anatomy teaching, it may not always be possible to create pre-recorded videos where an educator explains an anatomical concept using cadaveric specimens due to the donors' consent and ethical reasons. Therefore, educators can consider using cadaveric images from textbooks or other reliable sources that show the anatomical structures of interest to create recordings for students who might have missed their anatomy session with cadaveric specimens. In universities that use plastic models only for instance, educators can create pre-recorded videos using cadaveric and non-cadaveric images or their plastic anatomical models, so that students who could not attend their sessions and missed the opportunity to engage with the models can do so. In some institutions it might be possible for students to schedule "self-revision" sessions within the anatomy laboratory or classroom where they can use the prosected specimens (if available), real or plastic bones, and plastic anatomical models. These sessions might be useful for students who either missed an anatomy practical session or who were not well prepared when they attended their assigned session. Students who prefer revising alone or taking additional time while using the specimens and models might benefit as well. Major constraints associated with offering students these sessions include timing the session when students do not have any other university commitments, offering a booking system to limit the number of students within the learning space, and the need for a designated person (PD) to be present in case human tissue is used.

In addition, when uploading learning material or content on the learning management system, such as Canvas or Moodle, it is important to provide that material in various formats. For example, if using PowerPoint slides, along with providing slides in PPT format, they should also be provided in pdf version as certain devices do not support every media format and students often face technical issues that may hinder their

Fig. 6.15 Neurodiversity Celebration Week 2024 Events

Neurodiversity Celebration Week 2024 Events:

- An Introduction to Neurodiversity
- Neuro-inclusive language
- Neurodiversity: We don't all fit into one box!
- Neuro-inclusive language
- Neurodiversity in Higher & Further Education
- Neurodiversity at Work Q&A: Ask the experts
- Culture, Community and Class in Neurodiversity: Who gets left behind?
- The benefits of neurodiversity in the Workplace
- Creating a neuro-inclusive classroom
- Neurodiversity: Fuelling creative minds
- Neurodivergent LGBTQIA+: The 'Double Rainbow' Intersection
- Connecting Classrooms to Careers: Neuro-inclusive Tech
- Neurodivergence isn't a mental health problem
- Understanding Neurodiversity as a Family
- Neurodiversity for HR professionals
- Late Discovered Autism & Menopause
- Empowering yourself as a Neurodivergent Woman in the Workplace
- Neurodiversity for Parents & Carers
- The Experience of Neurodivergent Women and Girls
- Global Perspectives: Neurodiversity Around the World
- Good Practice for Neurodiversity Professionals

Table 6.4 Online resources on neurodiversity and neurodiversity-inclusive teaching practice

Source	Link and QR Code	Pertains to:
Lexxic	https://lexxic.com/resources	Neurodiversity in different work fields, neurodevelopmental conditions
Neurodiversity Education Academy	https://www.neurodiversityeducationacademy.org/	A neurodiversity toolkit for coaches, trainers, facilitators, teachers, school leaders, founders, and human resource managers
Neurodiversity at UCL: Guidance for Staff and Line Managers by University College London on Academy	https://www.ucl.ac.uk/equality-diversity-inclusion/equality-areas/disability-equality/neurodiversity-ucl-guidance-staff-and-line-managers	Neurodiversity key terms and associated terms, and types of neurodiversity
Supporting Neurodiversity in Education by University College London	https://www.ucl.ac.uk/teaching-learning/publications/2023/oct/supporting-neurodiversity-education	What neurodiversity is, why awareness of neurodiversity and neurodivergence is important, important terms, creating neurodiversity-friendly environments, supporting neurodivergent students and scholars, supporting group work, adjustments for postgraduate students
NeurodiversityHub	https://www.neurodiversityhub.org/resources-for-universities	Autism diagnosis, supporting students with autism, disability, and accessibility support services
Resources for Working with Students with Neurodiversities by Loyola University Chicago	https://www.luc.edu/media/lucedu/facultycenter/pdfs/Resources%20for%20Working%20with%20Students%20with%20Neurodiversities.pdf	Supporting ADHD learners, understanding autism, how to talk about autism respectfully, universal design for learning and neurodiversity, resources for supporting neurodiverse students

(continued)

Table 6.4 (continued)

Source	Link and QR Code	Pertains to:
Teaching Strategies for Neurodivergent Students by the Resilient Educator	https://resilienteducator.com/classroom-resources/teaching-strategies-neurodivergent-students/	Classroom set ups, teaching methods, lesson plans
Exploring Neurodiversity in the Workplace Online Course on FutureLearn	https://www.futurelearn.com/courses/exploring-neurodiversity-in-the-workplace?utm_source=google&utm_medium=paid:search&utm_campaign=fl_DSA_All_Products_Courses_UK/IRE&gad_source=1&gclid=CjwKCAjwjeuyBhBuEiwAJ3vuoaaZAsD9e6HeWYFc9pGGvM2LIIKMRYUmkKXr823MzGOHsNiZ8txk4RoCznoQAvD_BwE	Classroom set ups, teaching methods, lesson plans

Table 6.5 Five peer-reviewed articles that discuss neurodiversity in higher education and the adjustments that can be made to support students with neurodivergence

Article Title and Authors	Journal	Link
Neurodiversity in Medical Education: How can we improve postgraduate learning for neurodiverse doctors? Robinson (2022)	Medical teacher	https://doi.org/10.1080/0142159X.2022.2039383
Medical Education: Accommodating Students with ADHD Bailey, Grotowski and Bailey (2023)	Medical Teacher	https://doi.org/10.1080/0142159X.2023.2287984
Has the time come for Heutagogy? Supporting neurodivergent learners in higher education Friedman and Nash-Luckenbach (2023)	Higher Education	https://doi.org/10.1007/s10734-023-01097-7
I can be a "normal" student: the role of lecture capture in supporting disabled and neurodivergent students' participation in higher education Horlin, Hronska and Nordmann (2024)	Higher Education	https://doi.org/10.1007/s10734-024-01201-5
Compassionate pedagogy for neurodiversity in higher education: A conceptual analysis Hamilton and Petty (2023)	Frontiers in Psychology	https://doi.org/10.3389/fpsyg.2023.1093290

learning. Another example would be, converting a video-recorded lecture into an mp4 or mp3 format which is easily downloadable so students can view or listen to it offline. This may allow some students to listen to the recording whilst doing other activities, such as running, travelling, providing childcare or if they are in a remote location where they do not have access to the Internet. Some may argue that the learning might not be effective through this, but the emphasis here is on making learning accessible and not ensuring that learning is taking place.

It is important to consider accessibility in digital classrooms as well, for example, when providing activities online or worksheets to fill in a session, it is worth printing some copies for those who might not have a device or prefer learning with traditional pen and paper format. Despite this, it may not always be possible to print a large number of papers as it is important to consider sustainability, both financial and environmental.

In addition, it is important to ensure the creation of educational resources from an accessibility lens, such as using a dyslexia friendly guide when preparing teaching and learning material to ensure easily readable fonts, colours, writing style, etc. (BDA 2023).

Table 6.6 provides some online resources that educators may find useful.

6.7 Supporting Students with Hearing Difficulties

Rodrigues et al. reported that deaf or hard of hearing (DHH) students face significant physical and mental fatigue, particularly when learning online. This may impact their wellbeing and engagement; hence, the need for inclusive and accessible approaches is highlighted (Rodrigues et al. 2022). Another article highlights that the learning issues become significant when educators are wearing masks which makes it difficult for DHH learners to lip read. Hence, the need for stakeholders, including educators and information technologists to come together to address the learning needs of DHH students (Soucie et al. 2021). A qualitative study conducted by de Oliveira et al. reported educators not being aware of students' disabilities or additional special needs. Students with hearing impairments expressed difficulty hearing the educators when the microphone was not loud enough, and it was embarrassing to ask the educators about new words that they were not able to hear properly (de Oliveira et al. 2022). For students who are studying anatomy and happen to have a hearing impairment, the struggle might be even greater as the anatomical terminology is a new language for students meaning that if they are not familiar with an anatomical term already, they might not be able to pick it up quickly.

As educators, it is our responsibility to ensure students with such additional needs do not feel disadvantaged. Therefore, it is always worth encouraging students to speak to the educator before or after the session if they have additional learning needs. In addition, when teaching online, it is important to make sure that the webcam is working and students can lip-read the educator however, this may add cognitive overload for students, hence, following the multimedia learning principles would be essential to prevent that (Mayer 2008). In addition, in an educational environment, whether in person or online, audio-dominated resources may exclude students with a hearing impairment, including those students for whom English is not their first language and may struggle to understand different accents. The addition of text captions on such resources (or the availability of speech-to-text software or a written transcript) and the use of a microphone would make them much more inclusive. Video-capturing software can generate a transcript based on speech which can be used by students who require assistance with their hearing or while reading or viewing the recorded lecture. Despite this, in our experience and upon discussing with fellow anatomists, it can be noted that the generated transcripts for anatomy lectures are not always accurate as many anatomical terms are not captured. That means that the anatomy educators needed to manually edit the transcript to ensure its anatomical accuracy. It will be ideal if in the future, the video-capturing software is trained so that it can recognize anatomical terms

Table 6.6 Some online accessibility checker resources

Source	Content	Link/QR Code
Imperial College London	Text alternatives for images with examples, including use of headings and tables	https://www.imperial.ac.uk/staff/tools-and-reference/web-guide/training-and-events/materials/accessibility/checklist/
The University of Edinburgh	Creating accessible teaching material in formats such as Word, video, and audio	https://www.ed.ac.uk/information-services/learning-technology/accessibility/checklist/teaching-resources
University of Manchester	Top tips on creating accessible documents, presentations, and videos	https://elearning.bmh.manchester.ac.uk/quickchecks/quick-checks-accessibility/
Instructure	Accessibility checker for instructors who use Canvas LMS	https://community.canvaslms.com/t5/Instructor-Guide/How-do-I-use-the-Accessibility-Checker-in-the-Rich-Content/ta-p/820
Microsoft 365	Accessibility checker for MS Word or MS PowerPoint	https://support.microsoft.com/en-gb/office/improve-accessibility-with-the-accessibility-checker-a16f6de0-2f39-4a2b-8bd8-5ad801426c7f
Microsoft 365	Adding alt text to an image for those who are blind or might have low vision	https://learn.microsoft.com/en-us/training/modules/create-accessible-content/
Microsoft 365	Accessibility tools for vision	https://support.microsoft.com/en-gb/topic/accessibility-tools-for-vision-b3c57606-e0af-46d2-97b4-fa6b5fba4fa1

(continued)

Table 6.6 (continued)

Source	Content	Link/QR Code
Microsoft 365	Accessibility tools for hearing	https://support.microsoft.com/en-gb/topic/accessibility-tools-for-hearing-111d7360-cd47-4104-aef1-2a33297909ad
Microsoft 365	Accessibility tools for neurodiversity	https://support.microsoft.com/en-gb/topic/accessibility-tools-for-neurodiversity-6dbd8065-b543-4cf8-bdfb-7c84d9e8f74a

and transcribe them accordingly. Moreover, moving around while speaking can be distracting and can cause significant variations in voice projection to certain parts of the room and face visibility. Hence, facing students during teaching delivery will help project voice evenly and will help those who lip-read.

It is worth noting that what we provide as educators to our students now will eventually benefit them in their practice. For example, when treating patients who are deaf and hard of hearing (DHH). Rannefeld et al. (2023) highlight the need for, and importance of, training healthcare professionals with digital technologies and other communication-enhancing interventions to improve healthcare provision for DHH patients due to the communication barriers that lead to dissatisfaction and lack of engagement with health services.

6.8 Supporting Students with Visual Difficulties or Impairment

Students with visual difficulties or impairment may find it challenging to learn in an educational or clinical setting if extra support is not in place for them. Rowe et al. (2019) highlight that a support system needs to be in place for such students and they should be given the opportunity to participate despite their disability. However, there may be a need for more than just a support system. For example, de Lima et al. (2022) in their systematic review highlight the need for adapting teaching resources to make them more accessible and inclusive, particularly for the visually impaired as there is a clear gap of such resources.

There is limited literature reporting the challenges faced by students with visual difficulties when learning anatomy. Mendonça et al. (2021) report the emotional challenges a student with low vision and blindness faced due to not being able to visualize the anatomical details in cadavers. As an alternative, the student focuses on palpatory anatomy so they can use their hands to feel the anatomical structures to consolidate their understanding. The student with low vision and blindness also suggests providing good lighting in a classroom environment to improve the visualization of anatomical structures (Mendonça et al. 2021). However, this may be a challenge when teaching medical imaging or radiological anatomy.

It is important as educators to support students with visual difficulties or impairment. Some of the things to consider are:

- Providing PowerPoint slides in advance so students can review the material before class or have it with them during the session to make it easy for them to view or take notes.

Alternatively, students could be encouraged to sit closer to the screen if a lecture is being conducted in person.

- Avoiding the use of red and green colour for comparing anatomical structures as it may disadvantage students who are colour blind (Patera 2021).
- Making sure the images or educational resources used in lectures, practical sessions, and exams are not blurry.
- Certain resources that are visually dominated, such as screencasts or lecture recordings without any narration, could be enhanced with the addition of descriptions including explanations of images or animations.
- It is also important to ensure the sound quality is good and clear, and the colours and images are visible.
- Students with reading difficulty due to low vision may also benefit from bigger fonts (de Oliveira et al. 2022).
- A study by Diniz and Sita (2019) describes the use of low-cost tactile neuroanatomy models for students with visual disabilities or impairments. These models could easily be reproduced using materials like fabric, velvet, pins, etc. The authors also report that the tactile material was effective in engaging the student who was blind, and the use of gloves did not impair the tactile perception either.
- Students could be exposed to a variety of multi-sensory delivery mediums in an exciting, engaging, and interactive way (AdvanceHE 2020b). For example, giving students the opportunity to physically engage with plastic models and prosections (if available), while delivering a didactic lecture. However, there may be logistical constraints associated with design and delivery of such activities.
- The first author (MK) had a medical student with Keratoconus who wore special type of lenses to be able to perform day-to-day activities, including attending university and engaging in learning. The student had difficulty concentrating on their learning and revision at home as the student could not wear the lens all day long and the eye strain with wearing these lenses long term is common. This student had a support plan in place with reasonable adjustments, such as extra time for exams.

6.9 Supporting Students from Different Faiths and Cultures

Students who come from a variety of diverse backgrounds and cultures could often be faced with circumstances where religious or cultural commitments may impact their engagement with learning or assessment. At the University of Sunderland, School of Medicine, we have flexible leave of up to 6 half days for undergraduate medical students. Students can use this leave and apply for it at least 48 h in advance of any planned absence. The reasons could include, but are not limited to, attendance at religious celebrations or observance. Students are not required to provide a reason for requesting flexible leave. However, students are required to remediate if they miss any teaching or clinical sessions that they would have attended otherwise. The only exception is that they cannot apply for flexible leave for the dates when they have an assessment scheduled.

Whilst one may argue that it is not always possible to have an understanding or familiarization with other people's cultures, faiths, or beliefs, the use of faith calendars can help educators avoid clashes with major religious or cultural events when scheduling assessments (ECU 2009). This would ensure students who have any major religious or cultural events to celebrate with family can do so without having to worry about their exams. Equally, certain clinical sessions are resource-intensive and cost institutions and clinical trusts a lot of money; hence, making sure to avoid scheduling such sessions on certain religious days would be good to avoid rescheduling them for students who may not be able to attend to remediate. This could also be assessed by keeping a record of dates/days when a proportion of students apply for flexible leave. This can also act as an indicator for educators to not schedule assessments or sessions that cost a lot, such as Observed Structured Clinical Examinations

(OSCEs) (Cusimano et al. 1994; Brown et al. 2015).

In addition, some students may have dedicated prayer times which is often difficult to avoid when scheduling teaching, but as educators, being open to students leaving the sessions early or entering late due to religious observance or prayers can make them feel included. Educators can encourage students to let them know in advance to be aware of anyone leaving early or arriving late. Additionally, at Sunderland University, there are also prayer facilities available for students to pray on campus and it is always good to signpost students to these facilities. This also could apply when there are public transport strikes which makes it difficult for students to commute, particularly in areas where public transport is limited.

Fortepiani and Marsh (2023) emphasize the importance of supporting the needs of students by getting to know them first. They say: "*Educators should start with relationship-centered teaching, paying attention to "who" the students are and "how" they experience the classroom*". They also emphasize the importance of mixing students in groups based on their skills, gender, abilities, or ethnicity wherever possible to allow heterogeneity. This is also supported by Hassaskhah and Mozaffari (2015) who report that teacher-assigned groups are much more successful in terms of accomplishing group tasks than student-selected groups. This approach may also bridge the gap in cultural competence between students. Mixing up students from different backgrounds for group work can provide rich opportunities to help students understand the advantages and challenges of working in international contexts.

It is worth noting that medical students will be treating patients who come from different cultures and faiths therefore, educators must ensure that the clinical encounters students have during their clinical and communication skills sessions are with diverse groups of people and with case scenarios that represent a variety of cultures and faiths. Verbree et al. (2023) report that medical students did not feel prepared to serve a diverse patient population as they identified lack of diversity in the curricular content, such as a lack of communication with culturally diverse patients. This emphasizes the need for medical educators to take responsibility for preparing students to serve a diverse society. As recommended by AdvanceHE, institutions could consider inviting guest speakers from different cultural and social backgrounds, who may reflect the diversity of the student population in an institution (AdvanceHE 2020b). This would be particularly helpful when the staff diversity is not representative of the student diversity in an institution.

6.10 Ensuring Teaching Space and Environment Is Accessible

It is often easy to assume that someone who may look physically fit from the outside would not necessarily need to use a lift. One of our students had a heart condition that was not visible from their physical appearance. This condition prevented them from taking the stairs on campus and they relied on the lifts in the buildings to get to their teaching sessions. This often took this student time as compared to other students who could simply use the stairs. On one occasion, the lift in one of the buildings was non-operational and the student could not get to their teaching session. Upon identifying the barrier encountered by the student, this resulted in rescheduling and relocating certain teaching sessions. This made us think, that it is important to consider environmental accessibility when scheduling teaching sessions. It is easy to forget that some people may not be able to access certain areas due to their health or a disability. Similarly, a study reported that students who used a wheelchair found it difficult to move between lectures when they were taking place at different buildings (de Oliveira et al. 2022). Hence, it might be good for educators to let students go 10 min before their next class to allow them enough time to move between teaching spaces or buildings. It is important for educators to ensure that the content they aim to cover within their lecture and/or practical can be covered sufficiently within the time they have allocated for each session. In the past, the second author (EP) was due to deliver a lecture on the gross anatomy of the heart to first-year medical students. Students had a lecture on the physiology of the heart prior to their anatomy lecture. The physiology lecturer took 1 h and 15 min to finish their lecture, meaning that the anatomy educator started to deliver their lecture late. At some point, there were 20 min remaining for the 1-h session to be over and the anatomy educator wanted to allow students 10 min before the end of an hour. The anatomy educator asked the students if they had a session immediately after their anatomy lecture and one student highlighted that they did not, but they were supposed to have their lunch break before having 2 more hours of lectures. This was an important reminder for the anatomy educator that medical students have an extremely busy timetable, and they are entitled to have enough time to have lunch and rest. Because of that, the educator decided to cover the remaining content they had left by uploading it as a prerecorded lecture that students could view later. Technology-related issues can cause further disruptions during a lecture, which can cause further delays, thus, educators should make sure that their lecture content can be covered within 45–50 min if they have been allocated an hour for a lecture.

The second author (EP) used to work at an institution where students were expected to stand for 2 h during their anatomy practical in the anatomy laboratory. There were several students who needed to use a walking stick to aid their walking, and requested a chair as it was very difficult for them to be standing for the whole duration of the practical. There were a couple of chairs available in the laboratory; however, some were not fully functional as their caster wheel was not spinning. A couple of other chairs were too short to match the table's height where the specimens were meaning that the students would not be able to utilize the specimens while being seated. Lastly, some chairs were highchairs and did not have wheels which meant that if students had to use those they would need to stand and move the chair to rotate around the table where the specimens were. Hence, it is important for anatomy educators to ensure that there is equipment available to assist students who may experience mobility issues. Similarly, if a student uses a wheelchair, it is important to ensure there is space available in the teaching space for them to fit their wheelchair without being disadvantaged.

In a study by de Oliveira et al. (2022), one participant reported difficulty with wearing normal gloves in the anatomy lab as their hand was smaller and only had two fingers. The authors highlight the barriers to inclusive practice are lack of awareness of educators and lack of dialogue between students and educators about how adaptations to their needs could be made (de Oliveira et al. 2022). It might be that due to lack of sense of belonging, students do not feel encouraged to communicate with their educators about their needs. On this occasion, it is important to ensure various glove sizes are available in the lab for students to use, made of material that does not cause any allergic reaction.

Regarding the use of donated bodies, it is reported in the literature that students with fear of death or anxiety experience challenges, particularly during their initial encounters with donated bodies (Chang et al. 2018; Romo-Barrientos et al. 2020; Allison et al. 2021). This may make it difficult for them to engage with cadaveric specimens and may negatively impact their exam performance. Educators must ensure to support these students, for example, by setting expectations at the very beginning, familiarizing students with the Human Tissue Authority (UK) (HTA 2023) processes, donor dignity, cultivating respect during the sessions, reminding them of the donor's wish for donating their body, and being available to students to discuss their concerns. Alternatively, as reported by Arráez-Aybar et al. (2004) taking students to the cadaveric lab when no cadavers are out or introducing cadaveric anatomy by showing pictures or videos before showing real cadavers in person can reduce students' anxiety. Lastly, Memorial and Thanksgiving services can also create a sense of gratefulness and act as a reminder about the donors' gift to medical education (da Rocha et al. 2020).

Finally, when delivering sessions, it might be worth considering that there are gender-neutral toilets and sanitary products accessible in spaces or close by along with water taps/dispensers available nearby for those who may feel the need to use them.

6.11 Supporting Students with Non-academic Related Matters

Some students may not have enough funding to pay their tuition fees; hence, they may have to work part-time. Some of these students may also have families to look after. It is important to signpost these students to the university financial guidance team which can help them manage their finances or signpost them to alternative funding options.

Similarly, international students may experience a culture shock and can often find it challenging to settle in or adjust, particularly those where the cultural difference is the greatest (McGarvey et al. 2015). For example, when going out in groups, particularly when celebrating over weekends, drinking alcohol is common in the UK but in certain cultures, this may not be the case. Similarly, some institutions, such as the Sunderland School of Medicine, use body painting and ultrasound to teach anatomy; students from certain cultures and faiths may not feel comfortable exposing a particular body part. Hence, educators could consider making engagement with such sessions optional or making sure the room is dark and the student is covered with some dividers or whiteboards which would be beneficial for them to engage in the activity. In addition, a female student might prefer a female facilitator in an ultrasound setting, this may be a challenge for departments who might be understaffed or who may not have many female members of staff.

DeLisa and Lindenthal (2012) also highlight the importance of increased minority representation in the workforce including the need for, and importance of, cultural competence or ethnic diversity training. Such training is important in challenging preconceived unconscious biases and stereotypes that may compromise health outcomes. It is important in providing awareness of communities that make up the world. This is why it is important to offer support that some students may require, which educators may not be aware of. This could include and may not be limited to wellbeing support, academic writing support, or supporting extenuating circumstances for assignment deadlines.

6.12 Conclusion

Whilst there might be lack of awareness regarding making anatomy curricula inclusive, it is of utmost importance that educators strive to make anatomy education as inclusive as possible. This can be achieved by ensuring the representation of diverse student cohorts within their anatomy content along with ensuring the consistent use of inclusive language. Students with visible or hidden disabilities and/or impairments may face barriers to accessing education or educational resources. Hence, educators have the responsibility of minimizing such barriers by making adjustments to their teaching content, delivery, environment, and assessment methods. Such adjustments can improve students' overall anatomy learning experience. Higher education institutions should train their staff so that they can adequately support students who may or may not have additional accessibility needs alongside ensuring the provision of high-quality teaching.

References

3D4Medical from Elsevier (2024) The world's most advanced 3D anatomy platform. https://3d4medical.com/. Accessed 11 Jun 2024

AdvanceHE (2003) Aligning teaching for constructing learning. https://www.advance-he.ac.uk/knowledge-hub/aligning-teaching-constructing-learning. Accessed 9 Jun 2024

AdvanceHE (2020a) Inclusive learning and teaching. In: Advance Higher Education. https://www.advance-he.ac.uk/guidance/equality-diversity-and-inclusion/student-recruitment-retention-and-attainment/inclusive-learning-and-teaching. Accessed 9 Jun 2024

AdvanceHE (2020b) Inclusive curriculum. In: Advance HE. https://www.advance-he.ac.uk/inclusive-curriculum. Accessed 10 Jun 2024

Allison S, Notebaert A, Perkins E et al (2021) Fear of death and examination performance in a medical gross anatomy course with cadaveric dissection. Anat Sci Educ 14:764–773. https://doi.org/10.1002/ase.2092

Arráez-Aybar L-A, Casado-Morales MI, Castaño-Collado G (2004) Anxiety and dissection of the human cadaver: an unsolvable relationship? The Anatomical Record Part B: The New Anatomist 279B:16–23. https://doi.org/10.1002/ar.b.20022

Barradell S, Barrie S, Peseta T (2018) Ways of thinking and practising: highlighting the complexities of higher education curriculum. Innov Educ Teach Int 55:266–275. https://doi.org/10.1080/14703297.2017.1372299

BDA (2023) Creating a dyslexia friendly workplace. In: The British Dyslexia Association. https://www.bdadyslexia.org.uk/advice/employers/creating-a-dyslexia-friendly-workplace/dyslexia-friendly-style-guide. Accessed 10 Jun 2024

Beni R, Fisher L, Longhurst GJ (2022) The importance of diverse and accurate descriptions of genital anatomy in textbooks. Anat Sci Educ 15:985–988. https://doi.org/10.1002/ase.2200

Brennan PA, Susan S, Sam W (2019) Female reproductive system. In: Gray's surgical anatomy, 1st edn. Elsevier, pp 551–551

Brown C, Ross S, Cleland J, Walsh K (2015) Money makes the (medical assessment) world go round: the cost of components of a summative final year objective structured clinical examination (OSCE). Med Teach 37:653–659. https://doi.org/10.3109/0142159X.2015.1033389

Bruner JS (1960) The process of education. Harvard University Press, Cambridge, MA

Carless D (2015) Exploring learning-oriented assessment processes. High Educ (Dordr) 69:963–976. https://doi.org/10.1007/s10734-014-9816-z

Castellon S (2018) What is neurodiversity celebration week? In: Lexxic. https://www.neurodiversityweek.com/. Accessed 10 Jun 2024

Chang H-J, Kim HJ, Rhyu IJ et al (2018) Emotional experiences of medical students during cadaver dissection and the role of memorial ceremonies: a qualitative study. BMC Med Educ 18:255. https://doi.org/10.1186/s12909-018-1358-0

Cusimano MD, Cohen R, Tucker W et al (1994) A comparative analysis of the costs of administration of an OSCE (objective structured clinical examination). Acad Med 69:571–576. https://doi.org/10.1097/00001888-199407000-00014

da Rocha AO, Maués JL, Chies GAF, da Silva AP (2020) Assessing the impact of a ceremony in honor of the body donors in the development of ethical and humanistic attitudes among medical students. Anat Sci Educ 13:467–474. https://doi.org/10.1002/ase.1920

de Lima P, Souza e Silva R, Guedert DG et al (2022) Teaching human anatomy to the visually impaired: a systematic review. Clin Anat 35:660–665. https://doi.org/10.1002/ca.23887

de Oliveira E, Dantas RG, Amaral GA et al (2022) Experiences of disabled students in undergraduate medical education. Med Teach 44:294–299. https://doi.org/10.1080/0142159X.2021.1985098

DeLisa JA, Lindenthal JJ (2012) Commentary: reflections on diversity and inclusion in medical education. Acad Med 87:1461

Diniz GB, Sita L V (2019) ARTICLE development of low-cost tactile neuroanatomy learning tools for students with visual-impairment

Dominoni M, Gritti A, Bergante C et al (2022) Genital perception and vulvar appearance after childbirth: a cohort analysis of genital body image and sexuality. Arch Gynecol Obstet 307:813–819. https://doi.org/10.1007/s00404-022-06826-4

Drake LR, Vogl AW, Mitchell AWM (2023) Gray's anatomy for students, 5th edn. Elsevier

Drexel University (2024) How to promote inclusion in the classroom. In: Drexel University School of Education. https://drexel.edu/soe/resources/student-teaching/advice/Promote-Inclusion-in-the-Classroom/#:~:text=Students%20who%20may%20have%20experienced,value%20as%20a%20human%20being. Accessed 24 Jul 2024

Duong D, Vogel L (2022) Untapped potential: embracing neurodiversity in medicine. Can Med Assoc J 194:E951–E952. https://doi.org/10.1503/cmaj.1096006

Easterling L, Byram J (2022) Shifting language for shifting anatomy: using inclusive anatomical language to support transgender and nonbinary identities. Anat Rec 305:983–991. https://doi.org/10.1002/ar.24862

ECU (2009) Religious observance in higher education: institutional timetabling and work patterns. In: Equality Challenge Unit. https://www.advance-he.ac.uk/knowledge-hub/religious-observance-higher-education-institutional-timetabling-and-work-patterns. Accessed 10 Jun 2024

ECU (2014) Equality and diversity for teaching staff in colleges. In: Equality Challenge Unit. https://www.advance-he.ac.uk/knowledge-hub/equality-and-diversity-teaching-staff-colleges#:~:text=Teaching%20staff%20have%20a%20crucial,or%20otherwise%20engage%20with%20them. Accessed 10 Jun 2024

Elkington S (2020) Assessment: understanding the basics. In: Marshall S (ed) A handbook for teaching and learning in higher education: enhancing academic practice, 5th edn. Routledge, pp 72–81

Elsevier (2024) Gray's anatomy for students, 5th edn. Elsevier. https://evolve.elsevier.com/cs/product/9780323934237?role=student. Accessed 25 Jul 2024

Endres P, Ziring D, Papanagnou D (2022) Medical student sensitivity training on the differences in sex development. Int J Med Educ 13:259–260. https://doi.org/10.5116/ijme.631e.e8fd

Fernández-Batanero JM, Montenegro-Rueda M, Fernández-Cerero J (2022) Access and participation of students with disabilities: the challenge for higher education. Int J Environ Res Public Health 19:11918. https://doi.org/10.3390/ijerph191911918

Finn GM, Danquah A, Matthan J (2022) Colonization, cadavers, and color: considering decolonization of anatomy curricula. Anat Rec 305:938–951. https://doi.org/10.1002/ar.24855

Fortepiani LA, Marsh SA (2023) Innovative techniques for developing an inclusive teaching environment. Adv Physiol Educ 47:904–907. https://doi.org/10.1152/ADVAN.00014.2023

GMC (2024a) The importance of inclusion in medicine. In: The General Medical Council. https://www.gmc-uk.org/education/standards-guidance-and-curricula/guidance/welcomed-and-valued/health-and-disability-in-medicine/the-importance-of-inclusion-in-medicine. Accessed 9 Jun 2024

GMC (2024b) The importance of inclusion in medicine. In: General Medical Council. https://www.gmc-uk.org/education/standards-guidance-and-curricula/guidance/welcomed-and-valued/health-and-disability-in-medicine/the-importance-of-inclusion-in-medicine#:~:text=All%20medical%20students%20and%20doctors,the%20competences%20required%20of%20them. Accessed 10 Jun 2024

Golden RN, Petty EM (2022) Learners with disabilities: an important component of diversity, equity, and inclusion in medical education. Acad Med 97:328–330. https://doi.org/10.1097/ACM.0000000000004496

Gov.uk (2010) Equality act 2010. In: legislation.gov.uk. https://www.legislation.gov.uk/ukpga/2010/15/part/2. Accessed 9 Jun 2024

Habbal O (2009) The state of human anatomy teaching in the MedicalSchools of Gulf Cooperation Council Countries. Present and future perspectives. Sultan Qaboos Univ Med J 9:24–31

Hafferty FW, Finn GM (2015) The hidden curriculum and anatomy education. In: Teaching anatomy. Springer International Publishing, Cham, pp 339–349

Hassaskhah J, Mozaffari SH (2015) The impact of group formation method (student-selected vs. teacher-assigned) on group dynamics and group outcome in EFL creative writing. Language teaching research, vol. 6. https://doi.org/10.17507/jltr.0601.18

Hayes JA, Temple-Smith MJ (2022) New context, new content—rethinking genital anatomy in textbooks. Anat Sci Educ 15:943–956. https://doi.org/10.1002/ase.2173

Hsu JL, Rowland-Goldsmith M, Schwartz EB (2022) Student motivations and barriers toward online and in-person office hours in STEM courses. CBE—life sciences. Education 21:ar68. https://doi.org/10.1187/cbe.22-03-0048

HTA (2023) The Human Tissue Authority. In: The human tissue authority. https://www.hta.gov.uk/. Accessed 10 Jun 2024

Ibe C (2024) About Chidiebere Ibe. In: Chidiebere Ibe. https://www.chidiebereibe.com/about-me/?doing_wp_cron=1718133899.5383601188659667968750. Accessed 11 Jun 2024

Jönsson A, Prins F (2019) Editorial: transparency in assessment—exploring the influence of explicit assessment criteria. Front Educ (Lausanne) 3. https://doi.org/10.3389/feduc.2018.00119

Kelly T, Rodriguez SB (2022) Expanding under-represented in medicine to include lesbian, gay, bisexual, transgender, and queer individuals. Acad Med 97:1605–1609. https://doi.org/10.1097/ACM.0000000000004720

Lea Aisia (2024) Anatome. https://anatomeinclusivity.wixsite.com/anatome. Accessed 10 Jun 2024

Lerchenfeldt S, Mi M, Eng M (2019) The utilization of peer feedback during collaborative learning in undergraduate medical education: a systematic review. BMC Med Educ 19:321. https://doi.org/10.1186/s12909-019-1755-z

Louie P, Wilkes R (2018) Representations of race and skin tone in medical textbook imagery. Soc Sci Med 202:38–42. https://doi.org/10.1016/j.socscimed.2018.02.023

Marieb EN, Wilhelm PB, Mallatt JB (2016) Human anatomy (8th ed.). Pearson

Martin GC, Kirgis J, Sid E, Sabin JA (2016) Equitable imagery in the preclinical medical school curriculum: findings from one medical school. Acad Med 91:1002–1006. https://doi.org/10.1097/ACM.0000000000001105

Matthan J, Finn GM (2020) The hidden curriculum of utilisation of imaging and unregulated digital resources within clinical education, pp 145–163

Mayer RE (2008) Applying the science of learning: evidence-based principles for the design of multimedia instruction. Am Psychol 63:760–769. https://doi.org/10.1037/0003-066X.63.8.760

Mbaki Y, Todorova E, Hagan P (2021) Diversifying the medical curriculum as part of the wider decolonising effort: a proposed framework and self-assessment resource toolbox. Clin Teach 18:459–466. https://doi.org/10.1111/tct.13408

McGarvey A, Brugha R, Conroy RM et al (2015) International students' experience of a western medical school: a mixed methods study exploring the early years in the context of cultural and social adjustment compared to students from the host country. BMC Med Educ 15:111. https://doi.org/10.1186/s12909-015-0394-2

Mehta LA, Natrajan M, Kothari ML (1996) Understanding anatomical terms. Clin Anat 9:330–336. https://doi.org/10.1002/(SICI)1098-2353(1996)9:5<330::AID--CA6>3.0.CO;2-G

Mendonça CR, de Souza KTO, Arruda JT et al (2021) Human anatomy: teaching–learning experience of a support teacher and a student with low vision and blindness. Anat Sci Educ 14:682–692. https://doi.org/10.1002/ase.2058

Morris M, Cooper RL, Ramesh A et al (2019) Training to reduce LGBTQ-related bias among medical, nursing, and dental students and providers: a systematic review.

BMC Med Educ 19:325. https://doi.org/10.1186/s12909-019-1727-3

Mukhopadhyay DK, Choudhari SG (2024) Critical thinking and clinical reasoning in undergraduate medical course: a mixed-methods study in a Medical College in Kolkata, West Bengal, India. F1000Res 13:259. https://doi.org/10.12688/f1000research.146009.1

Nazar M, Kendall K, Day L, Nazar H (2015) Decolonising medical curricula through diversity education: lessons from students. Med Teach 37:385–393. https://doi.org/10.3109/0142159X.2014.947938

Newcastle University (2024) Gender-inclusive language. In: Newcastle University

Nicol D (2021) The power of internal feedback: exploiting natural comparison processes. Assess Eval High Educ 46:756–778. https://doi.org/10.1080/02602938.2020.1823314

O'Callaghan Tassia (2024) The reframing revolution: make medicine diverse. In: Peanut. https://www.peanut-app.io/blog/reframing-revolution. Accessed 11 Jun 2024

Parker R, Larkin T, Cockburn J (2017) A visual analysis of gender bias in contemporary anatomy textbooks. Soc Sci Med 180:106–113. https://doi.org/10.1016/j.socscimed.2017.03.032

Patera E (2021) Inclusive lectures: how often do we see them across anatomy and medical education? Med Sci Educ 31:2137. https://doi.org/10.1007/s40670-021-01411-x

Philippopoulos E (2023) More than just pronouns—gender-neutral and inclusive language in patient education materials: suggestions for patient education librarians. J Med Libr Assoc 111:734–739. https://doi.org/10.5195/jmla.2023.1723

QAA (2018) Assessment. In: The Quality Assurance Agency for Higher Education. https://www.qaa.ac.uk/the-quality-code/advice-and-guidance/assessment. Accessed 10 Jun 2024

Ramsden P (2003) Learning to teach in higher education. Routledge

Rannefeld J, O'Sullivan JL, Kuhlmey A, Zoellick JC (2023) Deaf and hard-of-hearing patients are unsatisfied with and avoid German health care: results from an online survey in German sign language. BMC Public Health 23:2026. https://doi.org/10.1186/s12889-023-16924-w

RCOG (2013) Skin conditions of the vulva. In: Royal College of Obstetricians and Gynaecologists. https://www.rcog.org.uk/for-the-public/browse-our-patient-information/skin-conditions-of-the-vulva/. Accessed 9 Jun 2024

Rodrigues FM, Abreu AM, Holmström I, Mineiro A (2022) E-learning is a burden for the deaf and hard of hearing. Sci Rep 12:9346. https://doi.org/10.1038/s41598-022-13542-1

Romo-Barrientos C, Criado-Álvarez JJ, González-González J et al (2020) Anxiety levels among health sciences students during their first visit to the dissection room. BMC Med Educ 20:109. https://doi.org/10.1186/s12909-020-02027-2

Rompolski K, Pascoe MA (2024) Does dissection influence weight bias among doctor of physical therapy students? Anat Sci Educ 17:1473. https://doi.org/10.1002/ase.2497

Rowe M, Steyl T, Phillips J, Rhoda A (2019) Supporting a student with visual impairment in the intensive care unit. South African Journal of Physiotherapy 75. https://doi.org/10.4102/sajp.v75i1.1324

Sambell K, McDowell L, Montgomery C (2012) Assessment for learning in higher education. Routledge

Servier Servier Medical Art. In: Servier Medical Art. https://smart.servier.com/. Accessed 10 Jun 2024

Soucie J, Vandervoort L, Adkins LE, Miller S (2021) Experiential education for the deaf or hard of hearing learner in a pharmacy curriculum. Curr Pharm Teach Learn 13:1102–1105. https://doi.org/10.1016/j.cptl.2021.06.025

UKSCQA (2018) The quality code. In: The Quality Assurance Agency for Higher Education. https://www.qaa.ac.uk/the-quality-code#. Accessed 10 Jun 2024

UNISON (2024) Why pronouns are important

Van Merriënboer JJG, Sweller J (2010) Cognitive load theory in health professional education: design principles and strategies. Med Educ 44:85–93. https://doi.org/10.1111/j.1365-2923.2009.03498.x

Verbree A-R, Isik U, Janssen J, Dilaver G (2023) Inclusion and diversity within medical education: a focus group study of students' experiences. BMC Med Educ 23:61. https://doi.org/10.1186/s12909-023-04036-3

Weeks SE, Harris EE, Kinzey WG (1995) Human gross anatomy: a crucial time to encourage respect and compassion in students. Clin Anat 8:69–79. https://doi.org/10.1002/ca.980080113

Williams J, Kane D, Sagu S, Smith E (2008) Exploring the National Student Survey Assessment and feedback issues

Women's Health Victoria (2024) Labia Gallery. https://www.labialibrary.org.au/labia_gallery/. Accessed 24 Jul 2024

Wong SHM, Gishen F, Lokugamage AU (2021) 'Decolonising the medical curriculum': Humanising medicine through epistemic pluralism, cultural safety and critical consciousness. Lond Rev Educ 19. https://doi.org/10.14324/LRE.19.1.16

Beyond the Answer: Exploring Question-Asking and Promoting Inclusivity

7

Shannon N. Curran, Rebekah Freed, and Amanda M. Root

Abstract

While asking and answering questions in class can be an anxiety inducing activity for students, it is a common educational tool used in higher education and medical education. Whether a student participates in asking questions in class can be influenced by a variety of factors, particularly pedagogical, psychological, and cultural. This guide outlines the literature to review major concepts of pedagogical, psychological, and cultural factors that influence question-asking, and then explores how to create an inclusive environment to foster success with question-asking in the classroom. Exploring pedagogical influences demonstrates that active learning, founded in constructivist learning methodologies, lays the framework for learners to ask questions in class. Psychological factors include motivation, interest, engagement, academic emotions, and growth mindset, all of which influence an individual's likelihood of participating in asking questions in an academic setting such as medical school. Cultural considerations must look at the broader culture surrounding question-asking and the culture of the classroom setting itself, as both impact whether learners will ask questions in class. In addition to pedagogy, psychology, and culture, we also explore specific challenges to consider when encouraging participation in class, such as imposter syndrome, and pose them as opportunities for instructors to be proactive and influence a learner's choice to ask questions in class. Finally, the factors that go into creating an inclusive environment will be discussed to allow educators to create an optimal learning environment for students in higher education and medical education.

Keywords

Question-asking · Inclusive teaching · Pedagogy · Psychology · Student culture education

7.1 Beyond the Answer: Exploring Question-Asking and Promoting Inclusivity

Question-asking in class can be a daunting task for students, even adult students in higher education and the medical school setting. Student engagement and participation in class is the first step to fostering a setting in which questions can be asked and answered by students. Trowler

S. N. Curran (✉) · A. M. Root
Kansas College of Osteopathic Medicine, Wichita, KS, USA
e-mail: scurran@kansashsc.org

R. Freed
University of Washington, Seattle, Seattle, WA, USA

R. Door (ed.), *Accessibility, Inclusivity and Diversity in Education and Beyond*, Biomedical Visualization 8, https://doi.org/10.1007/978-3-031-81795-3_7

(2010) reviews the historical coverage of student engagement in the literature and how "student engagement" has been a large focus in higher education. Many studies have demonstrated the connection between student engagement, improved performance and outcomes in class (Coates 2005; Trowler 2010). A key part of engagement is long thought to be oral participation, often in the form of question-asking and answering (Frymier and Houser 2016). Questioning is the most frequently used instructional tool, and asking questions is a key element of the learning process (Almeida 2012). Therefore, it might be said that it is in the best interest of students to participate in questioning and answering questions, yet it is known that learners of all educational levels generally avoid asking questions (Almeida 2012). Some educators are even changing the way they instruct to focus on a question-centered approach to promote insightful question-asking (Bowker 2010).

Many studies regarding question-asking and answering in the classroom have been conducted in the K-12 classroom, with a paucity of literature looking at students in higher education and medical education. It is the aim of this guide to provide a background and synthesis of the extant literature on student learning as it pertains to asking and answering questions in class, and to provide what the authors believe to be the necessary information to create a safe and inclusive environment to promote question-asking. This guide will delve into the pedagogical, psychological, and cultural factors that go into the decision-making process of students and why they choose to participate in question-asking and answering in higher education and medical school settings. Furthermore, we will explore the challenges and opportunities of asking questions in class and the methods of creating an inclusive learning environment to best promote question-asking in educational settings.

7.2 Pedagogy of Question-Asking

Traditional teaching methods in higher education and medical education often suggest the image of the "Sage on a Stage," disseminating content to a passive audience of students (Morrison 2014). This has long been the traditional way of teaching and follows a teacher-centered approach to learning. This approach does not provide opportunity for active student participation, which is why there has been a shift in education away from the "Sage on a Stage" to an instructor role of "Guide on the Side," which in turn has also required a shift in student roles (Morrison 2014). Learning sciences strive to better understand the cognitive and social processes that result in best effective learning. This knowledge is used to design the most effective classroom settings for learning to promote deep and effective learning (Sawyer 2014). This new approach to teaching is learner-centered and provides the learner with the opportunity for deeper learning. Deep Learning requires learners: (1) relate new ideas to prior knowledge, (2) integrate knowledge into interrelated conceptual systems, (3) look for patterns and underlying principles, (4) evaluate new ideas and relate to conclusions, (5) understand the process of dialogue which knowledge is created by, and examine logic of argument critically, and (6) reflect on personal understanding and learning processes (Sawyer 2014).

7.2.1 Active Learning Strategies

Effective classroom methodologies that promote deeper learning are active learning strategies and include those such as problem-based learning, inquiry-based learning, and flipped-classroom approaches. This is due to the nature of all these approaches promoting knowledge construction within the framework of constructivism. Constructivism is the leading framework of thought surrounding how learning occurs. This model of learning outlines learning as a process through which individuals actively construct their knowledge into mental frameworks via performing mental operations on the information to integrate it into one's memory and understanding (Cobern 1993). At its core, this is the reason why active pedagogies are as such, encouraging learner participation in their education. Here we outline the ways in which dominant active pedagogies encourage question-asking in medical education classrooms.

Problem-Based Learning Problem-based learning (PBL) is a pedagogical approach developed in the 1950s to address the criticism of traditional learning in medical education (Hung et al. 2008). PBL became a popular solution to demands to reduce lecture time and increase problem-solving skills and self-directed learning, which in turn create life-long learners. PBL immerses students in real-life ill-defined problems and requires them to work together to problem-solve and think critically to work through each problem. A faculty member or tutor is present to facilitate, not direct, and ensure students do not stray too far off topic, but student learning is otherwise self-directed, and question driven. In this scenario, it is essential that students can ask questions and ponder answers to critically evaluate the steps of working through a problem. The students must decide what steps they must take to evaluate the problem and explore those steps to develop their knowledge of the problem. At the end of the learning period, students will summarize and integrate the knowledge they have gained (Hung et al. 2008). In the initial steps, question-asking is essential to determine what knowledge they are lacking and must explore in their learning process. In this way, question-asking is essential for the practice of PBL.

Inquiry-Based Learning Inquiry-based learning is one pedagogical approach developed to encourage authentic participation in classes, which mirrors the inquiry approach in authentic science fields to prepare learners for participation in these fields (Constantinou et al. 2018; van Joolingen and Zacharia 2009). There is a general process to inquiry-based learning, and while there are several classifications of inquiry processes, they mostly differ in their granularity, ranging from very detailed to quite broad. This chapter will use the de Jong (2006) classification for this discussion, which uses the following inquiry process: (1) *orientation*—the learner makes a broad analysis of the domain, (2) *hypothesis generation*—specific statements about the domain are chosen for consideration, (3) *experimentation*—a test to investigate the validity of the hypothesis is designed and performed, predictions are made and outcomes of the experiment are interpreted, and (4) *conclusion*—a conclusion about the validity of the hypothesis is formed or new ideas are formed. This process is developed on question-asking, as is seen with step (2) *hypothesis generation*. Inquiry-based learning requires the learner to think critically and develop the questions they are going to investigate in the inquiry process through experimentation. In this way, they are constructing knowledge, supported by the constructivist approach to learning (Bada and Olusegun 2015).

Flipped-Classroom Learning A third pedagogical strategy within active learning pedagogy and constructivism frameworks is flipped-classroom models. In this model of education, *educators* provide instructional learning activities before students come to class, such that they engage in didactic elements independently, and interactive learning in the classroom (Schmidt and Ralph 2016). For example, instruction for an anatomy class might require an online learning module prior to class attendance so when students come to class, they are ready to engage in question-asking and other group learning activities. This is one educational strategy that demonstrates the "Guide on the Side" approach mentioned earlier in the introduction to this chapter (Morrison 2014). Activities students might be invited to engage in within a flipped classroom include (1) group activities where students arrange themselves into small groups to answer questions posed by the instructor, (2) students ask *educators* to clarify content from the pre-work and instructional activities, and (3) jigsaw activities where the groups all approach the subject through delving into complementary components of a problem that build on each other in order to solve the big picture together (Cochon Drouet et al. 2023). Though there are variances in the fidelity of implementation, as well as the proportion of instructional materials presented inside and outside of class, the concept of flipping the instruction from the professor to the learners is the general guiding principle of this approach (Schmidt and Ralph 2016). This approach fosters classroom time for students to engage with the material in a deeper way and to ask thoughtful questions about the material.

7.2.2 Importance of Educator–Student Interactions

While active learning strategies based in the constructivist approach to learning favor a "Guide on the Side" role for the educator, that does not mean the student–educator interactions are not important. In each active learning strategy outlined, the educator plays a vital role of guiding the learners toward success and helping them develop their question-asking skills. The educator fosters a safe space to learn and build knowledge through discovery and assigning meaning to new experiences.

In a constructivist classroom, it is the educator's role to encourage learners to ask questions of their peers and of their teachers. This process of question-asking is an essential part of the learning process and knowledge acquisition process according to the constructivist learning theory (Bada and Olusegun 2015). Problem-based learning, inquiry-based learning, and flipped-classroom methodologies all approach learning in slightly different ways, but they hold the key tenets of active learning and constructivism, and each relies on the role of the educator. The role of the educator in each instance as the guide or facilitator is to help develop the learner's ability to apply meaning to new experiences and information in the knowledge acquisition process (Bada and Olusegun 2015). The encouragement of question-asking is essential to achieve this and have successful learner outcomes.

It is important to stress the necessity of establishing a safe and supportive classroom environment when encouraging question-asking in the classroom. We have established the necessity of active learning and interaction while in medical school classrooms, but an environment that encourages class participation and question-asking is also necessary for all of these behaviors to happen. Setting up a psychologically safe environment that is student-centered encourages student participation (Almeida 2012) and provides an environment that is positive, where they can be safe being vulnerable and asking questions about what they don't understand (Pekrun and Linnenbrink-Garcia 2012). This is in part accomplished by making sure the environment motivates students to participate in class. It is important to create a positive environment in the classroom so students feel like they can ask questions, and so they can engage in interesting class content. Part of this environment is the one in the student's mind, so it is important that students have internalized a healthy mindset for vulnerability and question-asking (Dweck 2006). In the next section, we will delve into what classroom conditions encourage psychological safety and classroom question-asking.

7.3 Psychology of Question-Asking

Psychologically, how does an instructor set the stage for question-asking in their course? The elements identified in this section highlight the psychological conditions needed for classroom participation and question-asking a teacher needs to motivate students to participate, create interest, provide positive academic emotions, and promote growth mindset.

7.3.1 Motivation

Students' motivation for their work influences their likelihood of asking questions in class (Weiner 2010). Motivation is both a colloquial term and a field of scientific study in social psychology. In this section, we will discuss the definitions, theory, and research behind motivation science, as well as how motivation influences whether a medical student asks questions in class.

Atkinson, one of the original motivation theorists, described motivation as being a combination of one's motive behind something, their expectancy around it, and its associated value (Weiner 2010). Motivation is the energy to pursue a desired future and the direction that helps channel this drive (Oettingen et al. 2015). There are many different things that influence motivation, but research focuses on measuring desirability value (i.e., how much someone desires their goal) and feasibility expectations (e.g., expectations of reaching your goal, be it because one's ability or the goal's feasibility) as main themes (Oettingen et al. 2015). For example, if a student really wants to contribute to their class discussion, she may raise her hand in class, but if she does not believe her contribution is valuable, then she may not participate. Beliefs are also key to motivation. If someone believes that they are a low performing student, then they may believe that others know more than they do, and they will not raise their hand to ask a question out of fear of what their peers might think.

Though people may think of motivation as being something inborn and immutable, motivation is not a stable trait, and it can be influenced by others or regulated by the individual student (Lazowski and Hulleman 2016). Motivation can be influenced by the amount of energy or attention available by an individual at any given time (Baumeister et al. 2018). To influence their motivation, one has to monitor it (Miele and Scholer 2018). In the medical school classroom, students may show signs of motivation in their behavior by participating willingly with some topics more than others, raising their hand in certain lectures, and asking questions with certain educators. It is important to practice observing students to be able to monitor their motivation in the medical school classroom and adapt course instruction and activities accordingly.

Goals Motivation is also influenced by the desirability and feasibility expectations of a goal (Oettingen et al. 2015). For example, the more a student wants to learn about vascular surgery, the more they will spend time setting learning goals and studying about vascular surgery. If they believe they can grasp the concepts and understand enough to become a vascular surgeon, then they may be more willing to ask for help and reach out to their instructors for support in this.

A goal is generally defined as "the aim or object of one's actions," (p. 62, Cleary 2018). How motivated a student is to perform well on an assignment is influenced by what that assignment entails. The goal must represent the desired result of performing an activity. Setting goals prior to learning is, therefore, very important. In fact, the type of goals a learner sets can drastically affect how motivated they are to engage with learning the material, and in what way. Students that set and pursue goals consider different factors when setting those goals, whether consciously or unconsciously (Cleary 2018). Students consider choice, attention, effort, and persistence when promoting specific, task-related goals that support student learning. For example, students who have a strong sense of identity as medical students may set different goals about their learning in medical school, influencing whether they seek out clarification or engagement in the topics by asking for help.

Goals can be either specific or general, short term and immediate or long term and distal, they can be easy or difficult, or they can pertain to the outcome or the process. In the case of medical students in a learning environment such as a laboratory, when educators set learning goals, they can take these factors into account to encourage motivation and participation in their students. For example, are their learning objectives specific or general? Specific goals are clear and articulate, which feel to the learner accepting them that they can be attained. Goals that are closer to us that take less time to reach truly motivate us to engage in goal pursuit toward those goals. Additionally, are these objectives at a good level of difficulty—not too easy, not too difficult, but do-able? This is the level of "desirable difficulty," which is that the goals are written at the level where the learners can attain them at their current level of ability and knowledge (Dunlosky et al. 2013). Lastly, objectives or goals that focus on the process rather than the outcome are more motivating because people are more likely to be encouraged to engage in the process, which naturally leads to the outcome, but does not make that the only part of learning. Those that focus on the process of learning will also allow students to meet their goals, and they will also hone their process and strategies for future learning (Cleary 2018).

Achievement goal theorists research reasons why students engage, choose, and persist at different learning activities. What kind of learning goals students create for themselves influences their behavior in the classroom. There are two broad types of achievement goals in this context: performance goals and mastery goals (Meece et al. 2006). Performance goals are learning goals associated with performance outcomes or performance in relation to others, such as the goal of getting a 4.0 grade point average (GPA) or to perform at the top of the class. Performance goals correlate with surface level strategy use, self-handicapping, cheating, and lower grades in college. This is thought to be due to the underlying performance orientation and students not focusing on the content of the goals. The other kind of achievement goal is mastery goals. These goals pertain to content of the learning and relate to understanding over performance, such as wanting to understand the vasculature of the brain or really mastering sutures in surgery. These types of goals are associated with positive learning and motivational outcomes. In fact, students who set mastery goals are more likely to hold a deeper understanding of learning concepts, and thus perform better than other students who only set performance goals. They also have better quality of life outcomes, because they are not likely to make attributions to ability if they fail—they are likely to make attributions to effort and strategy use. These attributions encourage a growth mindset (Dweck 2006).

How a classroom environment is set up can either encourage students to adopt a mastery goal orientation or a performance goal orientation. The different elements of an environment that influence a student's internalized goal structure are called classroom goal structures (Meece et al. 2006). A performance-oriented classroom goal structure includes a classroom where their grades focus on students' performance relative to others, when performance outcomes are the focus of the class rather than the learning, and when their assignments and exams are higher stakes. A mastery-oriented classroom goal structure includes classroom elements that emphasize process and understanding over product and outcome, where learning outcomes are the focus over performance outcomes, and where cooperation is encouraged over competition. For example, a medical school that uses grading on a bell curve, encourages competition, and using high stake tests encourages performance-oriented classroom goal structures, which in turn encourage students to internalize individual performance goal orientations (e.g., they only care about their grade rather than their understanding). In sum, classroom goal structures influence students' goal development, and students' personal achievement goals shape their behavior and learning in educational environments. This is because achievement goals are related to student motivation, strategy use, and ultimately, performance. In this way, classroom goal structures set up by the educator is one-way educators can influence how students learn and participate in class. Consciously setting up an environment that promotes cooperation and mastery goals encourages students to integrate mastery goal structures, and thus sets the stage for positive class participation and asking questions.

Attribution Attributions students make about their learning influence motivation for future goals and learning (Weiner 2010). Attributions are the conscious or unconscious associations individuals make to determine a cause of a behavior (Cleary 2018). For example, a student that asks a question in class is likely to attribute their asking to their lack of knowledge in that area and not their lack of ability because in the end, they indeed were able to seek out such information, understand it, and integrate it just fine. Thus, attributions do influence classroom environments consciously and unconsciously.

Attributions are something that humans make naturally, but certain attributions are more helpful in life than others, and this is where we can encourage helpful attributions in medical students. Attributions are evaluated as being either stable or unstable, controllable or uncontrollable, and internal or external (Cleary 2018). Specifically, attributions that are unstable are desirable because the cause of the behavior can be changed. For example, getting a question in class wrong, a student may attribute their failure to ineffective study strategies, rather than to their lack of ability or intelligence. Attributions that are controllable are more beneficial than uncontrollable attributions. For example, a student who gets a quiz question wrong in class may attribute their failure to lack of sufficient time to study, or they might attribute their failure to low intelligence or ability. Attributing their failure to not having enough study time is something malleable, rather than an innate ability, so they might dedicate more study time when preparing for the next quiz. Lastly, attributions that are internal promote motivation and future positive behaviors because attributing their failure on a test to an internal cause, such as their lack of effort in studying for that topic, is something they can do something about for the next exam. Meanwhile, attributions that are external, such as attributing their failure on the test to their instructor disliking them, are not helpful because that is an external attribution, and there is not anything students can do about it. In sum, in the medical school classroom, encouraging effort and strategy during studying and learning encourages learners to create productive attributions, rather than thinking they got the question wrong because they were just unable, which allows students to focus on learning, class participation, and question-asking.

7.3.2 Interest

Educators know how important it is to get students interested in learning and instruction. Without an audience to instruct, what is an instructor? However, interest is a dual-edged sword: interest is situationally triggered, so doing something to catch learners' attention for instruction is one way to trigger interest in the classroom (Hidi and Ainley 2012). Indeed, the effort of regulating or orchestrating a situation to be interesting and catch students' attention is an ambivalent challenge: it can go either way. Specifically, learners may remember the thing that triggered their attention, but whether they retain interest in the subject matter or instructional content is hard to predict. Alternatively, learners may recall the content they are passionate about, which then triggers individual interest in the topic because of the content. In sum, instructors must orchestrate their classroom environment purposefully and with intentionality to set up a situation where students are motivated to participate, which includes creating an interesting setting for learning and instruction.

Interest is triggered by elements in the learning environment (Hidi and Ainley 2012). The stages of interest development are triggered or situational interest, maintained situational interest, emerging individual interest, and lastly, well-developed individual interest (Hidi and Ainley 2012). Interest can be a situational, individual, or an affective state. Interest influences attentional processes, quality and level of learning, learners' selection of learning strategies, achievement goals, and choices students make during performance. Educators willing to engage students in their class via nurturing interest can become powerful players in a medical student's development of individual interest in a research subject or a clinical specialty area. For example, working on case-based questions to apply instruction can be set up somewhat like a mystery waiting to be solved. This real-life application and its associated case-based storyline is one way educators can encourage students' participation through developing their interest in the course content.

Because interest is not a stagnant state of being, this means it can fluctuate (Hidi and Ainley 2012). In fact, individuals can influence their own development of interest. For example, self-regulation and interest have a reciprocal relationship. The more interest grows, the more developed the subject's content knowledge becomes. This process of accumulating content knowledge requires motivation, and increasing interest can provide additional motivation for the learner (Hidi and Ainley 2012). Additionally, all these elements happen in context, which critically shapes, supports, or constrains how interest develops (Thoman et al. 2017). That is why it is so important to support student interest via engagement, such as asking questions in the classroom.

Interest can be triggered, regulated, and it also has important outcomes associated with learning. In fact, interest can facilitate the development and maintenance of learning (Hidi and Ainley

2012; Pekrun and Linnenbrink-Garcia 2012). Students who are interested in a subject are more likely to become self-regulated learners (Hidi and Ainley 2012). Self-regulation of learning refers specifically to the efforts made by the student to focus on the acquisition of academically relevant knowledge and skills. Conversely, students who lack interest in a subject must rely on internal resources that are more effortful to persist through to their learning goal (Hidi and Ainley 2012). Interested students do not need as much effort to stay engaged in course content. Thus, interest facilitates learning and optional levels of performance, and is necessary for maximizing academic success in the difficult medical school classroom.

7.3.3 Engagement

One key outcome that classroom instructors use to gauge the efficacy of their instruction is student engagement. But what is meant by the term "engagement?" There are five types of engagement: cognitive, behavioral, motivational, cognitive-behavioral, and social-behavioral (Hidi and Ainley 2012). Motivational engagement is influenced by emotions and can shape other kinds of engagement.

Engagement is also related to student emotions (Efklides et al. 2017). Positive emotions promote task-related thoughts, and negative emotions promote task-irrelevant thoughts. Thus, learners should experience more positive emotions during class to promote task-related thoughts.

How learners attribute their successes and failures in the medical classroom influences how they engage in the class. This is because attributions can be conducive or not conducive to learning. For example, students who attribute their academic success to effort and hard work will be more likely to engage with their peers and professors to ask questions.

The mood a student is in while learning occurs is important because students are more likely to remember and recall information learned in an emotional state consistent with that which they are in during retrieval (Hidi and Ainley 2012). This is called mood-congruent memory recall, and it has two parts: retrieval-induced forgetting and retrieval-induced facilitation. Retrieval-induced forgetting occurs when students are in a negative mood state while learning in class. Students are more likely to forget content when learned in a state of negative emotions. Retrieval-induced facilitation is when students experience greater remembering when learning content in a positive emotional state. Thus, it is important to create a positive environment in the classroom so students feel like they can ask questions, and so they can experience positive emotions and mood, thus remembering the information contained in the answers to their questions.

7.3.4 Academic Emotions

Learning and class participation is influenced by affect, or academic emotions (Efklides et al. 2017). Though many educators traditionally considered emotions and cognition to be very separate processes, modern researchers have found that emotions are a critical component of cognition (Immordino-Yang and Damasio 2007). Emotions that relate to learning are called epistemic emotions (Pekrun and Linnenbrink-Garcia 2012). Academic emotions can also be classified by valence (i.e., positive and negative emotions), activation (i.e., whether the emotion is activating or deactivating for students), and by the object or focus of the emotion (i.e., whether the emotion the student experiences relates to the learning activity, the outcome of the learning activity, or past experiences related to the learning activity). For example, if a student feels confusion about the present learning material, then they may seek out answers to their questions causing this emotion. However, if the classroom environment becomes contentious, they may not decide to ask their question because they determine it is not amenable to conversation. In this way, academic emotions also influence other cognitive processes that contribute to learning, such as perception, attention, social judgment, cognitive problem-solving, decision-making, and memory processes

(Efklides et al. 2017). Therefore, it is important to provide a safe and positive environment for students to be able to seek help and ask questions in class.

Academic emotions are not inflexible (Efklides et al. 2017). Indeed, they can be automatized or consciously changed through the regulation of emotions and appraisals. Environment, goals, and appraisals students make of their situations can induce, prevent, or modulate their emotions. For example, a student who feels positively about a class, holds mastery-oriented goals, and appraises their performance in the subject area to be generally positive will have positive emotions about the given class. These emotions can shape the type of learning experience a student has. Students and educators can both influence their class-related emotions via regulating their appraisals of themselves and their environment, as well as via the design of tasks and environments. For example, educators who alter the design of their courses and classwork to provide positive experiences for their students can influence how their students think about their class, and thus, influence how a student learns in their classroom.

In sum, academic emotions have both individual (e.g., motivation) and contextual (e.g., classroom environment) antecedents (Efklides et al. 2017). Emotions are complex, but they can be regulated to make sure students enjoy generally positive, activating emotions, which allow students to engage deeper in their learning activities and ask questions as needed (Pekrun and Linnenbrink-Garcia 2012). This is important because emotions ultimately influence learning strategies and academic performance.

7.3.5 Growth Mindset

Peoples' views of the nature of intelligence and other personal attributes vary. People view intelligence as either stagnant or changing. These are referred to as implicit theories of intelligence (Dweck 2006). Those who view intelligence as an immutable trait that one is born with are considered to have an entity theory of intelligence. Those who view intelligence as a malleable trait that someone can grow through learning more and making new connections are considered to have an incremental theory of intelligence. Having an incremental implicit theory of intelligence leads individuals to make causal attributions of their behavior to effort. For example, if they raise their hand to ask a question in class, they will attribute their lack of knowing the answer to effort rather than being unintelligent. Having an entity implicit theory of intelligence leads individuals to make causal attributions of their behavior to innate ability. For example, if they raise their hand to ask a question in class, they would be more likely to attribute their lack of knowing the answer to being not intelligent, rather than simply needing to put in more effort. In sum, having an implicit theory of intelligence which is incremental is what we colloquially consider to be having a "growth mindset," and having an implicit theory of intelligence which is that intelligence is an entity is also called having a "fixed mindset." Those who have growth mindsets make more beneficial attributions for learning than those who have fixed mindsets.

It is, therefore, important to have a growth mindset because of the attributions someone with a growth mindset makes (Dweck 2006). When someone with a growth mindset takes a risk and makes a mistake while learning, they are more likely to attribute this mistake to effort and try again in a different manner. For example, a student who raises their hand in class might be asking a question for clarification on something they got wrong or do not understand. The student with the growth mindset is more likely to accept feedback about process, strategy, or effort while learning. They integrate the new information or feedback from the educator, and they move on to more learning. In contrast, when someone with a fixed mindset makes a mistake while learning, they are more likely to attribute this failure to ability, think that they are less smart than others who know the information, and not try as hard next time they don't understand something. So, when a student with a fixed mindset raises their hand in class, they take a risk of receiving feedback in front of the class that might be attributed to their ability (i.e., low ability), and they would be less likely to integrate this information or feedback into their work in the future in a way that is resilient and open to learning. In sum, taking chances, making mistakes, and getting feedback should be opportunities for learning, such as when students raise their hands in class for this purpose (Bjork et al. 2013). Students and classroom instructors can both play parts in cultivating an environment that is conducive to asking questions, which we will discuss in the next section on cultural contexts.

7.4 Culture and Question-Asking

Culture is an important lens one should consider when studying question-asking in higher education. Vygotsky, educator and socio-cultural researcher, declared that culture and cultural agents are mediators for individual learning. The complexity of information and the scaffolds that exist in society is the basis for the Vygotskian notion of the zone of proximal development—the difference between a student's current abilities and potential abilities is the supports and affordances in his or her environment that can help them learn and reach their full potential (Greeno and Engestrom 2014). Not only does Vygotsky's research show how important culture is as a mediator of the educational experience, it also outlines how important the role of the educator is in navigating that for the role of the student.

Research has found that the culture in which a learning community is situated influences a student's motivation and learning (Kumar et al. 2018). Cultural forms are cognitive processes based upon assumptions that underlie cultural thinking that can overlap with values (Stewart et al. 1998) such as assuming time, space, energy, and logical process of another defined upon a culture's characteristics. Common human problems that can be assigned certain assumptions and values are human activity, social relations, motivation, perception of the world, and perception of self. Stewart, Danielian, and Foster (1998) investigated how these specific cultural forms were measured against American values. A dominant mode for American society is the activity of "doing." A person is measured by the visible ways that they are working on something, such as learning or teaching, "Americans implicitly assume that learning is an active process requiring performance by the learner, whose incentive to learn is either further reward or the avoidance of punishment." (Stewart et al. 1998, p. 159). Therefore, question-asking in an American classroom is a primary measurement and evaluation of if a student is meeting academic expectations. If a student is not actively participating in the classroom, then they are perceived as not caring and unmotivated.

Many American college science courses are held in a lecture-based format with student mastery of concepts measured by question-and-answer participation (Nadile et al. 2021; Tanner 2009). Many faculty get frustrated by the lack of engagement in the classroom despite sharing that question-and-answer is how they measure excellence in material. Nadile et al. (2021) measured 417 undergraduates from large lecture-based science courses, evaluating how students felt asking and answering questions. It was found that students most frequently reported that they felt uncomfortable voluntarily asking instructors questions in front of the whole class if they did not feel confident about the course material (72.5%) and if they felt that other students would judge them (67.5%). Over half of students also selected that they felt uncomfortable if they felt like there were too many people in the class (55.8%) and if the instructor responded negatively to other students' questions (52.5%). Gender bias was found within participants, with women 2.9 times more likely to feel uncomfortable asking questions compared to men. Why such a discrepancy in gender? According to the Spring 2019 Health Assessment by the American College of Health Association (2019), 34.2% of undergraduate college students had indicated the top impediment to learning was stress, with depression (20.2%), anxiety (27.8%), and sleep issues (22.4%) following closely behind. It was found that females consistently rated higher than their male counterparts in coping with feelings of hopelessness, loneliness, exhaustion, anxiety, anger, and stress that directly tie back to the previous statements of what was impeding a student's performance.

These differences in classroom behavior by demographics highlight how individuals integrate different aspects of the broader culture into their identities through socialization. Cooley (1902) theorized that an individual's self-concept is developed by the active and passive reflections of others' perceptions. This process occurs: (1) imagined appearance as perceived by another person, (2) imagination of other persons' judgments of that appearance, and (3) self-feelings. Individuals constantly imagine what others think of them and how they are perceived (Franks and Gecas 1992). Self-feelings, such as pride or mortification, are affected by others' verbal and nonverbal responses, influencing individuals persistently to construe and appraise themselves based on such judgements in the process of developing self-concepts. This demonstrates that how individuals integrate culture into their self-concept affects their participation and engagement in the classroom. One cannot talk about participation and engagement with asking questions in class without also talking about the broader cultural influences on individuals in the classroom. Harrell (2000) found that race/ethnicity and stress was not a product of racial group membership but rather due to accumulated experiences with racism. Therefore, if a person was impacted by several experiences of perceived racism, then their self-esteem and wellbeing would be impacted. So, if a person is in a stressful environment (i.e., medical school) and has experienced perceived racism and felt targeted, they are more likely to have increased personal vulnerabilities that can in turn develop into mental health issues (Bowman 2013) impacting their performance in the classroom and willingness to ask questions.

7.4.1 Mental Health Considerations

TV shows like Grey's Anatomy or ER portray that during medical education, students are pummeled with complicated questions about clinical concepts and patient care. There is always the strict doctor that will belittle and shame the student into rising above and becoming the best. Barry et al. (2022) investigated how fear of repercussions drives behaviors on rotation wards and ultimately the students' learning experience. It was found that 75% of the students noted that at least sometimes they refrained from answering questions on rounds because they were afraid of being wrong or exposed for their lack of knowledge. How much of this is reality and how much is drama?

To reach the academic achievement of being accepted into medical school, one must obtain a reasonably high GPA within pre-medical courses, exceptional Medical College Admission Test (MCAT) scores, demonstrate support with recommendation letters from professors and healthcare professionals, and support their resume with volunteer and leadership experience. It takes an individual, continuous effort, studiousness, and drive to meet these conditions. A global analysis from Quek et al. (2019) analyzed data from 69 studies comprising of 40,348 medical students, finding that the overall global prevalence of anxiety among medical students to be around 33.8%, "discovering that around one in three medical students globally have anxiety, a prevalence rate which is substantially higher than the general population" (p. 2735). Generalized Anxiety Disorder otherwise known as GAD (American Psychiatric Association 2013) is defined as excessive anxiety and worry (apprehensive expectation), occurring more days than not for at least 6 months with symptoms of restlessness, fatigue, difficulty concentrating or going blank, irritability, muscle tension, and sleep disturbance. Posselt and Lipson (2016) found a positive relationship between high levels of perceived competition in classes and an increase in levels of depression and anxiety. Mental health influences academic achievement, with one study finding that students that were positively screened for symptoms of anxiety and depression displayed lower GPA levels than other students (Eisenberg et al. 2009).

7.4.2 Self-Efficacy

Bandura (1986) theorized that a person's effort and persistence would increase based on their belief in their self-efficacy of successfully completing a defined task. He did note that there would be times when a person's actual ability would not match one's believed ability, but believed that overall, a person would do better and try harder if they only believed themselves to be more efficacious. There are fears that high self-efficacy that does not match reality can be dangerous to patients, but the same could be argued for a person with low self-efficacy completing a task anxiously. Other learning theorists such as Pekrun (2006) agreed that beliefs of self-efficacy were valuable in the achievement of individuals. Determining that environmental factors, personal beliefs and emotions, and academic success interact as determinants of one another. For example, academic self-efficacy not only impacts on academic achievement but negative feelings, such as test anxiety, can also influence feelings of self-efficacy.

Ryan, Gheen, and Midgley (1998) researched how self-efficacy can affect the willingness of a student asking for help in the classroom. It was found that individuals asking for help tended to have a higher personal sense of academic efficacy whereas those who felt less efficacious were more likely to report avoidance of seeking help. Medical education research has found that an individual's self-efficacy belief is important for the development of medical knowledge, skills proficiency, and resiliency in the career (Artino et al. 2012). For example, if a medical student feels self-efficacy in anatomy and physiology, they would be more likely to participate and ask questions in their anatomy lab, but if they have poor self-efficacy and negative feelings about themselves, then they would be less likely to ask questions in anatomy lab because they would not feel confident in their abilities. Thus, those who feel higher self-efficacy would then develop deeper medical knowledge, skills, proficiency, and resiliency through their participation and self-confidence (Artino et al. 2012).

7.4.3 Imposter Syndrome

A psychological challenge that happens among high achieving individuals, known as Imposter Syndrome, is a pattern of behavior where people doubt their professional knowledge and abilities, resulting in a persistent fear of being exposed as a fraud regardless of their success otherwise (Mullangi and Jagsi 2019). It has been found that female osteopathic medical students experience imposter syndrome and at a higher rate than their male counterparts (Shill-Russell et al. 2022). Female osteopathic medical students also experience avoidance of incompetency (Balanoff et al. 2022).

The underlying reason for impostor syndrome, however, is rarely discussed in the established literature. This is because impostor syndrome derives from messages in the larger society about who fits in and who is othered (Tulshyan and Burey 2021). Members of society who experience overt or covert messages of belonging are more typically women and people of color, and in high-performing training programs (e.g., grad school) and positions of employment, these members of society are not exempt from societal messages. Those who are made to feel othered by society are those who have not historically been in positions of power and authority (i.e., members of low socioeconomic status, women, and people of color). These messages are internalized consciously or unconsciously and experienced by the individual as impostor syndrome. This is why it is broadly documented more frequently in women and people of color. But it is important to not place the blame of experiencing impostor syndrome on the individual who feels it. The cause is not a deficit within them, it is derived from a deficit in society, which is unfair and inequitable. This is why calling it a diagnosis is problematic. The concept was developed originally in the 1970s, though it was a phenomenon devoid of context. Studies did not acknowledge systemic racism, classism, xenophobia, and other biases when explaining how individuals receiving messages from those in power and authority experience impostor syndrome. In one popular article by the Harvard Business Review

(Tulshyan and Burey 2021), authors explain this, stating, "The answer to overcoming imposter syndrome is not to fix individuals but to create an environment that fosters a number of different leadership styles and where diversity of racial, ethnic, and gender identities is viewed as just as professional as the current model." Though broader cultural change and corporate cultural change are necessary to help solve the problem of impostor syndrome, there must be ways to aid individuals experiencing this in the present. Otherwise, women and people of color who feel impostor syndrome may not feel confident participating and asking questions in class, as this provides an avenue for unwelcome messages of non-belonging and vulnerability.

Indeed, there are some strategies to overcome the challenge of impostor syndrome. Researchers in one study (Banaloff et al. 2022) suggested that, instead of creating comparison schemes between residents, we can assist them in reaching their goals based on their own achievements and desire for success. This is an impostor-syndrome-fighting strategy that also benefits all learners because it encourages conscious mastery rather than competition. However, ultimately, impostor syndrome is a phenomenon that comes from the larger society, so beyond sweeping societal changes pertaining to overt or covert messages about who belongs in what institutions and contexts, impostor syndrome will still be experienced by members of society who experience marginalization.

7.5 Creating an Inclusive Environment

The benefits of having a safe and supportive learning environment have been discussed, but how does one achieve that? Inclusive learning environments are often regarded as being grounded in the framework of the Universal Design for Learning (UDL), which has three main principles, (1) provide multiple means of engagement, (2) provide multiple means of representation, and (3) provide multiple means of action and expression (Gronseth 2018). The use of active learning pedagogies, such as the ones discussed earlier in the chapter, aligns with this framework and supports the development of an inclusive learning environment. Discussion will continue about the additional influences of UDL and creating inclusive learning environments in this section.

7.5.1 Influence of Educator Experience

Several studies demonstrate the educator's individual experiences and other characteristics that might influence the use of inclusive practices in education. Lombardi et al. (2011) found that non-tenure track female faculty were more likely to favor and practice inclusive course instruction. Additionally, Lombardi et al. (2011) also found that individuals with prior disability-related training or disability experience were more likely to positively endorse attitudes toward inclusive instruction. Martins et al. (2018) performed semi-structured interviews of 18 educators to gain perspectives on the use of inclusive instruction. Participants varied in their roles and experience levels, as well as training with inclusion practices. Results suggest educator attitudes on using inclusive instruction were based on their individual experiences, but no formal correlation test was performed.

Reupert et al. (2010) invited inclusive educators to participate in their study to describe the teaching practices of inclusive educators. All participants self-identified as inclusive educators and advocates for inclusive instruction. This suggests the idea of "inclusive educator" as a distinct identity within education, which encompasses a collection of characteristics. Identification of key characteristics of inclusive instructors might support the development of targeted training to promote inclusive instruction in educators less supportive of the practice. This is essential to broaden the reach of inclusive learning environments and support student participation and question-asking in class.

7.5.2 Inclusive Course Design

In a review of the literature, Messiou (2016) found that what inclusion means in the context of inclusive course design is contested and varies depending on the author. For example, Nieminen (2021) defined inclusion in their educational study as a "radical, political quest to include disabled people in higher education," (p. 1) while Penner (2018) focused inclusive educational design on experiences that are accessible to all students in the class regardless of their demographic/social identities and prior learning experiences. Messiou (2016) suggested that focusing only on certain groups of students is contrary to the idea of inclusion, which they found most studies on inclusive education to do. Lombardi et al. (2011) discussed inclusion in the context of students with disabilities; however, they made the clear point that by utilizing inclusive course design practices, all students benefit. While there are several ideas regarding the meaning of inclusive course design, some authors and groups, including the Disabled Students Sector Leadership Group suggest the terms "inclusive approaches," "Universal Design for Learning," and "inclusive teaching and learning" may be used interchangeably (Draffen et al. 2017). Morgan and Houghton (2011) provided this definition for creating an inclusive course design suitable for all populations, "An inclusive approach has its foundation in a commitment to promoting equality and diversity by embedding the following principles in the design process: anticipatory, flexible, accountable, collaborative, transparent, and equitable." (pp. 12–13). By taking this definition and applying these concepts to active learning strategies such as problem-based learning, inquiry-based learning, and flipped classroom, a strong example of inclusive learning environments can be made. This establishes a setting in which students will feel supported to ask and answer questions as an important method of learning.

7.6 Conclusion

The act of questioning in educational settings, particularly in higher education and medical school, is pivotal for fostering student engagement and facilitating deeper learning experiences. Despite its recognized importance, there exists a reluctance among students to actively participate in questioning (Almeida 2012). However, the literature indicates a clear correlation between student engagement, improved academic performance, and outcomes in class, highlighting the significance of encouraging question-asking and answering. Through an exploration of pedagogical, psychological, and cultural factors influencing student participation, this guide endeavors to shed light on the complexities surrounding student questioning behaviors. By understanding these factors, educators can better create a safe and inclusive environment that nurtures a culture of inquiry and promotes active engagement in class activities. By adopting active learning pedagogies and implementing strategies to cultivate an inclusive learning environment, educators can empower students to overcome their apprehensions and actively participate in the questioning process, thereby enhancing their learning experiences and academic success.

References

Almeida PA (2012) Can I ask a question? The importance of classroom questioning. Procedia Soc Behav Sci 31:634–638. https://doi.org/10.1016/j.sbspro.2011.12.116

American College of Health Association-National College Health Assessment (2019) Reference Group Executive Summary

American Psychiatric Association (2013) Diagnostic and statistical manual of mental disorders, 5th edn. https://doi.org/10.1176/appi.books.9780890425596

Artino J, Dong T, DeZee KJ, Gilliland WR, Waechter DM, Cruess DF, Durning SJ (2012) Development and initial validation of a survey to assess students' self-efficacy in medical school. Mil Med 177(9 Suppl):31–37. https://doi.org/10.7205/MILMED-D-12-00240

Bada SO, Olusegun S (2015) Constructivism learning theory: a paradigm for teaching and learning. J Res Meth Educ 5(6):66–70

Balanoff C, Fatunmbi A, Aarons C, Protyniak B, Joshi A, Woelfel I, Hoffman R (2022) Relationship between gender, training level and goal orientation achievement motivation: implications for learners and faculty. J Surg Educ 79(6):e38–e47. https://doi.org/10.1016/j.jsurg.2022.07.015

Bandura, A., 1986. Social foundations of thought and action: a social cognitive theory

Barry K, Schiffman FJ, Collins BJ (2022) Assessing medical student fear and shame as barriers to active participation on the wards. Brown J Hosp Med 2(1). https://doi.org/10.56305/001c.40087

Baumeister RF, Tice DM, Vohs KD (2018) The strength model of self-regulation: conclusions from the second decade of willpower research. Perspect Psychol Sci 13(2):141–145

Bjork RA, Dunlosky J, Kornell N (2013) Self-regulated learning: beliefs, techniques, and illusions. Annu Rev Psychol 64:417–444

Bowker MH (2010) Teaching students to ask questions instead of answering them. Thought Action 26:127–134

Bowman PJ (2013) A strengths based social psychological approach to resiliency: cultural diversity, ecological, and lifespan issues. In: Resilience in children, adolescents, and adults: translating research into practice. Springer, New York, pp 299–324

Cleary TJ (2018) The self-regulated learning guide. Routledge, New York, NY

Coates H (2005) The value of student engagement for higher education quality assurance. Qual High Educ 11(1):25–36

Cobern WW (1993) Constructivism. J Educ Psychol Consult 4(1):105–112

Cochon Drouet O, Lentillon-Kaestner V, Margas N (2023) Effects of the Jigsaw method on student educational outcomes: systematic review and meta-analyses. Front Psychol 14:1216437

Constantinou CP, Tsivitanidou OE, Rybska E (2018) What is inquiry-based science teaching and learning? Springer International Publishing, pp 1–23

Cooley CH (1902) Human nature and the social order. Routledge

de Jong T (2006) Scaffolds for scientific discovery learning. Handl Complexity Learn Environ, Theory and Research, pp 107–128

Draffen EA, James A, Martin N (2017) Inclusive teaching and learning: what's next? J Incl Pract Further High Educ 9(1):1–12. https://eprints.soton.ac.uk/415140/

Dunlosky J, Rawson KA, Marsh EJ, Nathan MJ, Willingham DT (2013) Improving students' learning with effective learning techniques: promising directions from cognitive and educational psychology. Psychol Sci Public Interest 14(1):4–58

Dweck CS (2006) Mindset: the new psychology of success. Random house

Efklides A, Schwartz BL, Brown V (2017) Motivation and affect in self-regulated learning: does metacognition play a role?. In: Handbook of self-regulation of learning and performance. Routledge, pp 64–82

Eisenberg D, Golberstein E, Hunt JB (2009) Mental health and academic success in college. BE J Econ Anal Policy 9(1). https://doi.org/10.2202/1935-1682.2191

Franks D, Gecas V (1992) Autonomy and conformity in Cooley's self-theory: the looking-glass self and beyond. Symb Interact 15(1):49–68

Frymier AB, Houser ML (2016) The role of oral participation in student engagement. Communication Education 65(1):83–104. http://dx.doi.org/10.1080/03634523.2015.1066019

Greeno JG, Engeström Y (2014) Learning in activity. In R. K. Sawyer (Ed.), The Cambridge handbook of the learning sciences (2nd ed., pp. 128–147). Cambridge University Press. https://doi.org/10.1017/CBO9781139519526.009

Gronseth S (2018) Inclusive design for online and blended courses: connecting web content accessibility guidelines and universal design for learning. Educ Renaissance 7:14–22. https://eric.ed.gov/?id=EJ1218623

Harrell SP (2000) A multidimensional conceptualization of racism-related stress: implications for the well-being of people of color. Am J Orthopsychiatry 70:42–57

Hidi S, Ainley M (2012) Interest and self-regulation: relationships between two variables that influence learning. In: Schunk DH, Zimmerman BJ (eds) Motivation and self-regulated learning: theory, research, and applications. Routledge, New York, pp 77–110

Hung W, Jonassen DH, Liu R (2008) Problem-based learning. In Handbook of research on educational communications and technology. Routledge, pp 485–506

Immordino-Yang MH, Damasio A (2007) We feel, therefore we learn: the relevance of affective and social neuroscience to education. Mind Brain Educ 1:3–10

Kumar R, Zusho A, Bondie R (2018) Weaving cultural relevance and achievement motivation into inclusive classroom cultures. Educ Psychol 53(2):78–96

Lazowski RA, Hulleman CS (2016) Motivation interventions in education: a meta-analytic review. Rev Educ Res 86(2):602–640

Lombardi AR, Murray C, Gerdes H (2011) College faculty and inclusive instruction: self-reported attitudes and actions pertaining to universal design. J Divers High Educ 4(4):250–261. https://doi.org/10.1037/a0024961

Martins MH, Borges ML, Gonçalves T (2018) Attitudes towards inclusion in higher education in a Portuguese university. Int J Incl Educ 22(5):527–542. https://doi.org/10.1080/13603116.2017.1377299

Meece JL, Anderman EM, Anderman LH (2006) Classroom goal structures, student motivation, and academic achievement. Annu Rev Psychol 57:487–504

Messiou K (2016) Research in the field of inclusive education: time for a rethink? Int J Incl Educ 21(2):146–159. https://doi.org/10.1080/13603116.2016.1223184

Miele DB, Scholer AA (2018) The role of metamotivational monitoring in motivation regulation. Educ Psychol 53(1):1–21

Morgan H, Houghton A-M (2011) Inclusive curriculum design in higher education: considerations for effective practice across and within subject areas. High Educ Acad, pp 1–20. https://shorturl.at/6XeiX

Morrison CD (2014) From 'sage on the stage 'to 'guide on the side': a good start. https://scholars.wlu.ca/musi_faculty/23/
Mullangi S, Jagsi R (2019) Imposter syndrome: treat the cause, not the symptom. JAMA 322(5):403–404. https://doi.org/10.1001/jama.2019.9788
Nadile EM, Williams KD, Wiesenthal NJ, Stahlhut KN, Sinda KA, Sellas CF, Salcedo F, Rivera Camacho YI, Perez SG, King ML, Hutt AE, Heiden A, Gooding G, Gomez-Rosado JO, Ford SA, Ferreira I, Chin MR, Bevan-Thomas WD, Barreiros BM, Alfonso E, Zheng Y, Cooper KM (2021) Gender differences in student comfort voluntarily asking and answering questions in large-enrollment college science courses. J Microbiol Biol Educ 22(2). https://doi.org/10.1128/jmbe.00100-21
Nieminen JH (2021) Assessment for inclusion: Rethinking inclusive assessment in higher education. Teaching in Higher Education, 1–20. https://doi.org/10.1080/13562517.2021.2021395
Oettingen G, Schrage J, Gollwitzer PM (2015) Volition. In: Corno L, Anderman E (eds) Handbook of educational psychology, 3rd edn. Routledge, New York, NY, pp 104–117
Pekrun R (2006) The control-value theory of achievement emotions: assumptions, corollaries, and implications for educational research and practice. Educ Psychol Rev 18(4):315–341. https://doi.org/10.1007/s10648-006-9029-9
Pekrun R, Linnenbrink-Garcia L (2012) Academic emotions and student engagement. In: Christenson SL, Reschly AL, Wylie C (eds) Handbook of research on student engagement. Springer, Amsterdam, pp 259–282
Penner MR (2018) Building an inclusive classroom. J Undergrad Neurosci Educ 16(3):268–272. https://www.ncbi.nlm.nih.gov/pmc/articles/PMC6153021/
Posselt JR, Lipson SK (2016) Competition, anxiety, and depression in the college classroom: variations by student identity and field of study. J Coll Stud Dev 57(8):973–989
Quek TT, Tam WW, Tran BX, Zhang M, Zhang Z, Ho CS, Ho RC (2019) The global prevalence of anxiety among medical students: a meta-analysis. Int J Environ Res Public Health 16(15):2735
Reupert A, Hemmings B, Connors J (2010) Do we practice what we preach? The teaching practices of inclusive educators in tertiary settings. Inte J Teach Learn Higher Educ 22(2):120–132. https://eric.ed.gov/?id=EJ930141
Ryan AM, Gheen MH, Midgley C (1998) Why do some students avoid asking for help? An examination of the interplay among students' academic efficacy, teachers' social–emotional role, and the classroom goal structure. J Educ Psychol 90(3):528
Sawyer RK (2014) The new science of learning. In: Sawyer RK (ed) The Cambridge handbook of the learning sciences. Cambridge University Press, Cambridge, MA, pp 1–16
Schmidt SM, Ralph DL (2016) The flipped classroom: a twist on teaching. Contemp Issu Educ Res 9(1):1–6
Shill-Russell C, Russell RC, Daines B, Clement G, Carlson J, Zapata I, Henderson M (2022) Imposter syndrome relation to gender across osteopathic medical schools. Med Sci Educator 32(1):157–163. https://doi.org/10.1007/s40670-021-01489-3
Stewart CE, Danielian J, Foster JR (1998) Cultural assumption and values. In: Bennett JM (ed) Basic concepts of intercultural communication. Intercultural Press, pp 157–172
Tanner K (2009) Talking to learn: why biology students should be talking in classrooms and how to make it happen. CBE Life Sci Educ 8(2):89–94. https://doi.org/10.1187/cbe.09-03-0021
Thoman DB, Sansone C, Geerling D (2017) The dynamic nature of interest: embedding interest within self-regulation. In: O'Keefe PA, Harackeweicz JM (eds) The science of interest. Springer, Cham, pp 27–47
Trowler V (2010) Student engagement literature review. Higher Educat Acad 11(1):1–15
Tulshyan R, Burey JA (2021) Stop telling women they have imposter syndrome. Harv Bus Rev. https://short-url.at/ZN8ei
van Joolingen WR, Zacharia ZC (2009) Developments in inquiry learning. Technology-enhanced learning: principles and products, pp 21–37
Weiner B (2010) The development of an attribution-based theory of motivation: a history of ideas. Educ Psychol 45(1):28–36

Part III

Improving Diversity in Education and Beyond

8 Human Remains as Pedagogic Aids: In a World of Alternatives, Is the Use of Real Human Remains Still a Viable Option?

Georgina A. Goodison and Christopher Aris

Abstract

Human remains have been used for millennia as pedagogic resources, with their focus to enhance medical understanding and advancement in the treatment of the living. The last two-hundred years, however, have seen the use of human remains in the teaching of non-medical subjects, including archaeology and forensic science. Their use in education has peaked and plateaued over time, with numerous alternatives introduced due to advancements of skill and technology. In more recent years, the use of alternative pedagogic aids has begun to supersede the real thing, with plastic models, digital imaging, and interactive software coming to the fore in anatomical teaching. Providing access to such a diverse range of alternative resources as additional learning materials could aid in the education of students. Yet, if access to real human remains is replaced with alternatives, numerous issues could arise, including a lack of inclusivity, perhaps a lesser learning experience, and even the ability to empathise.

It could be argued that both legal and ethical policies surrounding the handling of human remains could be pushing institutions more towards using alternative resources for teaching. Within medicine, dissection is slowly being replaced with digital alternatives, whilst in forensic science and archaeology, replicated skeletal material is becoming more widely used. The question to be considered is if these are *viable* alternatives which produce the same levels of learning and understanding as real human remains. Alternatives may be more accessible, but only by looking at studies comparing the use of real and replicated material can conclusions on their educational viability be drawn, and thus the inclusivity provided by institutions.

In the context of anatomy education, institutions can improve the diversity of their learning opportunities through the provision of a wide range of pedagogic resources, including both real and replicated human remains. To ensure their learning environments are inclusive and accessible, it is essential that students are given access to a wide variety of resources so that they can use each resource to fulfil their own learning needs. Only then can universities be classed as fully diverse, inclusive, and accessible within the teaching of anatomy.

Keywords

Anatomy · Anatomical models · Bone · Body donors · Pedagogy

G. A. Goodison (✉) · C. Aris
School of Chemical and Physical Sciences, Keele University, Keele, UK
e-mail: g.goodison@keele.ac.uk

R. Door (ed.), *Accessibility, Inclusivity and Diversity in Education and Beyond*, Biomedical Visualization 8, https://doi.org/10.1007/978-3-031-81795-3_8

8.1 Introduction

The teaching of anatomy has been conducted certainly for centuries if not millennia and covers a spectrum of disciplines comprising the obvious biomedical sciences, but also the arts and humanities (Balta et al. 2016). Teaching extends to a multitude of subjects, including the expected biomedical science subjects such as medicine and physiotherapy, but also forensic science, archaeology, and heritage, among others (Deathridge 2017; Licata et al. 2020; Goodison and Aris 2023). A great many things can be learned from access to real human remains for the purposes of learning and education, from the obvious how to treat the living and identify the dead, but also how to accurately illustrate the human form and understand the human body in motion.

To provide students with the best educational support possible whilst learning anatomy, a course should provide resources and learning opportunities that are accessible, diverse, and inclusive. In this context, accessibility would allow for students to have access to sufficient pedagogic resources; diversity would mean a broad range of resources, including real and replicated human anatomy, digital, and physical material; and inclusivity would mean that all students have access to the same types of resources and thus the same opportunities to engage with the learning experience. Whether for biomedical sciences students, forensic students or for archaeology and heritage students, the best method of learning involves compliance with all three of these vital criteria.

8.2 A History of Human Remains as Pedagogic Resources

During the nineteenth century, when excavations in Egypt were rife, a series of papyrus texts was discovered which were found to date to 1700–1500 BC (Haas 1999). These papyri were determined to be copies of even older documents, estimated to date to around 3000 BC (Loukas et al. 2011; Barr 2014; Standring 2016). To date, these are the earliest documented examples not just of medicine and treating the living, but also act as the earliest pedagogic resources serving as both a reference and instruction on anatomical teaching. This, however, does not mean that the Egyptians used human remains specifically to obtain better anatomical understanding, rather that they learnt by using the living. Although human dissection was carried out in Ancient Egypt, it did not occur until the third century BC, under the influence of the Ptolemaic pharaohs (O'Malley 1964; Walker 2007; Standring 2016). In the quest to expand medical knowledge, and understand better how to treat the living, Greek physicians in Alexandria, including Herophilus of Chalcedon and Erasistratus of Chios (Singer 1957; Lloyd 1975; Walker 2007; Loukas et al. 2011; Ghosh 2015), requested the ability to dissect the dead. This, however, proved highly controversial due to religious beliefs on impurity and the afterlife and so was only permitted for a brief time (Edelstein 1935; Ghosh 2015; Pallocci et al. 2020).

With a surging population, and the need for greater medical understanding, the use of dissection as a pedagogic resource exploded in Renaissance Italy (Papa and Vaccarezza 2013). After the introduction of the first anatomy theatre in Salerno in 1235, numerous others began to appear, including at Bologna, which became well known in the anatomical world due to Andreas Vesalius, the 'Founder of modern anatomy' (O'Malley 1964; Bay and Bay 2010; Papa and Vaccarezza 2013; Lennox 2016). Determined to enhance anatomical study, and challenge the work of anatomists before him, Vesalius compared his own dissections against the previous literature and encouraged his students to take a more hands-on approach (Heseler 1959; Klestinec 2004). Even during the sixteenth century, Vesalius believed that the best way for his students to learn was by interaction—to touch and manipulate—and get a feel for the subject that they were learning from. His first publication 'De Humani Corporis Fabrica' (Sawday 1995; Carlino 1999; Richardson 2001; Mitchell 2012; Ghosh 2015) provided a complete overhaul of anatomical theory and was accompanied by multiple highly detailed illustrations (Buckland-Wright 1985; Phillipson 2022), which introduced a new pedagogical and more easily accessible resource—the two-dimensional image.

At the beginning of the sixteenth century, anatomisation was legalised in the United Kingdom, so that medical institutions could keep up with their European counterparts (Richardson 2001; Thomas 2006, 85; Cain 2017). Initially, this allowed for the bodies of four convicted criminals to be anatomised each year (Parker 1912; Buckland-Wright 1985; Burn 2001; Thomas 2006), which was later extended to six bodies, due to demand (Frank 1976; Richardson 2001). To begin with, legislation regarding human dissection applied solely to London, and so this low figure was sufficient. However, with the opening of medical schools across the United Kingdom, there was a greater need for bodies to be anatomised, leading to the introduction of the Murder Act of 1752 (Sawday 1995; MacDonald 2005; UK Parliament 2014; Grogan 2017; Tarlow and Battell Lowman 2018). This act only sufficed for a brief period, and soon freshly interred bodies began to be stolen from their graves by body snatchers, otherwise termed Resurrection Men, to supply medical schools with subjects to anatomise (Sawday 1995; Richardson 2001; Walker 2007; UK Parliament 2014). Once again, there was an update in legislation: the Anatomy Act was passed in 1832 to provide more bodies for anatomisation (Frank 1976; Mitchell 2012; UK Parliament 2014; Grogan 2017), thus stopping the illegal acquirement of corpses from their graves. This Act allowed for all unclaimed bodies from poor houses and hospitals to be given to medical institutions for educational purposes and remained in place until the update by the Anatomy Act 1984, which introduced donor programmes and the ability to choose to donate your body after death (Mahadevan 2006; UK Parliament 2014; Ghosh 2015; Legislation.gov.uk 2023). In 2004 (and subsequent revisions in Scotland in 2006), this legislation was altered and tightened, following the unauthorised retention and storage of tissues at several UK hospitals (discussed in greater depth below).

8.3 The Use of Replicated Material for Teaching Anatomy

The use of real human remains as pedagogic resources has proved problematic over the centuries, from more than just a legal perspective. One of the biggest issues is the rate of decomposition, and thus the actual ability to anatomise an individual. The human body begins the decomposition process immediately following death (Ferreira and Cunha 2013; Dash and Das 2020; Volk 2023), meaning that a body must be hastily dissected or preserved if possible. For centuries, there was no well documented way of preserving the dead, meaning anatomisation had to take place quickly. It usually took place during the winter months, when the temperatures were lower, and bodies would decompose at a slower rate (MacDonald 2005; Klestinec 2004; Ghosh 2015). To make anatomy more accessible during this time, many anatomists worked alongside artists to create accurate models for students to learn from.

Countless mediums have been used to create anatomical models, including wood, ivory, plaster, stone, glass, and in the present day, plastics (Narang et al. 2021). Some of the most detailed and realistic models were created from wax during the seventeenth and eighteenth centuries (Ballestriero 2010; Marković and Marković Živković 2010; Riva et al. 2010), under the supervision of anatomists and surgeons to make sure they were precise as pedagogic aids. It was hoped by some Renaissance anatomists, such as Felice Fontana, that anatomical waxes could be used to replace cadaver dissection as a teaching resource (Riva et al. 2010; Surgeon's Hall Museums 2022), due to their durability, permanence, and lack of odour (Landes 2008). One such anatomical wax can be seen in Fig. 8.1, which shows the life-size body of a female individual, with removable parts. Unfortunately, these anatomical waxes proved to be visual aids instead of the intended interactive ones, due to the materials made to create them—they were highly durable as long as they were kept in the right conditions, but not regularly touched or interacted with (Ballestriero 2010).

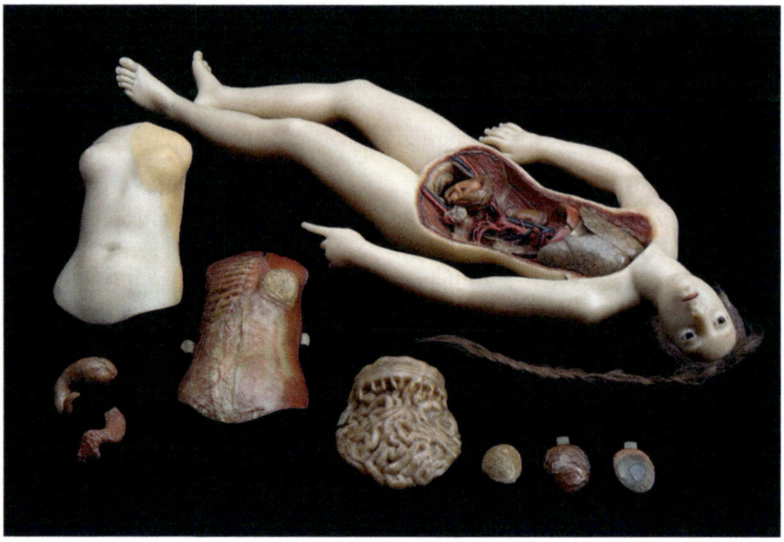

Fig. 8.1 The Anatomical Venus created by Clemente Susini during the eighteenth century. This anatomical model of the female form was manufactured from wax and can be moderately interacted with to show the internal organs and their positioning within the body when the exterior piece is removed. Italian female models created in wax were referred to as Venuses—taken from the Roman goddess of love and beauty—and were usually created in a similar reclining position, so they could be displayed delicately on silks and cushions. Many female models focused more on their reproductive traits, and their difference to the male body. This, like many other wax models, seems to be more a work of art than an interactive learning tool. Image courtesy of Wellcome Library (2014), London CC BY 4.0, Wikimedia Commons

During the nineteenth century, a new, more robust, and interactive medium was used for creating anatomical models: papier-mâché. Having already been used for a significant period to create children's toys and decorations (Valdecasas et al. 2009; Maerker 2014; Surgeon's Hall Museums 2022), there was an obvious durability to this material, allowing for models to be touched and manipulated and not just observed. Several methods were used to create these models, including the use of a real human skeleton as a foundation which provided a degree of interactivity, and later models which were made completely of papier-mâché (Maerker 2014), which allowed them to be fully interactive. One such piece, created by French doctor Louis Auzoux, represents a six-foot male, and comprises 2000 labels and 92 detachable parts (University of Aberdeen 2006; Surgeon's Hall Museums 2022). A similar Auzoux model can be seen in Fig. 8.2, with elements removed to show the extent of interactivity which could be achieved. Having been housed at the University of Aberdeen since 1882, the model was cleaned and conserved several decades ago. During the cleaning process, it was noted that the pieces of the model had accumulated a significant amount of dirt (University of Aberdeen 2006). This shows that the model had been interacted with quite heavily, more than likely by students who were learning anatomy, removing, and replacing the pieces to better familiarise themselves with how the human body fits together.

Within the last century, newer, more robust materials have been used to create anatomical models on a larger scale. By the 1950s, the properties of plastic were fully realised for the creation of interactive models (Hallam 2013). As a completely durable material, plastic could be repeatedly held and manipulated, and so could be used to create both visual and interactive anatomical models. The use of more durable materials for anatomical models only began to grow from this point, creating many anatomical structures to varying scales: from life-sized, to enlarged for easier structural visibility and understanding. An

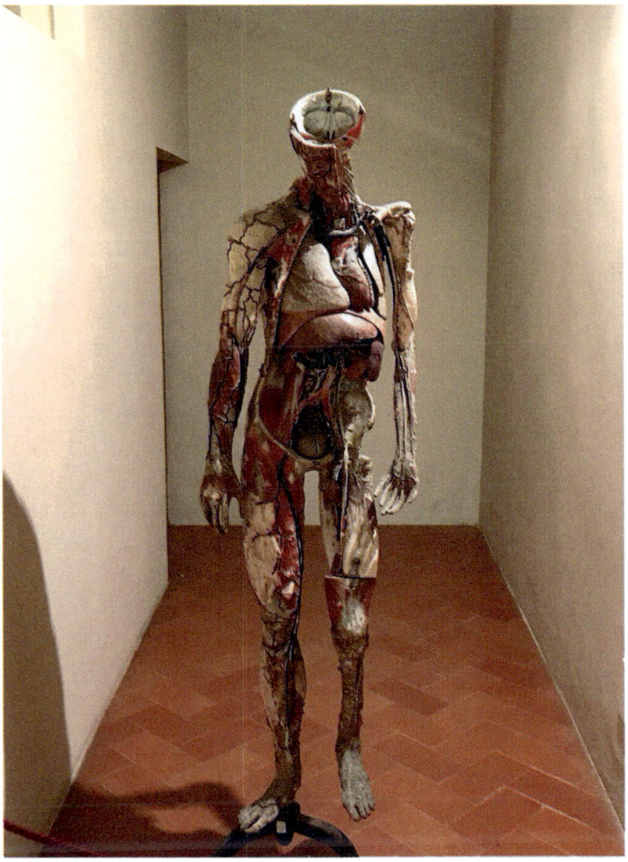

Fig. 8.2 Papier-mâché anatomical model created by Louis Auzoux. Auzoux began creating anatomical models from papier-mâché during the nineteenth century as it was a more robust material than wax and so could be more heavily interacted with. This particular model, made in France in 1861, is now housed in the anatomy museum at the University of Pisa. Unlike the wax model seen in Fig. 8.1, this anatomical is much more detailed, showing different tissues and their relationships. Image courtesy of Federigo Federighi (2019), CC BY 4.0, Wikimedia Commons

example of an enlarged plastic model can be seen in Fig. 8.3, which depicts the interior of the human eye. The 1980s saw the introduction of the first composite bone models, created using polyurethane foam as the trabecular core (Metzner et al. 2022) and an epoxy-reinforced braided glass sleeve as the exterior (Elfar et al. 2014). Although these models reflected the bones they were meant to be, they were not fully anatomically precise, and the interior did not match that of real bone. However, in the late 1990s after several more phases, the methods of manufacture altered, with models being cast from real human bones, creating a higher level of anatomical correctness (Elfar et al. 2014).

In 2004, after a decade of testing similar medical simulation products, the first synthetic cadaver was created and marketed by SynDaver Labs in Tampa, Florida (Syndaver 2020; Healthy Simulation 2020). Manufactured from polymer fibres, salts, and water, these synthetic cadavers are produced to replicate living tissue, from the texture to the colouration (Richardson et al. 2020) and have been used for a variety of reasons including anatomical dissection, but also surgical training and ballistics testing (Healthy Simulation 2020). Although the SynDaver, as can be seen in Fig. 8.4, cannot be dissected in the standard way using a scalpel, it can be fully interacted with, even by individuals with minimal dissection experience (Mitchell et al. 2016). Depending on the type of model being used, superficial muscles can be reflected and folded back to access deeper muscles and the internal organs. These models also showcase the delicate vascular and nervous systems, which can be interacted with on a much higher level than real donor bodies due to their robusticity. Although SynDavers are not cheap to purchase, providing they are handled and stored correctly, they can be used repeatedly (Syndaver 2020; Healthy Simulation 2020), making them more cost effective than real donor bodies in the

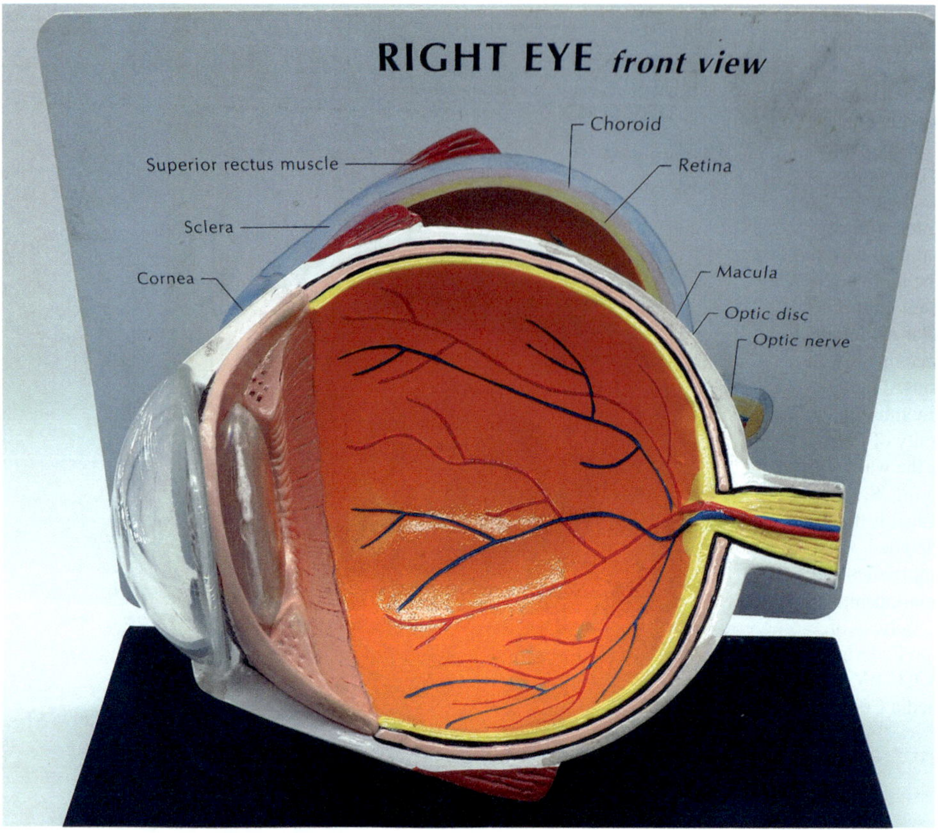

Fig. 8.3 Plastic model depicting the human eye. Although simple in its design, having been created at a much larger scale to the size of the real human eye allows the observer to differentiate between tissues. Image courtesy of YVC Biology Department and Netha Hussain (2020), CC BY 2.0, Wikimedia Commons

long term (this will be discussed further below). Synthetic cadavers also allow for a low-risk alternative to chemical-heavy, traditional, dissection, making anatomy more accessible and inclusive for individuals who should not be exposed to such harsh chemicals for health reasons.

One of the most recent advancements in this area is the use of digital screens and software, meaning that students can use programmes to learn anatomy without having to enter a dissection lab or even leave their homes (Preim and Saalfeld 2018; Laspro et al. 2023). Such software can be accessed online or downloaded as applications to allow for interactive study. The range of software available allows for a basic approach—simply naming a bone or organ of the body—to in-depth imagery—showing the intricate vascular and nervous systems. One such example can be seen in Fig. 8.5, which illustrates the lingual artery and its relation to nearby structures. The exterior skin and interior bone can also be seen, showing additional relationships with both superficial and deep tissues. Specifically tailored software can also prove helpful on specialised training programmes, such as the use of virtual reality (VR) in surgical training, where a variety of training procedures can be undertaken repeatedly, rather than singularly on a real donor (Seymour et al. 2002). Technology such as this can allow for greater inclusivity of students as they do not necessarily have to be in a teaching facility to access it, but it also means that they can choose how often they want to use it, enhancing their own learning. Building on this, in a classroom setting, is the use of the anatomy visualisation table: life-size interactive computer screens which have been programmed with CT scans from cadavers (Darras et al. 2018; Aland et al.

Fig. 8.4 A demonstration of SynDaver anatomy being given to students at Stanbridge University. The materials used to manufacture SynDavers allow for easy manipulation of tissues, whilst not having to use traditional dissection methods means that they can be interacted with time and time again, by multiple cohorts. Additionally, the lack of toxic chemicals means that SynDavers can be used without the need for regulated laboratories, or specialist air filtration systems in place. Image courtesy of Monirb (2016), CC BY 4.0, Wikimedia Commons

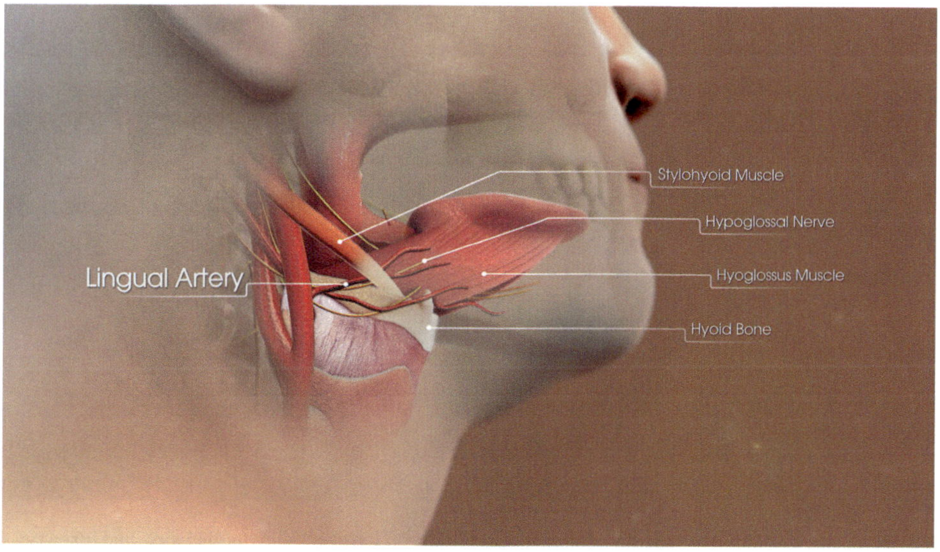

Fig. 8.5 3D medical illustration of the structure of the lingual artery, showing the hyoid bone and hypoglossal nerve. Software such as this can allow for a three-dimensional view of certain structures, with the viewer able to interact with and manipulate the images to achieve their best level of understanding. This type of resource can be highly beneficial when learning the relationships between structures and tissues. Images and software such as this can be a useful aid during self-study, when access to physical models may not be possible. Image courtesy of Scientific Animations and Manu5 (2020), CC BY 4.0, Wikimedia Commons

2023; Anatomage 2024), allowing students to see how structures should appear in real, living patients. In 2004, the American company Anatomage (https://anatomage.com/) began to create digital learning platforms for anatomy. Their most detailed and recent product being the Anatomage Table, which uses an extensive series of both CT and MRI scans to create digital cadavers which can be dissected over and over (Darras et al. 2018; Anatomage 2024). This type of resource also allows for an individual to be 'dissected' at speed, removing layers more easily, unlike a real donor body which requires a slower dissection process to allow for all tissues to be removed in each phase (Washmuth et al. 2020). The virtual anatomical model on the table, demonstrated in Fig. 8.6, allows for delicate structures to be viewed, including blood vessels, nerves, and fascial layers. Additionally, the image can be manipulated to see structures from different angles, perhaps allowing for a greater understanding of the relationships between structures and how specific structures work (Washmuth et al. 2020; Alasmari 2021). Although full-size anatomy visualisation tables can only be used in a laboratory/classroom setting, there is also the option to purchase Anatomage Tablets as a portable digital resource (Anatomage 2024).

The use of CT and MRI scans has also aided in the creation of three-dimensional pedagogic resources using 3D printers (Lauridsen et al. 2016; Otton et al. 2017; Narang et al. 2021). Depending on the quality of both the data used to create the three-dimensional images, and the printer, the results can prove to be highly accurate and create to-scale models which can be handled repeatedly (Smith et al. 2017; Collings and Brown 2020). The levels of accuracy with

Fig. 8.6 Medical students at Kharkiv International Medical University being taught using an Anatomage Table. Similar to anatomy software, Anatomage Tables can be used to manipulate and enhance images for students to gain the best understanding of tissues and structures. However, due to the size and expense of this resource, their use takes place mainly within an instructor-led, group setting. Image courtesy of Binil Zack (2021), CC BY 4.0, Wikimedia Commons

3D printing have also allowed for its inclusion within forensic and criminal cases (i.e. Baier et al. 2017), such as where skeletal material is small and/or highly fragmented, or where continued interaction with skeletal remains could cause degradation (Collings and Brown 2020). Bones can be printed with a high level of accuracy as they are monochromatic and made of hard tissue (AbouHashem et al. 2015). However, prints are not solely tied to the recreation of bone and have also been used to recreate other aspects of anatomy. These 3D prints can capture a range of complicated features which are unique to each individual and may not be seen in standard cadaver dissections, giving anatomy students access to a wider range of resources which may assist within their education. Websites such as the National Institute of Allergy and Infectious Disease (https://3d.nih.gov) allow for both viewing digital 3D material and access to printing. Sites such as these mean that universities can widen the material they have access to—providing they have the facilities to 3D print. If institutions were to collaborate, online databases could be created and shared between them, meaning that universities may have the ability to both digitally observe and 3D print material that was collected at another university. This would make more anatomical and pathological features available for students to learn from, making anatomical teaching more diverse and accessible.

With such a wide scope of anatomical resources, both physical and digital, the teaching of anatomy has never been so diverse. The broad range of materials meets several needs, including the ability to physically touch and manipulate specific anatomical areas using synthetic cadavers and interactive models, the ability to view varying angles and rotate anatomy through digital software and CT imagery, and the ability to understand relationships between structures through all types of anatomical resource. This diverse range of pedagogic resources can also allow for greater accessibility for students who may not have access to traditional dissection.

8.4 Stability Within Teaching: Should We Be Using Real or Replicated Human Remains?

8.4.1 Real Donors: Traditional Anatomy

There are currently a multitude of resources which can be used as pedagogic aids for teaching anatomy—not just the use of donor bodies, prosections, and real skeletal material. Many anatomists and educators are firm that the only way to truly learn human anatomy is from real, donor bodies, with full body dissection classed as the gold standard in anatomy education (Sugand et al. 2010; Smith et al. 2017; Richardson et al. 2020; Huri et al. 2021). However, numerous institutions are beginning to move away from the traditional resources in favour of alternative methods of anatomical teaching, opting for replicated anatomical models, rather than using real donors (Nichols 2014; Williams 2023). Several of these institutions, including Lancaster University (2024), state explicitly that they do not use traditional dissection, and instead favour other methods of teaching, including Anatomage Tables, posters, and 3D models (Smith et al. 2022). Other universities, such as the University of Sheffield, have recently removed traditional dissection from their syllabus (Hamilton 2023), but have provided no detail as to what resources will replace the use of human material (University of Sheffield 2024).

The benefits, both physical and psychological, of using donor bodes as learning resources are well established in the literature (Korf et al. 2008). Such studies frequently examine the use of human remains as teaching materials, assessing students' reactions to cadavers, and the knowledge they gain from dissection. Asante et al. (2021) noted that 90.1% of the 161 participants (medical students at the University of Health and Allied Sciences, Ghana) believed that access to dissection deepened their understanding, whilst 87% believed it helped them to recall what they were learning. A similar study, involving 174 undergraduate medical students, determined that dissection and the use of prosections were the most effective methods of learning, with 75% of participants believing that dissection was the 'single most useful method of learning anatomy' (Gogalniceanu et al. 2010; Papa and Vaccarezza 2013). Further to this, a study with 131 students from the Faculty of Medicine at the University of Ruhuna, determined that 97% of participants believed dissection to be essential in learning anatomy (Prabodha et al. 2016). Additionally, 85.5% of participants believed that it promoted teamwork, and that working together enhanced their learning of anatomy (Prabodha et al. 2016). A study conducted with first year medical students at the turn of the millennium, focused on the learning outcomes of cadaver dissection from both a hands-on and demonstrated perspective (Johnson 2002; Older 2004). The results showed that students who took part in this study preferred to dissect for themselves as opposed to watching someone else dissect, believing that their own interactions helped to develop a greater understanding of the structures present (Johnson 2002).

From a physical perspective, having access to real human donors and traditional dissection showcases the diversity of the human species. Variation in human anatomy can be vast, with no two donors anatomically the same, and many donors in one room all showing varying traits (Older 2004; Balta et al. 2016; Gholipour 2019). These traits can be both developmental, including the growth of extra ribs or blood vessels, reversed internal anatomy, et cetera or procedural including the need for amputations and/or transplants, the use of foreign objects such as pacemakers or plates and pins, and even the addition of breast implants. Instances such as these show students that anatomy, and the human body, is not perfect, and that things can always change after a procedure has begun. It is also important to mention how physical interaction helps with understanding. Tactile learning is fundamental within the education of anatomy. Although it has been determined that haptic sensation provides a lower response in the neocortex than visual sensation (11.5% haptic vs. 55% visual), it is still extremely important for understanding (Bremner and Spence 2017). While it is important to see structures and their corresponding relationships to other structures around them, it is also useful to be able to feel them and determine their textures.

Psychologically, students who interact with the once living as part of their education are exposed to a completely different type of experience, which promotes respect, empathy, and an understanding of ethics (Rizzolo and Stewart 2006). From a medical perspective, it has been stated that 'physician's own emotions help them attune to and empathetically understand patient's emotions' (Halpern 2001). The aforementioned study conducted by Asante et al. (2021) also touched on the perception and attitude of undergraduate medical students towards cadaveric dissection; 72% of participants noted an enhanced respect for the human body, and 96.3% recognised that the donor was once human. In which case, the use of a real donor body as part of a student's training will help them understand humility and allow them to form connections with real individuals, alongside granting them the ability to reflect on their own feelings of mortality (Saylam and Coskunol 2005; Gunderman and Wilson 2005; Rizzolo and Stewart 2006; Hallam 2016; Washmuth et al. 2020). The use of real donors has not just a physical impact on students' education and understanding, but also, perhaps equally as important, an emotional impact.

There are, however, challenges to using real human remains for teaching, including registering donors and later their transferral, storage and associated financial implications (Gholipour 2019). Regarding donor bodies, institutions within the United Kingdom must be informed by family members after the death of a donor, which may not happen for varying reasons (i.e. family are not aware of the individual's wishes or are finding the event too painful). In addition, depending on the circumstances surrounding death, such as the time of year, history of disease, and whether the individual needs a postmortem (although digital postmortems are now becoming more widely used) to name but a few, the donor may not be suitable for educational instruction (Hallam 2016). If the institution is contacted, arrangements must then be made for the transferral and preservation of the donor. The donor body will then be embalmed and stored until it is ready for use.

These instances all provide certain implications. Without contact after the death of a donor, an institution may miss out on an opportunity for the education of their students, limiting students' access to donor bodies. Alternatively, if the institution is contacted, there are then multiple financial considerations, including transferral and embalming and of course the institution having the correct facilities to both store and dissect the donor (Washmuth et al. 2020). Although the chemicals used for embalming are relatively low cost and easier to use than other preservation techniques (Balta et al. 2016), they are toxic (Deng et al. 2018). As of 2016, the laws surrounding the use of formaldehyde changed, classifying the chemical as a Category 1B carcinogen (GOV 2015). Formalin, which is used in the preservation of donors for traditional dissection, is a solution made up of 37% formaldehyde (GOV 2024). Unfortunately, there are several issues pertaining to the use of formalin as a preservative, including irritation of the respiratory system through inhalation, meaning areas used for both dissection and storage must be sufficiently ventilated. With this issue being regularly re-examined and defined in law, making any required changes can have further financial impacts on universities. Additionally physical contact with the chemical can cause skin irritation, and in extreme cases, cause blisters and hives (GOV 2024). The chemicals used for preservation also severely impact the tissues of the donor, altering both the colour and texture and making it almost impossible to reposition or manipulate digits or limbs (McLachlan et al. 2004; Balta et al. 2016; Gholipour 2019). Further to this, it has been stated that learning anatomy through dissection takes considerably longer than other methods (Washmuth et al. 2020), meaning more hours are required for dissection, and fewer hours are available for alternative resources. To counter the hazardous levels and tissue-altering chemicals, other methods of embalming are becoming more widely used, including the Thiel method which was first used in the 1990s (Liao and Wang 2019). This method uses a small amount of formalin, along with sodium nitrate, potassium nitrate, and sodium sulphide (Amrit et al. 2023), and the use of formalin-free N-vinyl-2-pyrrolidone (NVP) (Haizuka et al. 2018; Marayuma et al. 2019). Both methods provide flexibility of the body, so limbs can be easily repositioned, and help to keep a better colour and texture, allowing the donor to look and feel more lifelike (Niels 2022). Many argue that the cost of using real donor bodies for dissection is minimal and that it is the facilities that incur the initial high cost. Theoretically, this means that the initial outlay pays for itself over time. There may, however, be alternative, more diverse, and accessible options.

8.4.2 Synthetic Options: New Age Anatomy?

In 2016, a study explored the use of synthetic cadavers in anatomical teaching (Mitchell et al. 2016). Synthetic cadavers, predominantly the American SynDaver (https://syndaver.com/) are created using polymers, salts, and water to replicate human tissue, with many types of models available, such as anatomical and surgical models, individual organs and even veterinary models (SynDaver 2020). Anatomists in the School of Osteopathic Medicine compared the use of SynDavers and real donor dissections. Comprising 54 medical students, the study included the use of real dissection videos for participants to watch, and the ability to interact with a SynDaver in the lab. A survey completed by the students showed that although the synthetic cadaver did prove useful, it would be better served as an additional teaching aid, rather than instead of real donor dissection (Mitchell et al. 2016), perhaps highlighting the importance of diversity amongst pedagogic resources. With this in mind, it is important to understand what resources students already have access to, and whether they are deemed appropriate, or if they could be enhanced with other resources to further knowledge and education. A 2019 study investigated the educational value of synthetic cadavers as additional aids to plastic models (Richardson et al. 2020). Initially, students were taught using only plastic models, but halfway through the course, an experimental group of students was introduced to a SynDaver. Of the experimental group, 81% believed synthetic cadavers helped them to identify key structures, whilst 66% believed it aided in their learning experience. It is important to note that these students had no access to real donor bodies, only to plastic models before the introduction of the synthetic cadavers. In which case, having access to synthetic cadavers as an educational resource can prove highly valuable.

A significant challenge to using synthetic cadavers is their cost. The cost of a SynDaver human model is not publicly available, however as a comparison, the SynFrog Dissection Models start at US$202.50 (SynDaver 2020) and the Silicone Anatomy Model has been quoted at US$65,000. Due to the work that goes in to creating them, from creating the 'tissues', to piecing them all together into a human-looking model, the high cost is understandable. The fact that the SynDaver can be used repeatedly for multiple student groups does lower the overall cost in the long run. However, institutions will need to purchase a number of these models to facilitate their students, meaning a considerable investment akin to the outlay for the facilities needed for the use of traditional dissection.

8.4.3 Plastic Models

A study on the use of 3D printed models in anatomical pedagogy proved highly insightful (Smith et al. 2017). In total, 127 first-year medical students and 19 second-year students participated in the study which 3D printed 32 different models from a donor body—which was scanned before the embalming process took place. Students then had access to the printed anatomical models but not to the donor body as a comparative. The first-year students were allocated to focus groups, whilst the second-year students were given 3D printed models to take away with them. Of the second-year students, 83% stated they would use the models again, while 67% wanted to keep the models. Overarching positive comments also included the durability of the prints, and the ability to pass them around in large groups allowing for their interactivity. However, 63% of the second-year students believed that the scale of the models was too small for effective use, and it was also noted overall that identifying and using some of the models without any other material was challenging (Smith et al. 2017). This study shows both the positive and negative aspects of using 3D prints as pedagogic aids but does not use them as a comparative to dissection. It could have proved beneficial if the donor used to create the prints was then used for dissection, to compare points of interest on the models.

8.5 Screens and Software

Since their introduction, countless studies have been conducted on the utility and educational value of Anatomage Tables. One such study was undertaken with the use of Anatomage Tables as an accompaniment to dissection (Alasmari 2021). Of the 78 medical students who took part in the study, 73% reported that Anatomage Tables enhanced active learning, while 81% believed they aided as an additional method of learning to dissection (Alasmari 2021). Although this was a qualitative study, focussing on the opinions of student's as opposed to the grades they achieved while using specific pedagogic resources, the results of this research are highly valuable for understanding how students prefer to learn, and what resources benefit their learning processes. An interesting study was conducted by Anand and Singel on the utility of Anatomage Tables versus traditional dissection (Anand and Singel 2017). Two groups were created from 122 first-year medical students: Group A who had access to Anatomage Tables, and Group B who performed real dissections. Both groups were subjected to pre- and post-learning tests to determine the value of the resources they had access to. After the practical sessions and testing, the groups were swapped over, to gain an insight into what the other students had learnt from, and then all students were given a questionnaire to complete on the usefulness of the Anatomage Table. 51% believed that the Anatomage Table helped them to understand the topic better, whilst 55% enjoyed using it as a learning tool. Additionally, 79% of students agreed that the table enhanced their learning experience, and 84% believed the tables should act as regular teaching tools. The pre- and post-practical testing revealed that in both instances, student's scores improved from before to after the practical session, with no significant differences noted between the methods of teaching (Anand and Singel 2017).

There are numerous other studies, with similar student responses, which all seem to arrive at the same conclusion, that Anatomage Tables are indeed an extremely useful pedagogic resource, but that they should be used alongside dissection, not as a replacement (such as Brown et al. 2015; Kavvadia et al. 2023). As with every other method, there are setbacks to this type of learning. From the perspective of the student, not being able to feel differences in texture and manipulate a real donor body can limit tactile learning (Darras et al. 2018; Washmuth et al. 2020). From a technological angle, unfortunately, there can sometimes be glitches; when clicking on one structure, the screen can highlight a different one instead, perhaps making some smaller, more detailed areas, more difficult to 'dissect' (Washmuth et al. 2020). The screen can sometimes lag too, causing a delay (Alasmari 2021), which, like many other issues, is addressed in the user manual (Anatomage Table 10 2023). From an institutional point of view, Anatomage Tables prove to be a huge financial outlay that still require maintenance and updates. Prices for Anatomage Tables range from US$50,000–US$80,000, with the most popular table—the Table Convertible—purchased with a five-year warranty costing approximately US$100,000. Additionally, these estimates are without VAT or shipping costs, meaning the final prices would be significantly higher. Once again, the initial cost for multiple tables (to accommodate a full cohort) could equate to the same cost needed for traditional dissection facilities. Studies have also not been conducted for a significant timeframe yet to determine how well Anatomage Tables compare to traditional dissection on a cognitive level (Darras et al. 2018).

Of course, the above studies all refer to gross anatomy education, with a focus on teaching the biomedical sciences. The worlds of forensic science and even osteoarchaeological education are still considered relatively new subjects, and consequently have been subjected to fewer pedagogical anatomy studies. For both forensic science and archaeological cohorts, access to real skeletal material is essential to help students understand what they may encounter during their future careers. Access to dissection may also prove an important experience, whether that be through demonstration or personal dissection, but possibly not on the same scale that it is needed for medical students. In their future careers, archaeology students may be faced with human burials, and so need to understand the composition of bone and the context of the burial. Depending on the positioning of the skeletal remains, it may also prove useful to have an understanding of musculoskeletal anatomy, therefore access to dissection. This does not need to be the same as the training given to biomedical science students, relying purely on superficial, muscular anatomy. From a forensic science standpoint, it is essential that students have access to real human remains at some point in their education, whether that be through gross anatomy or skeletal anatomy, as it is highly likely that within the field they will encounter human remains. Even if they do not go on to specialise in forensic anthropology, it is important that students can identify bones and soft tissues and differentiate between human and animal. Additionally, it is important for students to experience variation between individuals, encountering instances such as supernumerary bones and/or teeth, thickening of areas of cortical bone, and anomalies within bone fusion (Christensen et al. 2014).

8.6 Diversity of Pedagogic Resources in Education

As previously mentioned, there is a myriad of resources, made from a great many different materials, that have been and/or are currently still used in the teaching of human anatomy. Although certain materials, such as bronze, ivory, and stone are no longer used in their creation, anatomical models are still widely used as pedagogic resources. For centuries, the teaching of anatomy has been supplemented using pedagogic aids—from the initial hand drawn anatomical illustrations and wax models of the Renaissance, to photographic textbooks and atlases, interactive plastic models, and the use of ultrasound imaging in the present day (Smith et al. 2017). These technological advances have allowed for a diverse range of products with varying qualities, tailored to specific subjects and levels of skill. They have also allowed for anatomy to become more accessible in both education and for members of the public wanting to learn. For example, a quick search of the Internet for replicated human bones brings up a number of options at varying prices, with the cheapest being of the lowest quality, and accordingly, the most expensive being the best quality and most anatomically correct. Initially, a student may simply be able to use the cheapest replications, to obtain a basic understanding of skeletal anatomy, such as identifying specific bones, how to side and position each bone, and so on. However, when it comes to learning specific anatomical features such as facets, ridges, and foramina, having access to an anatomically correct model is highly beneficial (Bone Clones 2024). One such anatomically correct model can be seen in Fig. 8.7, which depicts a Bone Clones® cast of a robust Asian male used for teaching on Keele University's forensic anthropology module. In this instance, replicated skeletal material could act as a highly beneficial precursor to the use of real human bones, giving students a deeper understanding of skeletal anatomy before they begin to handle donated material.

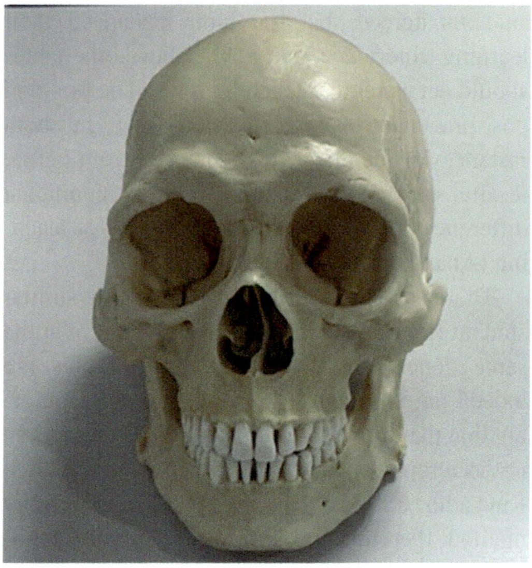

Fig. 8.7 Bone Clones® Asian Male Robust Skull (Catalogue Number BC-287). The use of casting from real skeletal remains provides highly detailed but durable replications, which can be handled more frequently. Unlike a generic skull model, this cast shows the exact traits of the individual it was taken from, including a large supraorbital ridge, a 90-degree gonial angle, and shovel-shaped incisors. Because of the skill and detail involved, reproductions such as this are not cheap to purchase, perhaps limiting the audience. Personal Image

This scalable approach could have a similar impact in the teaching of gross anatomy, with a diverse range of products and stages of quality introduced throughout study. For example, lower grade models could be used to create the foundations of knowledge which can be built on further with higher quality models, synthetic cadavers and then real human remains at a later stage. Access to full or part-body anatomical models at the beginning of their studies would allow medical students to appreciate basic structures and the relationships between structures in the human body and aid in interaction, before moving onto 'dissection' of synthetic cadavers. An introduction to the lab and dissection could prove beneficial with the use of a synthetic cadaver, allowing students to become familiar with their environment, whilst also giving them the opportunity to understand human anatomy in a more lifelike form. These stages of training could also provide a useful solution to one key issue regarding the use of donor bodies used for dissection; a donor body can only be dissected once (Washmuth et al. 2020). Although there is the full body to dissect including each of the limbs, the trunk, and the head (Hallam 2016), it can only be cut in to once, marking both the first and last time that each incision can be made. This of course means that each donor body has an educational lifespan, and due to the ever-increasing number of students attending universities, access to more than one donor per student group might not be possible. Theoretically, this could suggest that access to real human remains would be highly beneficial at a later stage, after initial anatomical training and understanding is already gained.

Similar can be said of digital resources, including virtual reality such as Anatomage Tables, and other e-learning resources. By introducing students first to virtual learning, it may help them to identify structures and their relationships before moving on to the next level. For example, first-year students could have access to solely e-learning platforms, second-year students could have access to synthetic cadavers, and third/fourth-year students could have access to traditional dissection. Of course, this is just a loose suggestion, and as previously mentioned studies have noted, multiple methods are better used in conjunction with one another. The crux of the issue is what materials students have access to and when. Ultimately, access to real human remains, synthetic cadavers, Anatomage Tables, et cetera, can only be given in a supervised, laboratory setting. One of the key factors is not necessarily the diversity of pedagogic resources available, it is the staffing levels and amount of time allocated to students to access them.

Fig. 8.8 Basic disarticulated teaching skeleton purchased through Amazon. This was bought in 2020 for less than £100 as an introduction to skeletal anatomy. The complete skeleton is present—including rubberised costal cartilage surrounding the sternum—but only to a basic standard. Although some foramina are present on the skull, no other details, such as textures on the auricular and symphyseal surfaces are present. This type of resource can prove highly useful for learning the basics of skeletal anatomy, but not when learning how to estimate biological sex or create an age profile. Personal image

8.7 Accessibility of Resources

During self-study, a student may only have access to textbooks and digital material (i.e. supplied interactive software and anatomical websites), unless they can afford to purchase physical teaching aids themselves. There are numerous online retailers who sell replicated anatomical models (such as Bone Clones® 2024; SOMSO 2024; Evotech Scientific 2024, etc.), but these can be very expensive, especially when purchasing an anatomically correct model. There are, of course, more basic alternatives which can be purchased for a fraction of the cost of the expensive replications. These include both basic articulated or disarticulated medical skeletons (as can be seen in Fig. 8.8), individual and grouped bones (such as an articulated pelvic bones and articulated hands or feet, etc.), and interactive models which can be manipulated to show specific tissues (as can be seen in Fig. 8.9). Items such as these can be bought from online outlets such as Amazon and eBay. These are not always the most anatomically precise models, but they can still be used to grasp a basic understanding and allow for an interactive experience. Even so, this still relies on the student having to purchase their own learning materials to enhance their study. This could potentially lead to the division of students who can and cannot afford to purchase their own additional learning resources, creating an additional level to inclusivity and accessibility issues. As students have limited time in supervised teaching areas, it is essential that during contact hours they have access to a diverse range of pedagogic resources that they can handle and sufficiently interact with. The study by Prabodha et al. (2016) found that students used textbooks as supplementary material for learning—accounting for 70.2% of participants—rather than use of the Internet or

Fig. 8.9 Interactive model of the human foot purchased through Amazon. This model was purchased in April 2024 for just over £90. Branded with the Evotech Scientific logo, it was cheaper to purchase the item through Amazon than to deal directly with the company which is based in America. The model, which uses both pins and magnets to hold it together, can be manipulated to remove singular and grouped muscles to understand how the soft tissue of the foot is comprised. Numerous types of tissues can be seen because of the different colours used, and numbers have been added which correspond to a paper hand out, so that each structure can be easily identified. Personal image

3D models. This could be a perfect example of the issue of access.

Theoretically, students always have access to textbooks, which are usually purchased by the student themself or borrowed from the university library, and so could take them to lectures, practical classes and have access during their personal study time. A study published in 2020 determined that 52.9% of the 120 students included preferred to use textbooks as supplementary material for understanding anatomy (Zargaran et al. 2020). Sadly, textbooks are not interactive and show only two-dimensional pictures which cannot be manipulated or viewed from other angles to get a better concept of structures. Further to this, images of full body anatomy tend to reflect that of males, with images of specific female anatomy only when deemed necessary. In fact, it was not until 2021 that the full digital female body was released through Complete Anatomy (3D4Medical 2021). In which case, the use of e-learning through online programmes could provide additional benefit (Preim and Saalfeld 2018; Leung et al. 2020). There are countless websites which provide information on anatomy; however, choosing the right website is not always the easiest task. Simply typing into a search engine the type of structure to be studied can be riddled with issues as the Internet is not always the most reliable source. There are, however, many accredited websites which can be accessed for free such as Anatomical Society (https://www.anatsoc.org.uk/home) and Kenhub (https://www.kenhub.com)—which also has the option to purchase additional content. Websites such as these use a multitude of digital materials to enhance the anatomical learning experience, including flash cards, quizzes, and videos, with useful links to validated external websites. One study undertaken by Leung et al. (2020) determined that 30% of the student participants favoured anatomy websites as additional anatomy resources for private learning. Nowadays individuals do not even need a computer to access this type of material, it can all be done on a mobile phone or a tablet, with anatomy applications even available to download. The study by Zargaran et al. (2020) also determined that 63.7% of students also used apps to enhance their learning (specifically surgery related). One issue that could arise here is access to the Internet. Universities generally provide free Internet service to their students, and typically students have home access also. It is possible to have Internet issues for short periods of time, but if individuals have prolonged periods of Internet disturbance for whatever reason, their learning experience could be compromised. It is also important to mention that not all digital material is accurate. Depending on what is typed into a search engine, and what website/image is studied by students, the results may not prove to be of any educational value or may in fact provide students with incorrect information. A digital study compiled by Wink et al. (2022) investigated the use of digital images through Google Image search and documented the top ten results for each of the 39 anatomical structures. The study determined that out of 390 gathered images, 363 were relevant (showed the structure in humans), and of those 339 were accurate, whilst 24 showed errors, including mislabelled structures and errors in morphology and location. Although overall a small percentage, this shows that by not providing access to specific, correct digital material, students may find the wrong information during their independent searches. If institutions provide guided independent learning, it will aid in student understanding and make them better self-learners (Cukurova et al. 2017).

In the present day, especially during the covid pandemic, many students did not have access to physical teaching aids, both real and replicated, which meant they had to rely on textbooks and online platforms. As of April 2020, universities in 191 countries closed their physical doors to students, meaning that measures had to be quickly taken to move every programme of study online (Evans et al. 2020). Like other subjects, anatomical learning was moved online, where lectures and seminars could be attended from home without the risk of contracting or spreading the virus (Evans et al. 2020; Huri et al. 2021; Potu et al. 2021). Although this meant that students could still 'attend' lectures and seminars, unfortunately it stopped physical access to pedagogic aids. Potu et al. (2021) surveyed undergraduate medical students who had begun their tuition before the pandemic and thus experienced learning in both physical and digital environments. The overarching response was that students preferred face-to-face demonstrations as they believed it helped them better understand what they were learning (Potu et al. 2021).

Sadly, with the precedent set by the pandemic, and with the already debated financial and health related aspects, many institutions have begun to question whether dissection and access to real human remains is completely necessary (Williams 2023). With such a wealth of alternatives, including synthetic cadavers and anatomy visualisation tables, many teaching units are having decisions made for them on the materials they should use to educate students.

8.8 Inclusivity of Resources for Students

Although it is not possible to determine explicitly what every student across every anatomy related subject and institution has access to, without an extremely intensive study, by looking at what information is provided on university course listings, it might be possible to determine what materials students will have access to as part of their study.

When reviewing biomedical courses, the lead author of this paper used the search term 'Anatomy' on a university course search engine (whatuni.com), returning eight courses across six universities. Of course, there are a great deal of more biomedical courses that do not include that keyword—for example, using 'Biomedical Sciences' as the key term instead brings up 303 courses—but that would be a study within itself (similar can be said of skeletal anatomy with the keyword 'Archaeology', which brings up 296 courses). As a brief insight into the teaching of anatomy, this relatively small number of universities could still prove useful. For example, the

Applied Anatomy BSc at the University of Bristol introduces students to dissection during their first year of study (University of Bristol 2024). The University of Liverpool begins dissection during the second year of their Anatomy and Human Biology BSc (University of Liverpool 2024). Yet the Biomedical Sciences BSc at the University of Aberdeen, and the Anatomy BSc at the University of Glasgow do not begin using cadaver dissection until their third years (University of Aberdeen 2024; University of Glasgow 2024). Additionally, the Anatomy, Developmental and Human Biology BSc at King's College London does not state at any point what physical materials will be used for anatomical teaching (King's College London 2024). The same can be said of the Anatomy and Development BSc at the University of Edinburgh, where, once again, there is no mention of the physical resources used for learning (University of Edinburgh 2024).

Within the realms of forensic science, such a study has recently been conducted (Goodison and Aris 2024). Focussing on the issue of transparency surrounding the information provided on online university course listings, 39 universities in the United Kingdom were found to offer 64 courses in forensic science with specialist training in forensic anthropology and the identification of human remains. Of these course listings, only 12 courses across 6 universities specifically mentioned that they use real human remains as pedagogic resources; 4 universities suggest the use of real human remains across their seven courses but do not state this explicitly; and the remaining 29 universities provide no information whatsoever about the physical resources used to teach 45 courses—with one course being purely digital (Goodison and Aris 2024). This does not necessarily mean that there is no access to real human remains, or that they have been replaced with models, it merely shows what information the university is happy to promote. This could be for several reasons, with the most likely being that of the numerous ethical polices surrounding the storage and use of human remains.

Both ethical policies and legislation can evolve and change dramatically over short periods of time. For example, the Human Tissue Authority, which regulate all recently harvested human tissues in the UK (HTA 2024), came into being because of a significant revelation on the procurement and storage of human tissues during the 1990s and early 2000s. This incident, often referred to as the Alder Hey Affair, saw several hospitals removing and storing human tissues which they did not acquire consent for (Bauchner and Vinci 2001; MacDonald 2005). The repercussions of this discovery completely changed everything regarding human tissues which are less than 100 years old within the United Kingdom, including the use of donor bodies in dissection.

After the introduction of the Human Tissue Act in 2004 (which covers England, Wales, and Northern Ireland, but not Scotland), individuals must give their signed consent, and have it witnessed if they want to donate their bodies for education after their death (Anatomical Society 2016; HTA 2024). This obviously provides institutions with a considerable amount of paperwork and the subsequent collection, embalming, and storage of their donors. As these processes all conform to the legislation set out by the Human Tissue Authority, institutions possibly feel comfortable to advertise their use of traditional dissection within their course listings. However, the skeletal material used in archaeological and forensic science education typically does not conform to the Human Tissue Act as it is over their threshold. When creating the legislation, an end date of 100 years was created, meaning that the remains of any individual who died within the last 100 years must be registered and authorised under the Human Tissue Act, however, the remains of any individual deemed to have died before the 100-year marker do not have to be registered (HTA 2021). This could create issues surrounding the use of archaeological human remains more of an ethical, rather than a legal issue. The rise of the Internet has made it possible to buy skeletal material online. Although the likes of eBay, Amazon, and other selling sites have prohibited their sale (Hugo 2016), a simple search on social media can connect buyers and sellers very easily, with multiple groups set up simply for that function (Graham and Huffer

2022; Pester 2022; Huffer and Graham 2023). Further to this, a recent television auction show highlighted a bidding war for a partially articulated set of human remains (Curve Media 2023). These instances show that human skeletal materials can be easily bought and sold, meaning that universities are less inclined to mention their use as pedagogic resources within course listings so they do not feel obliged to justify their origins, or simply it may be to cover themselves should any legal or ethical policies change suddenly.

The university resource searches above highlight just how differently subjects are being taught. The 'Anatomy' keyword search brought up three different years in which students are introduced to donor dissection, showing there is no consistency across all universities in how students are taught. Students who can interact with donor bodies throughout their degree programme may have a greater opportunity to obtain a deeper understanding of the human body and its workings than students who only have access during their final years. This could of course be a way for a university to seem superior, and to attract more students because of the resources it provides access to. It would be incredibly useful to see a study conducted on the thoughts and grades of students across all universities offering biomedical science subjects with access to dissection to determine at what stage in their degree programme they begin using donor bodies and how this affects their learning outcomes. It could be suggested that universities promoting their use of real skeletal material perhaps want to attract greater applicant numbers, whilst those not stating what resources they use might be holding back for ethical reasons. Again, a study across all universities that provide forensic anthropology modules would be highly beneficial, to determine what students have access to, at what stage in their programme, and for how long.

One extremely useful study to contemplate is that undertaken by Biasutto et al. (2005). In total, 1043 students across two universities were divided into three groupings and surveyed. Group A (comprising 698 students), from Cordoba National University, had access to real donor bodies which they dissected in detail to observe all regions and structures, Group B (comprising 330 students), from La Rioja National University, only used digital material as there were no dissection facilities available during class time, and Group C (comprising 145 students), also from Cordoba National University, had access to both real donor bodies and digital material. This study ran over the course of 5 years, with the data obtained from Groups A and B in 1999, and the data from Group C obtained in 2004. The two groups from Cordoba National University were provided with the same lecture content, taught by the same professor. The data was obtained from two sets of exams—the midterms and end of term, pass rates for which proved highly enlightening. For the midterm exam, the pass rate for Group A (real donor bodies) was 92.61%, Group B (digital resources) was 53.41%, and Group C (both real and digital) was 94.24%. Additionally, pass rates for the final exam showed Group A (real donor bodies) passing by 82.30%, Group B (digital resources) passing by 60.41%, and Group C (both real and digital) passing by 99.23% (Biasutto et al. 2005). This study highlights how including a more diverse range of resources, in this case digital material and real donor bodies, affects students and shows that access to a variety of pedagogic aids allows for a greater understanding and learning experience than only having access to one resource. In an instance like this, results could be used to propagate the information that one university is better than another, with Cordoba National University receiving the highest exam results for both Groups A and C potentially being seen as a better university than La Rioja National University. However, numerous arguments could be made about facilities, resources and also sample size, with perhaps the best argument being that Group C was a much smaller cohort—less than half the size of Group B and much less than a quarter or the size of Group A. This could suggest that a smaller group size promoted better access to group resources—in this case, the donor body and subsequent laboratory time—allowing students to get a better understanding of the structures they were looking at.

It would be more beneficial for students if universities provided the same diverse materials, so that rather than vying for best university, department, and so on, they all allow their students to achieve the same level of training, using the same types of resources. Another similar study was conducted by Deng et al. (2018). For this study, 120 fourth-year medical students were split into four separate groups—two control groups and two experimental groups. Students in the control groups began straight away in the dissection lab with textbooks and atlases as aids, whilst students in the experimental group were taught first using digital anatomy and then given access to the dissection lab. For the survey, 60 students were chosen—30 from the control group and 30 from the experimental group. The overall results showed that the students who had access to both virtual learning and then dissection scored higher than those in the control group who only had access to dissection. Additionally, 88% of students believed that it aided in their scientific thinking, while 93% believed it helped with their self-learning (Deng et al. 2018). The results of both studies highlight the increased understanding gained by students because of using multiple methods of teaching, rather than just a singular type of resource. If all universities adopted this policy of inclusivity, incorporating as many pedagogic resources as possible, introduced at the same times, graduates would be on an equal footing when applying for jobs.

Further to this, the use of alternative methods to traditional dissection can in fact prove more inclusive in certain situations. The study conducted by Washmuth et al. (2020) came about as a response to a specific individual on an anatomy course. The individual in question, who was training in nurse anaesthesia, became pregnant during her studies, and due to the harmful chemicals used to preserve donor bodies, needed an alternative to traditional dissection. The other 24 members of her cohort continued with traditional dissection, while this student was instead given access to an Anatomage Table. In this instance, the university providing access to another form or anatomical teaching aid promoted inclusivity and allowed for the student in question to learn the same things as the rest of her cohort, just using a different medium. This can also be beneficial for inclusivity regarding multifaith cohorts, where individuals may not feel comfortable using a real donor body because of their religious beliefs. Having access to alternative pedagogic resources can prove more inclusive to students who feel they cannot undertake traditional anatomy as part of their religious beliefs (Dental Schools Council 2023).

8.9 Discussion

There are numerous advantages and disadvantages for the use of each type of anatomical resource—far more than have been noted here—but that does not mean that any singular resource is more or less valuable than another, just that they have their own important qualities. As stated by Washmuth et al. (2020), 'the best way to teach anatomy may be combining multiple pedagogical resources to complement one another'. Whilst having access to real human remains, including donor bodies, skeletal material, and prosections is incredibly useful, it is also important for students to have access to replicated materials. Over the last three decades, countless studies have been undertaken on the use of donor dissection as an anatomical teaching method, with increasing numbers of studies looking at dissection versus other teaching methods. Generally (but not always) these studies have shown traditional dissection to be the superior method of learning. However, when looking at studies which use multiple methods of learning, these provide even better student results than dissection alone, showing that the best way to instruct students is to provide them with access to a diverse range of pedagogic resources. These studies generally focus on dissection and one or two alternatives, but it would be highly beneficial if a study was conducted using as many methods of learning anatomy as possible, to better understand the student experience and the learning outcomes.

Unfortunately, access to resources may always be an issue. Regarding timetables, giving students enough supervised access to pedagogic aids is not always the easiest task. With growing cohort numbers, but stagnant staffing numbers, departments are stretched further in trying to allocate enough time for students. This can have a serious impact on students' learning abilities. Investing in other resources may not necessarily help in this instance either. Rather than employing new staff to cover certain materials and methods of teaching, this could potentially stretch staff even more. Further to this is the cost to increase the range of resources. There is indeed a financial outlay for the use of real donor dissection, as is the same for purchasing synthetic cadavers and for anatomy visualisation tables. If an institution wanted to incorporate as many pedagogic resources as possible to aid in teaching, it would incur a significant initial cost, one which many universities can simply not afford to do, and so they must pick a specific method which suits them. Potentially, this could be aided by the sharing of information between institutions (i.e. CT/MRI scan data), which can be digitally observed by students and potentially 3D printed by the university to lessen costs.

This all, however, revolves around laboratory based hours for teaching, which are supervised. As mentioned previously, limited access to the laboratory and taught environment means students must develop their own learning process. If guidance and resources are provided for students, they can learn how best to study alone and develop critical thinking skills (Cukurova et al. 2017). By providing links to websites, uploaded content such as videos and quizzes to online learning environments, and even having open access laboratories with plastic models, replicated bones, and so on, students can choose to further supplement their learning, helping them to develop a better understanding through multiple additional methods. Theoretically, independent learning through solo study, and group learning in a laboratory setting allow for a more well-rounded student at the end of their programme of study.

8.10 Conclusion

Real human remains have been used as pedagogic resources for centuries. Although there is now a multitude of replicated teaching aids which can be used in anatomical education, this does not mean that they should be used instead of real human remains, but rather as supplementary resources. The use of real human remains as pedagogic aids also helps students develop other skills, including the ability to work as part of a team (McLachlan et al. 2004; Prabodha et al. 2016), the promotion of respect and empathy towards patients and other individuals (Rizzolo and Stewart 2006), and the chance to come to terms with their own mortality (Gunderman and Wilson 2005; Hallam 2016; Washmuth et al. 2020). It has been stated that 'knowledge acquisition is the most efficient when involving as many senses as possible' (Korf et al. 2008). In that case, surely the most efficient way of using as many senses as possible is also by using as many varied materials as possible. For universities to be inclusive for their students they must have both a diverse range of pedagogic resources for anatomical education, and be able to provide access to said resources.

References

3D4Medical (2021) The wait is over: full female model. https://3d4medical.com/blog/female-model-released. Accessed 25 May

AbouHashem Y, Dayal M, Savanah S, Štrkalj (2015) The application of 3D printing in anatomy education. Med Educ Online 20

Aland CR, Hugo HJ, Battle A, Donkin R, McDonald A, McGowan H, Nealon JR, Ritchie H, Stirling A, Tentrisanna M, Valter K, Xiao J (2023) A plethora of choices: an anatomists' practical perspectives for the selection of digital anatomy resources. Smart Learn Environ 10(66)

Alasmari WA (2021) Medical student's feedback of applying the virtual dissection table (Anatomage) in learning anatomy: a cross-sectional descriptive study. Adv Med Educ Pract 12:1303–1307

Amrit G, Ray S, Mohapatra S (2023) Preparation of soft embalmed cadavers by the modified Thiel embalming technique for surgical skill training and the development of a universal quantitative scoring system to assess the suitability of soft embalmed cadavers for such training purposes. Cureus 15(8)

Anand MK, Singel TC (2017) A comparative study of learning with "anatomage" virtual dissection table versus traditional dissection method in neuroanatomy. Indian J Clin Anat Physiol 4(2):177–180
Anatomage (2024) Anatomage. https://eu.anatomage.com. Accessed 21 Mar 2024
Anatomage Table 10 (2023) Anatomage Table EDU 10.0 User's manual. https://hsasf.hsa.washington.edu/wp-content/uploads/2023/12/Table_EDU_10.0_Application_User_s_Manual__Rev_A_.pdf. Accessed 26 Mar 2024
Anatomical Society (2016) Body Bequeathal. https://www.anatsoc.org.uk/home/bequests/information-about-body-donation Accessed 20 Apr 2023
Asante EA, Maalman RS, Ali MA, Donkor YO, Korpisha JK (2021) Perception and attitude of medical students towards cadaveric dissection in anatomical science education. Ethiopian J Health Sci 31(4):867–874
Baier W, Warnett JM, Payne M, Williams MA (2017) Introducing 3D printed models as demonstrative evidence at criminal trials. J Forensic Sci 63(4):1298–1302
Ballestriero R (2010) Anatomical models and wax Venuses: art masterpieces or scientific craft works? J Anat 216(2):223–234
Balta JY, Cronin M, Cryan JF, O'Mahoney SM (2016) The utility of cadaver-based approaches for the teaching of human anatomy: a survey of British and Irish anatomy teachers. Anat Sci Educ 10(2):137–143
Barr J (2014) Vascular medicine and surgery in ancient Egypt. J Vasc Surg 60(1):260–263
Bauchner H, Vinci RJ (2001) What have we learnt from the Alder Hey affair? BMJ. Br Med J 322:309–310
Bay NS, Bay BH (2010) Greek anatomist Herophilus: the father of anatomy. Anat Cell Biol 43(4):280–283
Biasutto SN, Caussa LI, Criado del Rio LE (2005) Teaching anatomy: cadavers vs computers? Ann Anat 188:187–190
Binil Zack (2021) Anatomage Table at Kharkiv International Medical University, CC BY 4.0, Wikimedia Commons. https://commons.wikimedia.org/wiki/File:Anatomage_Table_at_Kharkiv_International_Medical_University.jpg. Accessed 23 Apr 2024
Bone Clones (2024) Human anatomy. https://boneclones.com/category/human-anatomy. Accessed 29 Feb 2024
Bremner AJ, Spence C (2017) The development of tactile perception. Adv Child Dev Bahav 52:227–268
Brown J, Stonelake S, Anderson W, Abdulla M, Toms C, Farfus A, Wilton J (2015) Medical student perception of Anatomage – a 3D interactive anatomy dissection table. Int J Surg 23:17–18
Buckland-Wright JC (1985) Readers of anatomy at the barber-surgeons company in the Tudor period. J R Soc Med 78:802–811
Burn I (2001) The company of barbers and surgeons. Farrand Press, London
Cain T (2017) Getting under the skin. https://wellcomecollection.org/articles/WXXZUyoAAJpKfPOq. Accessed 8 Mar 2023
Carlino A (1999) Books of the body: anatomical ritual and renaissance learning. The University of Chicago Press, Chicago
Christensen AM, Passalacqua NV, Bartelink EJ (2014) Individual skeletal variation. Forens Anthropol: Curr Methods Pract 301–339
Collings AJ, Brown K (2020) Reconstruction and physical fit analysis of fragmented skeletal remains using 3D imaging and printing. Foren Sci Int: Rep 2:1–6
Cukurova M, Bennett J, Abrahams I (2017) Students' knowledge acquisition and ability to apply knowledge into different science contexts in two different independent learning settings. Res Sci Technol Educ 36(1):17–34
Darras KE, de Bruin ABH, Nicolaou S, Dahlström N, Persson A, van Merriënboer J, Forster BB (2018) Is there a superior simulator for human anatomy education? How virtual dissection can overcome the anatomic and pedagogic limitations of cadaveric dissection. Med Teach 40(7):752–753
Dash HR, Das S (2020) Thanatomicrobiome and epinecrotic community signatures for estimation of post-mortem time interval in human cadaver. Appl Microbiol Biotechnol 104:9497–9512
Deathridge JR (2017) Should human remains be displayed in museums? https://blogs.ucl.ac.uk/researchers-in-museums/2017/03/28/should-human-remains-be-displayed-in-museums/#:~:text=Human%20remains%20are%20an%20important,the%20study%20of%20medical%20sciences. Accessed 16 Apr 2024
Deng X, Zhou G, Xiao B, Zhao Z, He Y and Chen C (2018). Effectiveness evaluation of digital virtual simulation application in teaching of gross anatomy. Annals of Anatomy 218: 276–282.
Dental Schools Council (2023) Active inclusion, challenging exclusions: supporting students of different faiths. https://www.medschools.ac.uk/media/3049/active-inclusion-challenging-exclusions-supporting-students-of-faith.pdf. Accessed 25 May 2024
Edelstein L (1935) The development of Greek anatomy. Bull Instit Hist Med 3(4):235–248
Elfar JJ, Menorca RMG, Reed JD, Stanbury S (2014) Composite bone models in orthopaedic surgery research and education. J Am Acad Orthop Surg 22:111–120
Evans DJR, Bay BH, Wilson TD, Smith CF, Lachman N, Wojciech P (2020) Going virtual to support anatomy education: a stopgap in the midst of the Covid-19 pandemic. Anatomical Sci Educ 13(3):279–283
Evotech Scientific (2024) Welcome to Evotech Scientific. https://evotechscientific.com. Accessed 23 Apr 2024
Federigo Federighi (2019) Preparato di Auzoux at human anatomy museum (Pisa university), CC BY 4.0, URL: https://commons.wikimedia.org/wiki/File:Preparato_di_Auzoux_at_human_anatomy_museum_(Pisa_university).jpg. Accessed 23 Apr 2024
Ferreira MT, Cunha E (2013) Can we infer post mortem interval on the basis of decomposition rate? A case from a Portuguese cemetery. Forensic Sci Int 226(1–3):298

Frank JB (1976) Body snatching: a grave medical problem. Yale J Biol Med 49:399–410
Gholipour B (2019) Med school without cadavers? Scientific American. https://www.scientificamerican.com/article/med-school-without-cadavers/. Accessed 25 Mar 2024
Ghosh SK (2015) Human cadaveric dissection: a historical account from ancient Greece to the modern era. Anatomy Cell Biol 48(3):153–169
Gogalniceanu P, Palman J, Madani H, Sheena Y, Birch W, Paraskeva P, Douek M (2010) Traditional undergraduate anatomy education—a contemporary taboo? ANZ J Surg 80(1–2):6–7
Goodison GA, Aris C (2023) Analysis of the use of human remains as teaching materials in higher education: examining differing opinions between population subgroups. Eur J Anat 27(4):503–516
Goodison GA, Aris C (2024) The potential instability of forensic anthropology pedagogy in forensic science. Excursions 11:2
GOV (2015) Changes to Formaldehyde (Formalin) rules. https://www.gov.uk/government/news/changes-to-formaldehyde-formalin-rules#:~:text=From%20 1%20January%202016%20formaldehyde,market%20 by%20a%20registered%20supplier. Accessed 25 May 2024
GOV (2024) Formaldehyde: general information. https://www.gov.uk/government/publications/formaldehyde-properties-incident-management-and-toxicology/formaldehyde-general-information. Accessed 25 May 2024
Graham S, Huffer D (2022) When TikTok discovered the human remains trade: a case study. Open Archaeol 8(1):196–219
Greatest Auction (2023) Curve Media. Available at: Channel 4. Accessed 1st Apr 2024
Grogan S (2017) Death, dissection and disease: the life of a surgeon-apothecary 1750–1850. Pen & Sword History, Barnsley
Gunderman RB, Wilson PK (2005) Viewpoint: exploring the human interior: the roles of cadaver dissection and radiologic imaging in teaching anatomy. Acad Med 80(8):745–749
Haas LF (1999) Papyrus of Ebers and Smith. J Neurol Neurosurg Psychiatry 67(5):578
Haizuka Y, Nagase M, Takashino S, Kobayashi Y, Fujikura Y, Matsumura G (2018) A new substitute for formalin: application to embalming cadavers. Clin Anat 31:90–98
Hallam E (2013) Anatomical design: making and using three-dimensional models of the human body. In: Gunn W, Otto T, Smith RC (eds) Design anthropology: theory and practice. Bloomsbury Academic, Oxon, pp 100–116
Hallam E (2016) Anatomy museum: death and the body displayed. Reaktion Books Ltd, London
Halpern J (2001) From detached concern to empathy: humanizing medical practice. Oxford University Press, New York
Hamilton K (2023) Family tells of trauma as dad's body not accepted by Sheffield medical school after 'unforgivable' error. https://www.thestar.co.uk/news/people/family-tells-of-trauma-as-dads-body-not-accepted-by-sheffield-medical-school-after-unforgivable-error-4394307. Accessed 18 Apr 2024
Healthy Simulation (2020) SynDaver. https://www.healthysimulation.com/syndaver/#:~:text=SynDaver%20Labs%20products%20are%20made,against%20the%20relevant%20living%20tissue. Accessed 12 Mar 2024
Heseler B (1959) Andreas Vesalius' first public anatomy at Bologna: an eyewitness report by Baldassar Heseler, ed. and trans. Ruben Ericksson. Almquist and Wiksells, Uppsala
HTA (2021) Public Display: Code of Practice and Standards. https://content.hta.gov.uk/sites/default/files/2021-07/HTA%20Code%20of%20Practice%20D%20Public%20Display_1.pdf. Accessed 20th Sept 2024
HTA (2024) The human tissue authority. https://www.hta.gov.uk. Accessed 31 Mar 2024
Huffer D, Graham S (2023) These were people once: the online trade in human remains and why it matters. Berghahn, New York
Hugo K (2016) Human skulls are being sold online, but is it legal? https://www.nationalgeographic.com/science/article/human-skulls-sale-legal-ebay-forensics-science. Accessed 23 Apr 2024
Huri G, Gülsen MR, Karmis EB, Karagüven D (2021) Cadaver versus simulator based arthroscopic training in shoulder surgery. Turk J Med Sci 51(3):1179–1190
Johnson JH (2002) Importance of dissection in learning anatomy: personal dissection versus peer-teaching. Clin Anat 15(1):38–44
Kavvadia E-M, Katsoul I, Angelis S, Filippou D (2023) The Anatomage Table: a promising alternative in anatomy education. Cureus 15(8)
King's College London (2024) Anatomy, Developmental & Human Biology BSc. https://www.kcl.ac.uk/study/undergraduate/courses/anatomy-developmental-and-human-biology-bsc. Accessed 25 Mar 2024
Klestinec C (2004) A history of anatomy theatres in sixteenth century Padua. J Hist Med Allied Sci 59(3):375–412
Korf HW, Wicht H, Snipes RL, Timmermans JP, Paulsen F, Rune G, Baumgart-Vogt E (2008) The dissection course – necessary and indispensable for teaching anatomy to medical students. Ann Anat 190:16–22
Lancaster University (2024) Choosing Lancaster for medicine. https://www.lancaster.ac.uk/lms/medicine/medicine-at-lancaster/. Accessed 18 Apr 2024
Landes JB (2008) Wax fibers, wax bodies, and moving figures; artifice and nature in eighteenth-century anatomy. In: Panzanella R (ed) Ephemeral Bosides: wax sculpture and the human figure. Los Angeles, Getty Research Institute, pp 41–66
Laspro M, Groysman L, Verzella AN, Kimberly LL, Flores RL (2023) The use of virtual reality in surgical

training: implications for education, patient safety and global health equity. Surgeries 4(4):635–646

Lauridsen H, Hansen K, Nørgård MO, Wang T, Pedersen M (2016) From tissue to silicon to plastic: three-dimensional printing in comparative anatomy and physiology. R Soc Open Sci 3

Legislation.gov.uk (2023) Anatomy Act 1984. https://www.legislation.gov.uk/ukpga/1984/14/introduction. Accessed 20 Apr 2023

Lenox S (2016) Bodysnatchers: digging up the untold stories of Britain's resurrection men. Pen & Sword History, Barnsley

Leung BC, Williams M, Horton C, Cosker TDA (2020) Modernising anatomy teaching: which resources do students rely on? J Med Educat Curri Develop 7:1–7

Liao P, Wang Z (2019) Thiel-embalming technique: investigation of possible modification in embalming tissues as evaluation model for radiofrequency ablation. J Biomed Res 33(4):280–288

Licata M, Bonsignore A, Boana R, Monza F, Fulcheri E, Ciliberti R (2020) Study, conservation and exhibition of human remains: the needs of a bioethical perspective. Acta Biomed 91(4):1–10

Lloyd G (1975) Alcmaeon and the early history of dissection. Sudhoffs Arch 59(2):113–147

Loukas M, Hanna M, Alsaiegh N, Shoja MM, Tubbs RS (2011) Clinical anatomy as practiced by ancient Egyptians. Clin Anat 24(4):409–415

MacDonald H (2005) Human remains: dissection and its histories. Yale University Press, London

Maerker A (2014) Between profession and performance: Displays of anatomical models in London, 1831–32. Histoire Médecine et Santé 5(2):47–59

Mahadevan V (2006) The human tissue act 2004: implications for anatomical work at the college. Ann R Coll Surg Engl 88:264–265

Marayuma K, Yokoi H, Nagase M, Yoshida H, Noguchi A, Matsumura G, Saito K, Shiokawa Y (2019) Usefulness of N-vinyl-2-pyrrolidone embalming for endoscopic transnasal skull base approach in cadaver dissection. Neurol Med Chir 59(10):379–383

Marković D, Marković Živković B (2010) Development of anatomical models-chronology. Acta Medica Medianae 49(2)

McLachlan JC, Bligh J, Bradley P, Searle J (2004) Teaching anatomy without cadavers. Med Educ 38(4):418–424

Metzner F, Neupetsch C, Carabello A, Pietsch M, Wendler T, Drossel WG (2022) Biomechanical validation of additively manufactured artificial femoral bones. BMC Biomed Eng 4(6):1–12

Mitchell PD (2012) There's more to dissection than burke and hare: unknowns in the teaching of anatomy and pathology from the enlightenment to the early twentieth century in England. In: Mitchel PD (ed) Anatomical dissection in enlightenment England and beyond: autopsy, pathology, and display. Ashgate, Farnham

Mitchell T, Newton BW, Terreberry RR, VanCura MX (2016) Video dissections to determine the usefulness of synthetic cadavers vs real cadavers. J Hum Anat Physiol Soc 20(3):115–118

Monirb (2016) CC BY 4.0, Wikimedia commons. https://commons.wikimedia.org/wiki/File:MSOT_005-NEW.jpg. Accessed 25 May 2024

Narang P, Raju B, Jumah F, Konar SK, Nagaraj A, Gupta G, Nanda A (2021) The evolution of 3D anatomical models: a brief historical overview. World Neurosurg 155:135–143

Nichols H (2014) The digital age of medicine: cadavers still best choice for learning anatomy. https://www.medicalnewstoday.com/articles/284057. Accessed 18 April 2024

Niels H (2022) Thirty years of Thiel embalming—a systematic review on its utility in medical research. Clin Anat 35(7):987–997

O'Malley CD (1964) Andreas Vesalius of Brussels 1514–1564. University of California Press, Berkeley

Older J (2004) Anatomy: a must for teaching the next generation. R Coll Surg Edinburgh Ireland 2(2):79–90

Otton JM, Birbara NS, Hussain T, Greil G, Foley TA, Pather N (2017) 3D printing from cardiovascular CT: a practical guide and review. Cardiovasc Diagn Ther 7(5):507–526

Pallocci M, Petroni G, Treglia M, Giammatteo J, Marella GL, Arcangeli M (2020) Law proposal "provisions on the post-mortem body donation and the use of bodies for the purposes of study, scientific research and training": comment and analysis of the bill and the historical-juridical-ethical aspects of cadaveric dissection and practice of the donation of a corpse for scientific and medical training purposes. Acta Medica Mediterranea 36:999–1005

Papa V, Vaccarezza M (2013) Teaching anatomy in the XXI century: new aspects and pitfalls. Scientific World 2013:310348–310345

Parker G (1912) The barber surgeons. Bristol Med Chir J (1883) 30(118):327–338

Pester P (2022) Desecrated human skulls are being sold on social media in UK's unregulated bone trade. https://www.livescience.com/human-skulls-desecrated-uk-human-remains-trade. Accessed 23 Apr 2024

Phillipson T (2022) Anatomy: a matter of life and death. NMS Enterprises Ltd, Scotland

Potu BK, Atwa H, Nasr El-Din WA, Othman MA, Sarwani NA, Fatima A, Deifalla A, Fadel RA (2021) Learning anatomy before and during COVID-19 pandemic: students' perceptions and exam performance. Morphologie 106:188–194

Prabodha LBL, Nanayakkara BG, Ilayperuma I, Samarawickrama MB, Johnson P (2016) Is the cadaver dead? Learning anatomy beyond the horizon. Int J Adv Sci Eng Technol 4(1):57–61

Preim B and Saalfed P 2018. A survey of virtual human anatomy education systems. Computers and Graphis 71: 135–153

Richardson R (2001) Death, dissection and the destitute, 2nd edn. Phoenix Press, London

Richardson NS, Zwambag D, McFall K, Andrews DM, Gregory DE (2020) Exploring the utility and student perceptions of synthetic cadavers in an undergraduate human anatomy course. Anat Sci Educ 14(5):605–614

Riva A, Conti G, Solinas P, Loy F (2010) The evolution of anatomical illustration and wax modelling in Italy from the 16th to early 19th centuries. J Anat 216(2):209–222

Rizzolo LJ, Stewart WB (2006) Should we continue teaching anatomy by dissection when…? Anat Rec 289B:215–218

Sawday J (1995) The body emblazoned: dissection and the human body in renaissance culture. Routledge, Oxon

Saylam C, Coskunol H (2005) Orientation lesson in anatomy education. Surg Radiol Anat 27(1):74–77

Scientific Animations and Manu5 (2020) 3D medical animation still shot structure lingual artery. CC BY 4.0, Wikimedia Commons. https://commons.wikimedia.org/wiki/File:3D_Medical_Animation_still_shot_structure_Lingual_Artery.jpg. Accessed 25 Apr 2024

Seymour NE, Gallagher AG, Roman SA, O'Brien MK, Bansal VK, Andersen DK, Satava RM (2002) Virtual reality training improves operating room performance. Ann Surg 236(4):458–464

Singer C (1957) A short history of anatomy & physiology from the Greeks to Harvey. Dover Publications, New York

Smith CF, Tollemache N, Covill D, Johnston M (2017) Take away body parts! An investigation into the use of 3D-printed anatomical models in undergraduate anatomy education. Anatomical Sci Educ 11(1):44–53

Smith CF, Freemand SK, Heylings D, Finn GM, Davies C (2022) Anatomy education for medical students in the United Kingdom and Republic of Ireland in 2019: a 20-year follow up. Anat Sci Educ 15(6):993–1006

SOMSO (2024) Original SOMSO Modelle. https://www.somso.de/en/. Accessed 25 Feb 2024

Standring S (2016) A brief history of topographical anatomy. J Anat 229:32–62

Sugand K, Abrahams P, Khurana A (2010) The anatomy of anatomy: a review for its modernization. Anat Sci Educ 3(2):83–93

Surgeon's Hall Museums (2022) A model education. https://surgeonshallmuseums.wordpress.com/2022/05/25/a--model-education-2/. Accessed 25 Feb 2024

SynDaver (2020) SynDaver: beyond human. https://syndaver.com. Accessed 24 Mar 2024

Tarlow S, Battell Lowman E (2018) Harnessing the power of the criminal corpse. Palgrave Macmillan, Cham

Thomas DP 2006. Thomas Vicary, barber-surgeon. Journal of Medical Biography, 14(2): 84–89

UK Parliament (2014) Body snatching. https://www.parliament.uk/about/living-heritage/transformingsociety/private-lives/death-dying/dying-and-death/bodysnatching/. Accessed 26 Mar 2023

University of Aberdeen (2006) Papier mache man stars in exhibition. https://www.abdn.ac.uk/news/17538/. Accessed 21 May 2024

University of Aberdeen (2024) Biomedical sciences (anatomy) BSc. https://www.abdn.ac.uk/study/undergraduate/degree-programmes/462/B9BC/biomedical-sciences-anatomy/#year_3. Accessed 25 Mar 2024

University of Bristol (2024) BSc Applied Anatomy (B112). https://www.bristol.ac.uk/study/undergraduate/2025/anatomy/bsc-applied-anatomy/. Accessed 25 Mar 2024

University of Edinburgh (2024) BSc Anatomy and Development. https://www.ed.ac.uk/studying/undergraduate/degrees/index.php?action=programme&code=C183. Accessed 25 Mar 2024

University of Glasgow (2024) Anatomy BSc/MSci. https://www.gla.ac.uk/undergraduate/degrees/anatomy/#programmestructure. Accessed 25 Mar 2024

University of Liverpool (2024) Anatomy and human biology BSc (Hons). https://www.liverpool.ac.uk/courses/2024/anatomy-and-human-biology-bsc-hons. Accessed 25 Mar 2024

University of Sheffield (2024) Biomedical Science BSc. https://www.sheffield.ac.uk/undergraduate/courses/2025/biomedical-science-bsc#modules. Accessed 18 Apr 2024

Valdecasas AG, Correas AM, Guerrero CR, Juez J (2009) Understanding complex systems: lessons from Auzoux's and von Hagens's anatomical models. J Biosci 34(6):835–843

Volk D (2023) How long does it take for a body to decompose? https://www.livescience.com/how-long-bodies-take-to-decompose Accessed 8 Feb 2023

Walker P (2007) Bioarchaeological ethics: a historical perspective on the value of human remains. In: Katzenberg MA, Saunders SR (eds) Biological anthropology of the human skeleton. Wiley, Hoboken, pp 3–39

Washmuth NB, Cahoon T, Tuggle K, Hunsinger RN (2020) Visual dissection: alternative to cadaveric dissection for a pregnant nurse anesthesia student. Health Professions Educ 6(2):247–255

Wellcome library, Anatomical Venus. Wax figure of reclining woman, London (2014) CC BY 4.0, Florence. https://commons.wikimedia.org/wiki/File:Anatomical_Venus._Wax_figure_of_reclining_woman,_Florence._Wellcome_L0058207.jpg Accessed 23 Apr 2024

Williams T (2023) Is dissection dying out in UK medical schools? https://www.timeshighereducation.com/news/dissection-dying-out-uk-medical-schools#:~:text=The%20University%20of%20Oxford%20has,used%20only%20on%20certain%20courses. Accessed 18 Apr 2024

Wink AE, Telfer AN, Pascoe MA (2022) Google images search results as a resource in the anatomy laboratory: rating of educational value. JMIR Med Educ 8(4):e37730

YVC Biology Department and Netha Hussain (2020) Human eye model, CC BY 20, Wikimedia Commons. https://commons.wikimedia.org/wiki/File:Human_Eye_Model_(50693611856).jpg. Accessed 25 May 2024

Zargaran A, Turki MA, Bhaskar J, Spiers HVM, Zargaran D (2020) The role of technology in anatomy teaching: striking the right balance. Adv Med Educ Pract 11:259–266

9 "You Want Us to Consent to This? I Don't Even Know What You're Talking About!": Exploring Science Communication Using Cinemeducation and House M.D.

Andrew S. Cale, Hailey Wilkinson, Carissa Birznieks, and Rebecca L. Wisner

Abstract

Communicating complex medical information in a manner accessible to non-experts is an essential skill for medical professionals. Cinemeducation, a method of medical education that leverages the many benefits of cinema for educational purposes, may be ideal for cultivating this communication skill in medical professionals. Though originally intended as entertainment, medical movies and television can feature realistic portrayals of the medical field that engage and resonate with learners. The medical drama *House M.D.* is one such television show that can be used to teach communication skills to medical professionals. In a thematic analysis of the first two seasons of the show ($n = 46$ episodes), several scenes were identified as depicting either effective or ineffective communication of complex medical information to non-experts. Effective communication strategies that increased the accessibility of medical information included elaboration on jargon, use of analogies, and integration of visual aids. These scenes can be used in conjunction with additional pieces of medical film and television in a cinemeducation curriculum to teach medical professionals how to make medical information more accessible for non-experts.

Keywords

Cinemeducation · Cinemedicine · Communication · Medical television · House M.D

9.1 The Importance of Communication

Communication is a bioactive process, meaning it can directly impact an individual's biological processes (Parrish-Sprowl et al. 2023). For example, a positive interaction between a patient and their doctor can alleviate anxiety related to their illness, thereby reducing the concentration of stress hormones circulating in their body (Sumioka et al. 2013). Beyond this direct physiological impact, effective communication has

A. S. Cale (✉) · R. L. Wisner
Department of Anatomy, Cell Biology, & Physiology, Indiana University School of Medicine, Indianapolis, IN, USA
e-mail: ascale@iu.edu

H. Wilkinson
Modern Human Anatomy Program, University of Colorado Anschutz Medical Campus, Aurora, CO, USA

C. Birznieks
Integrative Physiology Program, University of Colorado Anschutz Medical Campus, Aurora, CO, USA

R. Door (ed.), *Accessibility, Inclusivity and Diversity in Education and Beyond*, Biomedical Visualization 8, https://doi.org/10.1007/978-3-031-81795-3_9

also been linked to several improved outcomes for patients, including increased satisfaction, adherence to treatment plans, sharing of diagnostically salient information, and self-efficacy (Ha and Longnecker 2010). Improved doctor–patient communication has also been positively correlated with patient pain tolerance, recovery, daily functioning, and mental health, as well as negatively correlated with patient rehospitalization frequency and duration, and even tumor growth (Ha and Longnecker 2010). Unfortunately, doctors often overestimate their communication skills, causing them to fail at properly informing their patients and capitalizing on the benefits of effective communication (Ha and Longnecker 2010; Ranjan et al. 2015). As such, it is essential for doctors to develop their communication skills. Cinemeducation, a method of medical education that leverages the many benefits of film and television for educational purposes, may be ideal for cultivating communication skills in medical professionals.

9.2 Cinemeducation: A Narrative Review of Film and Television Usage in Medical Education

While technical knowledge and skills can be developed through traditional didactic methods, interpersonal skills and humanistic values are more challenging to cultivate in medical professionals (Blasco et al. 2006). Development of these aspects requires learners to self-reflect on their behaviors, attitudes, values, and overall worldview (Blasco et al. 2006). One potential method for developing these skills and qualities in medical professionals is through cinemeducation, an educational technique that leverages the unique advantages of cinema (i.e., movies and television) to facilitate learning in medical education. This method of medical education was first pioneered by Fritz and Poe in 1979 with a seminar series of contemporary films designed to supplement a psychiatry residency curriculum (Fritz and Poe 1979). Later, in 1994, Alexander et al. (1994) first coined the term "cinemeducation" (a portmanteau of the words "cinema," "medicine," and "education") to describe this medical education technique. Since its introduction, numerous additional studies have further expanded what is known about cinemeducation and how it can be implemented in medical education (Darbyshire and Baker 2012; Hoffman et al. 2017).

9.2.1 General Educational Advantages

Cinema as a medium has a long history of not only entertaining but also educating audiences. Though it may not be their primary purpose, movies such as It's a Wonderful Life (1946) or Forrest Gump (1994) are well-known as modern parables that convey significant life lessons to their audiences (Blasco et al. 2011). Several unique attributes of cinema aid in effectively connecting with audiences and conveying these lessons. For one, cinema is an audio-visual medium, making it more engaging than either audio recordings or visual media alone (Lumlertgul et al. 2009; De Souza and Adams 2023). Cinema also uses storytelling techniques to create compelling narratives that evoke emotions in the audiences, resonating with them in ways that traditional lectures cannot (Blasco et al. 2011). Lastly, these narratives are also explored through the lens of characters that audiences can relate to, empathize with, and share in their experiences (Hirt et al. 2013; Weaver et al. 2014). These intrinsic qualities of cinema make it a powerful tool that not only teaches audiences but entertains and engages them as well.

9.2.2 Medical Cinema and Cinemeducation

Cinemeducation leverages these unique aspects of cinema to train learners on topics specific to the medical field. This technique can incorporate a wide range of cinema, but most applications include movies and television shows related to the medical field. In 1951, the first medical TV program, City Hospital (1951), premiered on network television and depicted the professional and personal life of a physician working in a large metropolitan hospital (Alismail et al. 2018). Since its premiere, medical cinema has increased dramatically in popularity.

9.2.2.1 Popularity and Viewership

Nowadays, every major television network airs at least one medical show in their programming, often in the coveted primetime slots when viewership is highest (Hirt et al. 2013). As a testament to their popularity, many of these television shows have aired for multiple seasons, sometimes even decades. Some examples include ER (1994) which aired for 15 seasons from 1994 to 2009 and Grey's Anatomy (2005) which is currently airing its 20th season. Over the years, several movies centered around medical issues such as the films Philadelphia (1993) and One Flew Over the Cuckoo's Nest (1975) have also garnered significant critical and commercial acclaim. Medical cinema has also expanded into a variety of genres such as drama, comedy, thriller, documentary, and reality, further increasing the appeal and viewership of these programs. Thanks to their popularity, medical cinema has left a significant impact on society, culture, and the public's perception of the medical field (Alismail et al. 2018). As more medical film and television titles become available, cinemeducation can expand accordingly as well, further increasing its effectiveness and impact on medical education.

Medical cinema is not only popular with general audiences, but also amongst medical professionals and learners in the medical field. Though the exact rate varies between studies, the number of medical professionals who report watching medical cinema remains consistently high. Czarny et al. (2008) determined that 84% of medical students and 81% of nursing students at a U.S.-based institution watched medical television shows at least once per week. Similarly, Weaver and Wilson (2011) found that 93.7% of their population of Australian medical students watched medical shows on a regular basis, while Williams et al. (2014) discovered that 100% of their U.K. medical student population viewed medical television. Amongst these students, roughly 49% of medical students and 55% of nursing students reported discussing the medical scenarios from medical cinema with colleagues (Czarny et al. 2008). Similarly, 41% of medical students and 47% of nursing students were asked by their family and friends in non-medical fields to discuss the medical cases depicted in these shows (Czarny et al. 2008). In another study by Haboubi et al. (2015), 65% of surveyed doctors reported watching medical dramas, with junior doctors being more regular viewers than their more senior colleagues. The majority (69%) of these doctors reported viewing these television programs for entertainment purposes, though 8% also reported watching for educational purposes. The high viewership of medical cinema by those in the medical field suggests that viewers have an interest in these programs and see value in them in some capacity.

9.2.2.2 Accuracy and Realism

Although intended as entertainment, medical cinema can be highly accurate and realistic. This is particularly true for movies or television shows where complex medical cases and scientific diagnoses are at the forefront of the plot. Despite *House M.D.* (2004) being a fictional television series, Katherine Lingenfelter, the supervising producer of the show has stated that she considers it a social responsibility to keep the scientific details of the show as accurate as possible (Jerrentrup et al. 2018). As such, the writers of *House M.D.* regularly based the medical cases portrayed in the series on real cases reported in notable medical journals such as The New England Journal of Medicine (NEJM), The Lancet, and the British Medical Journal (BMJ) (Jerrentrup et al. 2018). To further enhance realism, film studios often hire consultants to review the medical content of the show for accuracy and plausibility. During the years that *House M.D.* was airing on television, Dr. Lisa Sanders, an associate lecturer at the Yale School of Medicine, served as the show's medical consultant and reviewed scripts for medical inaccuracies (Baños et al. 2019). As a result, the medical content of *House M.D.* demonstrated a high degree of accuracy and realism, sufficient enough to even aid in the diagnosis and treatment of a real-life case of cobalt intoxication (Dahms et al. 2014).

However, this degree of realism does not fully capture every aspect of the medical field, nor is it consistent across all medical cinemas. Depending on the priorities and premise of the specific show, filmmakers may forgo some medical realism to increase entertainment value (Jain and Slater 2013). For example, Lapostolle et al. (2013) found that many of the treatment strategies in *House M.D.* were unrealistic in that they focused on the novel and unusual rather than the typical. Routine examinations were underused, invasive and expensive procedures were treated as first-line options, and patient populations were not representative of reality in terms of gender and age (Lapostolle et al. 2013). Baños et al. (2019) also determined that while *House M.D.* consistently portrayed the correct indications for pharmaceutical drugs, it did not include the depictions of the safety checks typically performed when those drugs are administered. Some medical events also tended to be over-dramatized (Rattani et al. 2020). For example, Cape (2009) found that depictions of drug withdrawal, overdose, and violence by individuals experiencing addiction are often heavily exaggerated for dramatic effect. Similarly, televised portrayals of cardiac arrest and resuscitation have also perpetuated the misconception that an automated external defibrillator (AED) can restart the heart from asystole, a non-shockable heart rhythm (Alismail et al. 2018). Fortunately, individuals who are familiar with the medical field are aware of these inaccuracies and proficient at identifying errors when they occur, whereas lay audiences are less able to do so (Czarny et al. 2008). Therefore, while efforts are made to enhance the realism of medical cinema, its depictions of the medical field must be taken with a grain of salt as some details may have been adjusted for entertainment purposes, leaving the public with a limited, skewed view of the medical field (Czarny et al. 2008).

While medical cinema may not be completely accurate or realistic, its fictional portrayals of the medical field and medical professionals are often representative enough to have educational merit. As noted previously, viewership of medical cinema is high amongst members of the medical community. This high rate can be attributed to an interest in medical topics and opportunities for the medical community to process their emotions and achieve a cathartic release. Many scenarios portrayed in medical movies and television shows can simulate real-life situations that individuals in the medical field may have encountered in their day-to-day practice (Aryanty and Raudhoh 2023). Certain scenes, particularly ones involving emotionally difficult topics like interpersonal conflicts, social issues, or death and morbidity, can resonate and evoke emotions that must be processed by the viewer (Baños 2007; Blasco and Moreto 2012; Hirt et al. 2013; Zeppegno et al. 2015; Kadivar et al. 2018; Parsi and Elster 2020; Alahmari 2023). Unlike in the clinic, in these moments, the viewers are free from clinical responsibility or judgment, allowing them to focus and critically reflect on their emotions and experiences (White 2008). This phenomenon is known as the *Don Quixote* effect (Shapiro and Rucker 2004). In the Spanish novel by Miguel de Cervantes, Sancho Panza, Don Quixote's companion, is better able to endure, process, and learn from unpleasant experiences of his reality by occasionally embracing Don Quixote's idealized, imaginary view of the world. Similarly, by watching medical cinema and embracing an idealized version of the medical field, the viewers can better process and learn from the complicated, stressful situations they encountered in the clinic (Shapiro and Rucker 2004). In some instances, even egregious portrayals of the medical field can serve an educational purpose. Scenes portraying unprofessional or unethical behavior, poor communication, or sociocultural conflict can serve as examples of actions or attitudes unbefitting of medical professionals. For example, Dr. Gregory House, the titular character of *House M.D.*, has been used as a case study for all manner of inappropriate behavior, ranging from basic medical history taking to unethical and illegal actions (Hirt et al. 2013). A study by Stanek et al. (2015) also identified multiple scenes from ER (1994), Grey's Anatomy (2005), and Scrubs (2001) that depicted unprofessional behavior and patient dehumanization that could be used as teachable moments in medical education. Since these inappropriate actions are performed by fictional characters, learners tend to feel more inclined to identify and discuss their faults and errors as opposed to using real-life examples that involve peers, colleagues, or supervisors (Klemenc-Ketis and Kersnik 2011; Stanek et al. 2015). Therefore, although medical cinema may not be perfectly representative of the medical field, the fictionalization of certain elements can actually enhance its educational value in some areas.

Medical cinema also allows viewers to experience the medical field from a perspective other than their own, making it ideal for teaching the psychosocial and humanistic aspects of practicing medicine. In some cases, medical professionals and learners may sympathize with a patient but struggle to empathize with them because they have not had the same experiences as the patient. Cinemeducation can bridge this gap between empathy and sympathy by providing viewers with insight into patient experiences and emotions (Zeppegno et al. 2015). For example, in a study by Kadivar et al. (2018), 82.8% of medical students participating in a medical film series agreed that it provided them with improved awareness of the humanistic side of doctoring. Modules featuring films centered around mental illness and other psychiatric topics provided medical students with a greater appreciation for the social, cultural, and historical contexts of mental illness (Datta 2009) and increased their empathy toward patients experiencing mental illness (Petkari 2017). Furthermore, in a study by Klemenc-Ketis and Kersnik (2011), medical students participating in a 4-month medical film elective course involving lectures, group discussions, and individual work reported greater appreciation and respect for effective communication, empathy, and palliative care that prioritizes the patient. Since film and television are also cultural products, they are excellent reflections of the societies and time periods in which they were created. As such, they can serve as lenses through which learners can explore sociocultural issues such as diversity, equity, inclusion, and accessibility in the medical field (Datta 2009; Alahmari 2023). Interestingly, the characters depicted in modern medical television generally represent a range of diverse identities, though certain groups still remain underrepresented. These shows tend to portray diversity in that they include more characters that are Black, women, or young (Painter et al. 2020). However, they include less representation of characters who are Asian or Hispanic, international, and older (Painter et al. 2020). Several titles such as Grey's Anatomy (2005), ER (1994), The Resident (2018), New Amsterdam (2018), and Scrubs (2001) have touched on topics related to race and racism such as stigmatizing language, health disparities, microaggressions, bias, and tokenism in their episodes (Hoffman et al. 2023). These episodes have been used in an anti-racism curriculum for medical residents to great effect (Hoffman et al. 2023). Gender and sexuality can also be explored with these medical television programs. While many titles include well-crafted and complex female characters, gender stereotypes at the individual and profession levels (e.g., nursing and surgery) are still commonly featured (Jubas 2013). The juxtaposition of depictions of both empowered and repressed character traits in a single television show can serve as excellent discussion topics on gender, sexuality, and misogyny in the medical field.

9.2.3 Application Strategies and Considerations

Given the many potential benefits, several studies have developed different strategies for applying cinema to medical education and evaluating its educational impact. Typically, these strategies have taken the form of discussion-based class sessions, ranging from semester-long formal courses (Datta 2009) to weeks-long modules (Gallagher et al. 2011) to hour-long workshops and informal film clubs (Blasco and Moreto 2012; Jerrentrup et al. 2018). Each of these variations includes both advantages and disadvantages that educators must carefully consider when implementing cinemeducation into their curricula. Extended courses allow for viewing of multiple works of medical cinema, increasing the variety of topics and perspectives that can be addressed. However, they require a greater time commitment from both the educator and the learners. Conversely, shorter workshops can be more easily slotted into busy schedules but are limited in the breadth and depth of what can be covered.

While cinema is the central focus of all these strategies, the movies and television shows selected for screening are highly dependent upon the desired educational outcomes. Cinemeducation has been used to teach a plethora of both technical and humanistic topics, including clinical diagnosis and treatment (Jerrentrup et al. 2018), bioethics (Aryanty and Raudhoh 2023), mental illness and stigma (Datta 2009; Tobia et al. 2015; Zeppegno et al. 2015; Petkari 2017), public health (Gallagher et al. 2011), anti-racism (Hoffman et al. 2023), cultural sensitivity (Kelly 2012), professionalism (Klemenc-Ketis and Kersnik 2011), and communication skills (Wong et al. 2009). Generally, movies and television shows with a medical focus are selected for these educational programs. However, media from outside the medical genre can be used to great effect as well. For example, Blasco et al. (2011) included a diverse selection of non-medical movies ranging from the The Lion King (1994) to teach reflective attitude, Spartacus (1960) to illustrate teamwork, and The Last Samurai (2003) to demonstrate leadership. Tobia et al. (2015) even used the sitcom Seinfeld (1989) to illustrate different types of delusions for medical students as preparation for their standardized shelf exams, which consist of expired exam questions taken from the United States Medical Licensing Examination (USMLE). This inclusion of media from outside the medical field can add both novelty and diversity of perspective, making cinemeducation more engaging for the learner. Ideally, the selected titles have not been viewed by the learners previously to increase their impact and initial responses to the media, though this is not always possible (Shankar 2019).

Several authors also consider debriefing and robust discussion amongst the participants as an essential aspect of cinemeducation (Datta 2009; Lumlertgul et al. 2009; Blasco and Moreto 2012; Williams et al. 2014; De Souza and Adams 2023; Hoffman et al. 2023). After viewing the selected titles, open discussion encourages active learning by giving the learners an opportunity to engage with and process their thoughts, emotions, and ideas that surfaced when viewing the program (Blasco et al. 2011). Learners can also ask questions and/or describe their own experiences related to the work of cinema, and then have their viewpoint potentially reinforced or challenged by their peers and colleagues. They are then encouraged to reflect on their own experiences and perspectives, leading to more authentic learning. As a result, students have greatly appreciated the opportunity to interact with and discuss the cinematic titles like this (Datta 2009). To ensure a rich, focused discussion that touches on the appropriate topics and achieves the desired outcome, facilitators of these sessions should come prepared with a variety of discussion questions (De Souza and Adams 2023; Hoffman et al. 2023). However, even when the discussion deviates from what the facilitator expects, it may still contribute to the diversity of opinion on the topic (Blasco and Moreto 2012; Law et al. 2015). For example, in the study by Blasco and Moreto (2012), one medical student stated, "You must make your comments while the movie is going on… Do you ask if I agree with you? No, not at all…But your comments push me to reflect." To encourage learner participation in the discussion, facilitators must establish a safe and supportive environment in which the students can share their views without fear of judgment or repercussions. As participants share sensitive personal experiences and become more comfortable being emotionally vulnerable, a connection is formed between them, encouraging further disclosure and learning (Kelly 2012).

Ultimately, cinemeducation has been incredibly well-received by both learners and facilitators. Learners consider cinemeducation to be a fun, engaging, and motivational alternative to traditional lectures (Blasco et al. 2011; Gallagher et al. 2011; Shankar et al. 2016; Jerrentrup et al. 2018; Shrivastava and Shrivastava 2022; Aryanty and Raudhoh 2023). After participating in a public health-focused session, 92.9% of medical students felt that movies helped them better understand social issues (Gallagher et al. 2011). In another study focusing on the psychosocial aspects of medicine, medical students thought that this application of cinemeducation was useful to them as physicians and would recommend it to their colleagues (Kadivar et al. 2018). Medical students who participated in the *Seinfeld*-based workshop on delusion types also felt that it enhanced their learning and prepared them for their standardized shelf exams (Tobia et al. 2015). Lastly, after a patient communication workshop that incorporated cinemeducation, residents felt more comfortable with the core principles of communication and with handling challenging clinical scenarios (Wong et al. 2009). These examples illustrate how cinemeducation can be successfully applied to a variety of educational topics relevant to learners in the medical field.

9.2.4 Limitations and Critiques

Although cinemeducation offers multiple benefits, it is not without limitations. Multiple studies have noted that the length of full feature films can be a significant drawback of cinemeducation due to time constraints (Darbyshire and Baker 2012; Aryanty and Raudhoh 2023). On average, feature films can range from 80 to 180 min in length, which is restrictive considering that class periods are often under 120 min in duration. While in-class screening of the film is possible, it does not leave much time for robust discussion, which is a key aspect of cinemeducation. To work around this limitation, several studies recommend using curated clips taken from the full-length film to save time (Blasco and Moreto 2012; Jerrentrup et al. 2018; Hoffman et al. 2023). However, this reduces the time available for the characters and narrative to resonate with the viewers, which may diminish the emotional impact of the film overall. Another alternative to screening full feature films is by using television episodes instead. Typically, episodes are 20–45 min in length, making them much more easily slotted into a class period while still leaving ample time for rich discussion. These alternatives allow for learners to engage with cinemeducation while still fitting into the constraints of the curriculum.

Another known limitation of cinemeducation is how and when the works of medical cinema are viewed by the learners. While most examples of cinemeducation involve synchronous, in-class screening of the selected titles, some educators have opted to assign learners to watch them asynchronously, in their own time (Gallagher et al. 2011). This allows the learner some flexibility when they view the assigned materials, but present obstacles related to the accessibility of the titles. Some films and television shows may require learners to subscribe to a paid streaming service or purchase them outright on digital or physical media to view the assigned materials. This also requires the learner to have stable Internet access and devices on which to play the media (e.g., television, computer, DVD, or Blu-ray player). To increase accessibility, educators may want to consider using films and shows available on free streaming services (such as Freevee, Crackle, or Vudu in the USA, Box of Broadcasts in the U.K., or Tubi and Popcornflix worldwide), curating relevant clips on YouTube, or screening the media in class for all learners to view together. An asynchronous approach also increases the amount of time between viewing the film and discussion, which may reduce the impact of the film and group discussion as details may be forgotten by this time. Additionally, the impact of experiencing a film as a communal event is difficult to replicate when viewing independently at home (Shankar 2019). In-class viewing allows learners to immerse themselves in the film and decreases the likelihood of distractions from phones, laptops, and other people. Like watching in a movie theater or attending a concert or sports game in person, collective reactions can heighten the emotional impact of a film, creating an experience unique to that group of people.

A common critique of cinemeducation also stems from the selection of movies and television shows. Most of the titles that have been used in cinemeducation were written, filmed, and produced by major Hollywood studios in the United States. As such, they represent a limited, U.S.-centric view of the medical field and do not accurately represent medical practice in remote areas of the USA not typically featured on film or internationally. It is a stretch to say that U.S medical programs would not be applicable or useful to rural or international learners, but educators should be aware of how portrayals of the medical field in these titles can differ from the structure, practices, and culture of the environment in which they teach. For example, programs like *House M.D.* often include scenes where doctors must make decisions based on the treatments or procedures a patient's private insurance would cover. In the British Broadcasting Company (BBC)-produced miniseries This is Going to Hurt (2022) which takes place in a U.K. hospital run by the National Health Service (NHS), a publicly funded healthcare system, the same scenarios would not make sense. In addition, the rank and title of doctors in the U.K. and other countries, as well as the educational trajectory to obtain licensure varies significantly from the USA. To address this limitation, some educators have supplemented their filmography with independent or foreign media written and produced internationally and in non-English languages. For example, the Korean drama series Hospital Playlist (2020) is a portrayal of the lives of five doctors at a fictional medical center in South Korea. The program has achieved international attention and can be found on Netflix with subtitles in several languages. The practice of including a diversity of media sources ensures that learners develop a more holistic and multicultural perspective on the medical field that is more representative of their own environment.

Lastly, the existing literature on cinemeducation is limited in evidence on its long-term impact on learners (Hoffman et al. 2017). Numerous studies have demonstrated both facilitators and learners are supportive of integrating cinemeducation into the traditional medical education curriculum (Blasco et al. 2011; Gallagher et al. 2011; Shankar et al. 2016; Jerrentrup et al. 2018; Aryanty and Raudhoh 2023). However, this perception data only satisfies the first level of Kirkpartick's Model of Training Evaluation (Level 1: Reaction) (Lumlertgul et al. 2009). Few studies have explored how cinemeducation impacts trainee knowledge (Level 2: Learning), and even to our knowledge, no studies have determined its impact on performance (Level 3: Behavior) and patient outcome (Level 4: Results). These aspects represent a significant gap in the cinemeducation literature that should be addressed by future studies.

Ultimately, film and television are unique resources that have the potential to greatly enhance medical education. As an audio-visual medium that is built around compelling narratives and characters, cinema can evoke emotions and resonate with the learner, resulting in more meaningful, lasting, and entertaining learning experiences. Though medical film and television can vary in its realism, several studies have successfully used cinemeducation to teach learners a variety of technical and humanistic topics. As new films and television shows are produced and cinema evolves, cinemeducation has the potential to grow alongside incorporating new media and exploring under-addressed topics.

9.3 Communicating with Dr. Gregory House M.D.

Communication skills are one such topic that is essential for medical professionals but has yet to be fully explored using cinemeducation. Due to the unique premise of the show and the way character dialogue is crafted, the television show, *House M.D.*, may serve as an ideal piece of medical cinema to explore science and patient communication.

House M.D. is a medical drama that aired on television for eight seasons from November 16, 2004 to May 21, 2012, totaling 177 episodes (IMDB 2024). While it aired, *House M.D.* was one of the most watched and critically acclaimed shows on network television, garnering millions of viewers each episode, winning multiple awards, and ranking in the top 10 television shows during its 2nd–4th seasons (IMDB 2024). Each episode of the show follows the titular Dr. Gregory House M.D. (played by Hugh Laurie) and his team of doctors (Dr. Eric Foreman, Dr. Robert Chase, and Dr. Allison Cameron) as they diagnose and treat complex and unusual medical cases (IMDB 2024). Dr. House also regularly interacts with series regulars, Dr. James Wilson, a fellow doctor and confidant, and Dr. Lisa Cuddy, the Dean of Medicine with whom he has an antagonistic relationship (IMDB 2024). As a "misanthropic medical genius" who believes "everybody lies," House's communication style often leads him to become embroiled in interpersonal conflict with fellow doctors, patients, and hospital administrators, which makes the show a great example of both effective and ineffective patient and science communication (IMDB 2024). Since the central focus of the show involves highly complex and unusual medical cases, the writers of the show must craft the dialogue of the characters in a manner that maintains the scientific complexity and accuracy of the medical case, but is still comprehensible to a general audience, which also makes the show useful for exploring science communication.

Therefore, the goals of this study are to 1. identify examples of both effective and ineffective science and patient communication from the first two seasons of *House M.D.* and 2. determine how character dialogue is crafted to make the complex scientific topics of the show more accessible to the general audience. Given the qualitative nature of this study, a priori hypotheses were not generated at the start of this study. The results of this study will further expand the applications of cinemeducation in the field of communication and serve as a foundation for future educational studies that seek to use *House M.D.* to enhance communication skills.

9.3.1 Methods

9.3.1.1 Episode Selection and Viewing

Although House M.D. aired for eight seasons, only the first two seasons of the show were viewed for this study, which totaled 46 episodes ($n_{\text{season 1}} = 22$ episodes and $n_{\text{season 2}} = 24$ episodes) and 2024 min overall (~44 min/episode). Both seasons originally aired on the Fox Television Network between 2004 and 2006. Only a portion of the full series was selected for viewing to remain feasible for the researchers while still capturing a representative sample of the show's characters, writing, and overall style. Streaming services such as Amazon Prime Video, Hulu, or Peacock were used to view the American medical drama during data collection as they hosted *House M.D.* at the time this study was performed. Subtitles were turned on during viewing to allow the researchers to closely follow the dialogue and identify relevant instances of either effective or ineffective communication.

9.3.1.2 Data Extraction

Prior to viewing each episode of *House M.D.*, the researchers recorded basic information about the episode using a data extraction tool hosted on Qualtrics (Qualtrics 2024). This included the season of the show, episode number, and episode length. As the researchers viewed the episodes, they then identified and extracted data from any scenes that depicted either effective or ineffective communication. Initially, three members of the research team (A.S.C., H.W., and C.B.) independently viewed the first three episodes of season one. During their viewings, the researchers noted any scenes that involved the communication of scientific or medical information between characters. The details of the scenes, including the timepoint, characters involved, direct transcript of the dialogue, and descriptions of any visual aids used (e.g., computer-generated animations and medical imaging) were then recorded using the data extraction tool. If readily available, investigators also recorded the speaking characters' names, age ranges, genders, race/ethnicities, and roles in the episode (patient, family member, etc.). The research team then reconvened to compare findings, discuss any discrepancies to consensus, and to revise the data extraction tool. Only minor revisions were made to the data extraction tool to increase usability. Afterwards, the research team viewed the remaining 43 episodes of seasons one and two, with at least two researchers viewing each episode to improve reliability.

9.3.1.3 Thematic Analysis

All extracted data was then exported from Qualtrics to Dedoose, a cross-platform application for qualitative and mixed-methods analysis (SocioCultural Research Consultants, LLC 2024). Thematic analysis of the character dialogue extracted from the identified scenes was performed using Braun and Clarke's method of thematic analysis (Braun and Clarke 2006; Clarke and Braun 2013; Braun et al. 2022). This method of thematic analysis involves six primary phases in an iterative process: (Alahmari 2023) familiarization, (Alexander et al. 1994) coding, (Alismail et al. 2018) searching for themes, (Allen et al. 2023) reviewing themes, (Aryanty and Raudhoh 2023) defining and naming themes, and (Baños 2007) reporting. First, the research team (A.S.C. and H.W) reviewed all the extracted dialogue to develop familiarity with items of interest related to the research questions. Following familiarization, the researchers inductively coded the extracted dialogue, allowing the data to drive the coding process from the bottom up. Afterwards, codes were organized into provisional themes based on their relationships to one another. Lastly, the provisional themes were reviewed, defined, and named by the researchers.

9.3.1.4 Reflexivity Statement

In the interest of transparency, the authors have provided details on their academic and professional backgrounds. All four authors are proficient with the English language and familiar with American culture and popular media. Additionally, all authors are currently pursuing or have completed graduate degrees (M.S. and/or Ph.D.) in disciplines related to biological sciences. As such, their knowledge and familiarity with some of the medical concepts and terminology used in *House M.D.* is more robust when compared to the general public. This advanced knowledge may have caused the researchers to determine a scene to portray effective, easily understood science communication when it would be difficult for an average member of the public to understand. This was potentially mitigated by the authors' experience and training in science communication. The primary author (A.S.C.) and fourth author (R.L.W.) have completed minors in Communicating Science and Education as part of their doctoral programs and have extensive experience participating in and developing successful science communication projects. Additionally, A.S.C directs an elective course for undergraduate medical students that uses popular media (e.g., film, television, and comics) to teach the principles of science communication. This expertise may have allowed these authors to more accurately determine what constituted effective or ineffective science communication during the data collection and analysis phases of this study. Though they have not received formal training in science communication, the second (H.W.) and third (C.B.) authors also have experience in science communication through public outreach projects. This experience likely aided them in their assessment of the communication portrayed in *House M.D.* too.

9.3.2 Results

Numerous scenes were identified from the first two seasons of *House M.D.* as depicting effective or ineffective science communication by the researchers. Most of these scenes involved Dr. House himself and his diagnostic team of doctors either discussing a case amongst themselves or with a patient or caregiver.

9.3.2.1 Thematic Analysis

Several key themes regarding effective communication were identified through thematic analysis of the dialogue extracted from the identified scenes, including *Immediate Elaboration on Jargon, Analogies, Visual Aids, and Ineffective Examples* (Table 9.1). Some of the following representative quotes drawn from the show were edited for clarity or brevity.

Table 9.1 Communication themes identified in *House M.D*

Theme	Definition
Immediate elaboration on jargon	After using medical jargon, the speaker or another character immediately used a more common term and/or elaborated on the meaning of the jargon in simpler language
Analogies	The speaker compared a medical structure or concept to a commonplace or famous comparator to clarify and make it more accessible
Visual aid	The scene included a visual aid (e.g., computer-generated animations, medical imaging, and whiteboard drawings) to make a concept more accessible
Ineffective examples	The scene depicted an example of poor science communication between characters. This could involve the speaker using excessive jargon, being rude or condescending, or not providing an adequate explanation

Immediate Elaboration on Jargon

One of the simplest and most consistently used techniques the writers of *House M.D.* employed to make complex scientific concepts more accessible to audiences was by having characters immediately define and elaborate on any jargon that is spoken. Here, jargon is defined as "the technical terminology or characteristic idiom of a special activity or group" (Merriam-Webster 2024b). These statements typically occurred when the show's doctors explained concepts to patients or their caregivers but sometimes arose in conversations between doctors.

> House: "*I had an infarction.*"
> Patient: "*A heart attack?*"
> House: "*It's what happens when the blood flow is obstructed. If it's in the heart, it's a heart attack. If it's in the lungs, it's a pulmonary embolism. If it's in the brain, it's a stroke. I had it in my thigh muscles.*" (Season 1, Episode 1)

Analogies

The writers of *House M.D.* also made complex scientific concepts more accessible to general audiences by regularly including analogies in the dialogue between characters. In this study, an analogy is defined as "a comparison of two otherwise unlike things based on resemblance of a particular aspect" and includes figurative language such as metaphors and similes (Merriam-Webster 2024a). Across the 46 episodes of seasons 1 and 2, several scenes portrayed a character using an analogy to communicate a complex scientific topic to another character. The specific comparator against which the scientific concept or structure was compared varied from analogy to analogy. In some cases, the comparator was a commonplace item or concept that was easily recognizable to most American audience members (e.g., household items, locations and structures):

> House: "*A small percentage of plague cases present with sleep disturbance. Imagine, an idyllic river of bacteria. Okay, it's not idyllic for her, but it serves my purposes. The steroids and the immuno-suppressants acted like a big hunk of dam across the river. Physics 101, put a dam up in front of a raging river, the river rises. By stopping the immuno-suppressants, we blew up the dam, and a hundred-foot wall of bacteria flooded her lymph nodes.*" (Season 2, Episode 18)

To illustrate complex scientific topics, the characters also use analogies that reference well-known landmarks or entities from pop culture, recent events, or sports:

> House: "*Take a close look at the corpus callosum.*"
> Chase: "*It looks ok.*"
> House: "*Are we all looking at the same thing? 200 million interhemispheric nerve fibers, the George Washington Bridge between the left and right side of the brain. It's subtle.*" (Season 1, Episode 2)

Visual Aids

In addition to dialogue, *House M.D.* often used visual aids to better illustrate scientific concepts to the audience. These could take the form of medical imaging (e.g., X-rays, ultrasound, computed tomography, etc.) viewed and discussed by the characters as part of the scene. The show also employed computer-generated imagery (CGI) of the patient's body at the cellular or organ level to illustrate a normal or pathophysiological event. These visual aids often accompanied a more detailed verbal explanation from the characters (Table 9.2).

Table 9.2 Example computer-generated (CG) animations from *House M.D*

Season	Episode	Subject	Description
1	2	Mutation of measles virus	The CG scene begins with a rotating view of a measles virus particle within the patient. The camera then zooms into the virus to show a double-helix strand of DNA. One strand of the double helix then breaks
1	9	Arterial clot removal	The camera pans from Drs. Chase and Cameron to a scan of the patient's head. A device can be seen moving superiorly through the patient's arteries toward their brain. The scene then switches to a CG view of an artery within the patient's brain in which blood flow is obstructed by a blood clot. The device then comes into view and pulls the blood clot out of the artery, restoring flow
1	12	Adverse leprosy drug reaction	The camera zooms into an IV containing a yellow medication, then switches to a CG view of the patient's nerves. Electrical activity can be seen traveling down the yellow nerve cells. Small white bacteria then detach from the nerve cells. Antibodies begin attaching to both the dead bacteria and the nerve cells, which begin to shrink and turn brown
2	1	Bowel perforation	The camera zooms into the patient's abdomen to show a CG animation of intestines that appears dark and inflamed. The camera zooms in further to show a brown substance oozing out from the wall of the bowel

Ineffective Examples

The show also included several scenes that could be used as examples of ineffective science communication. These scenes often involved a doctor character overusing technical jargon, acronyms, or numerical values to explain either a patient's condition or a medical concept or procedure. In some cases, the doctor is using this advanced language when conversing with another medical professional, whereas others involve the doctor overexplaining to a patient or caregiver without addressing their concerns. Often in these doctor–patient interactions, the doctor or a colleague catches the ineffective attempt at communication and corrects themselves either immediately or in a later scene.

> Foreman: "*So we've confirmed that the problem is this mutated virus. The treatment for SSPE is intraventricular interferon. We implant an Ommaya reservoir under the scalp, which is connected to a ventricular catheter that delivers the antiviral directly to the left hemisphere...*"
> Father: "*Look, you want us to consent to this? I don't even understand what you're talking about!*"
> Foreman: "*Well, the antiviral...Look, I'm sorry, I can explain this as best I can, but the notion that you're going to fully understand your son's treatment and make an informed decision, is, it's kind of insane. Now, here's what you need to know, it's dangerous, it could kill him, you should do it.*" (Season 1, Episode 2)

9.3.3 Discussion

Overall, this study demonstrates that the medical drama, *House M.D.*, can serve as a useful resource for teaching patient and science communication. Through thematic analysis of the first two seasons of the show, several scenes were identified as depicting examples of either effective or ineffective science communication. In scenes depicting effective communication, the characters used several different strategies to communicate complex medical information to other characters, and by extension, to the audience as well. These strategies include immediately elaborating on medical jargon, using analogies with recognizable comparators, and integrating visual aids. Multiple scenes that depicted ineffective communication between doctors and their patients or colleagues were also identified, which could be equally useful for teaching communication skills. Therefore, medical educators seeking to improve the communication skills of medical professionals should consider integrating cinemeducation and *House M.D.* into their curricula.

9.3.3.1 Jargon

Unsurprisingly, language heavily influences how quickly and easily complex medical concepts can be grasped by non-expert audiences. In the case of *House M.D.*, the writers increased the accessibility of medical concepts by crafting the show's language with multiple strategies. For one, whenever medical jargon is used in the show (e.g., "infarction"), the speaker or another character immediately follows up with a more common term (e.g., "heart attack") or an explanation of the jargon in more simplified language (e.g., "blood flow is obstructed"). This occurs even in conversation between characters who are medical experts, which suggests that the elaboration is for the benefit of the audience rather than the characters.

Jargon is widely considered to be a significant barrier to understanding medicine and science for the public (Thon and Jucks 2017; Baram-Tsabari et al. 2020; Riggs et al. 2022; Allen et al. 2023). According to Allen et al. (2023), 9 out of 10 surveyed adults preferred doctors who used jargon-free language over ones who speak with heavy jargon. Doctors who used jargon-heavy language were viewed as more confusing, too technical, and uncaring, whereas doctors who used simpler, jargon-free language were viewed as good communicators, caring, and more approachable (Allen et al. 2023). Heavy jargon use has also been associated with difficulty processing information, message resistance, decreased perception of credibility, and impaired judgment and decision-making (Riggs et al. 2022). Since jargon gives the impression of complexity and mental exertion, the public is less inclined to engage with the information and more likely to disregard the persuasive messaging (Shulman and Bullock 2020). Nevertheless, some studies suggest that jargon can be beneficial under certain circumstances (Shulman and Bullock 2020; Allen et al. 2023). For example, some surveyed adults consider doctors who use jargon as more competent, knowledgeable, and thorough (Allen et al. 2023). In high urgency situations such as during the COVID-19 pandemic, some members of the public also appreciated jargon since it conveys specialized information precisely and efficiently and reduces the likelihood of miscommunication (Shulman and Bullock 2020. In the context of a television show, characters have a limited time frame to establish their credibility and expertise while still conveying information to the audience in an accessible manner to progress the plot and keep them engaged. By having characters use medical jargon to demonstrate the character's authority and expertise, but then immediately following up with a simpler, more recognizable term or an explanation, the writers have achieved a balance between credibility and understandability. Since medical professionals must also quickly establish credibility while maintaining accessibility, this strategy would translate well to real patient interactions.

9.3.3.2 Analogies

Analogies are another strategy that the *House M.D.* show relies on to improve the accessibility of complex medical concepts for its audience. In their dialogue, the characters often compare medical concepts to objects, processes, or events that are more recognizable and easily understood. The advantages of analogies and metaphors have been well-documented in the patient and science communication literature (Casarett et al. 2010; Galesic and Garcia-Retamero 2013; Hildenbrand and Perrault 2022). Analogies have been associated with improved objective understanding and are viewed by patients as useful in explaining complex health issues (Hildenbrand and Perrault 2022). In patient evaluations, doctors who used more analogies and metaphors with patients were rated higher as communicators compared to those who did not (Casarett et al. 2010). However, multiple factors can diminish the effectiveness of analogies in communicating information to different audiences such as similarity to the comparator, familiarity with the spoken language, personal experiences, and cultural context (Galesic and Garcia-Retamero 2013). If used improperly, they can also contribute to public misunderstandings or reinforce negative stereotypes (Taylor and Dewsbury 2018). As such, it is recommended that doctors use analogies and metaphors when they are familiar with the patient and aware of their background (Galesic and Garcia-Retamero 2013).

In *House M.D.*'s case, the show effectively uses analogies to convey medical information but also sometimes depicts problematic usage. In many instances, the characters use analogies that include universally known comparators such as balloons or dams, which increases their effectiveness with broader audiences. However, since the show is written for an American audience, many of the comparators are only recognizable to those who live in the United States or are familiar with American landmarks and culture from the time in which the show aired (e.g., George Washington Bridge), which limits their effectiveness with international audiences. Some analogies are culturally insensitive/discriminating and may be potentially triggering to some patients such as Dr. House's comparison of a tumor to Al-Qaeda terrorists with explosive vests in Season 2, Episode 2. This reference is harmful in several ways. It implies that Afghanistan as a whole is part of the Al-Qaeda terrorist organization, and that the Afghani people are collectively suicidal terrorists. Moreover, making light of suicide and terrorism may be emotionally distressing for some viewers who have been affected by these actions. As such, *House M.D.* can serve as an example of both effective and ineffective use of analogies for communication in cinemeducation.

9.3.3.3 Visual Aids

In addition to careful crafting of the show's dialogue, *House M.D.* also regularly includes visual aids to make complex medical information more accessible for its audience. In many episodes, the show cuts to computer-generated animations of the patient's internal organs that illustrate how a medication or disease is affecting the patient (e.g., internal bleeding, neuronal misfiring, bone fracture, intestinal perforation, etc.). Some scenes also feature zoomed in views of medical imaging taken from the patient such as radiographs, MRI, and CT. Although these visual aids may not accurately portray the exact mechanism of a disease or treatment, they do quickly convey general information about the diseased organ and the state of the patient. Previous studies have shown that visual elements can support understanding of complex medical concepts (Lee et al. 2022; Riggs et al. 2022). For example, in a study by Lee et al., 90% of surveyed social media users believed complicated COVID-19 science was easier to navigate when presented as an infographic rather than a written article (Lee et al. 2022). Additionally, 65% of participants believed they retained more information from an infographic, while 75% found pictures, diagrams, and graphical elements useful in understanding science (Lee et al. 2022). Visual elements can also mitigate the negative effects of jargon on understanding of scientific concepts when presented together (Riggs et al. 2022). However, to maximize their effectiveness, visual aids must be carefully integrated alongside text or audio elements and tailored toward a specific audience (Rodríguez Estrada and Davis 2015). *House M.D.* makes effective use of the show's visual elements by having characters either narrate the physiological and pathophysiological events that occur in an animation or describe what they see on medical imaging.

9.3.3.4 Potential Applications

Given that *House M.D.* includes several scenes that depict both effective and ineffective communication, the next steps are to use the show to teach medical professionals how to make complex medical information more accessible to the public. As described in the literature review, this could take several different forms depending on the needs and constraints of the learners. Either full episodes or curated clips of the show could be easily slotted into a cinemeducation curriculum to illustrate communication skills while still leaving ample time for focused discussion. At the time of writing though, *House M.D.* cannot be legally streamed for free on any platform in the United States, making it less accessible than some other movie or television options. As a major Hollywood produced show, *House M.D* also presents an American-centric perspective on the medical field and may not be representative of communication in other cultures or medical contexts. Therefore, educators may want to supplement their use of *House M.D.* with additional resources to provide a more holistic view.

9.3.3.5 Limitations

This study is not without limitations. As noted previously, the researchers who conducted this study have all completed graduate degrees in the biological sciences and likely have a better understanding of complex medical concepts than the average member of the public. This advanced knowledge may have caused the researchers to overestimate the accessibility of medical concepts portrayed in the show. Their familiarity with American culture and proficiency with the English language may have also caused them to overestimate the effectiveness of some communication strategies for audiences with different cultural backgrounds or language proficiencies. However, the researchers' expertise in science communication may have helped mitigate this issue. Another limitation is related to the communication strategies described in this study. While evidence supports the elimination of jargon, use of analogies, and integration of visual aids to make medical information more accessible to a general audience, these strategies may not be equally effective across all audiences. Some of these strategies are highly dependent on the listener's language proficiency, culture, and sensory impairments. Therefore, learners should be forewarned to carefully consider their audience as they use these strategies learned from *House M.D.*

In conclusion, the medical drama *House M.D.* can serve as a useful cinemeducation resource for teaching communication skills that increase accessibility of complex medical information. Several scenes throughout the first two seasons of the show portrayed both effective and ineffective use of jargon, analogies, and visual aids, which are well-documented strategies of communicating with non-expert audiences. To provide learners with a more holistic perspective on communication and the medical field, educators using cinemeducation should include movies and television shows beyond *House M.D.* in their curricula. Given the amount of compatible teaching strategies and medical cinema currently available and in production, medical educators have a great deal of options in building their cinemeducation curricula to meet the needs of their learners and improve the communication skills of medical professionals.

Acknowledgments The authors would like to thank Dr. Elizabeth Agosto for her time and effort in reviewing and proofreading multiple iterations of this manuscript.

References

Alahmari AA (2023) Professionalism, ethics, and realism of television medical dramas as perceived by Saudi medical students. J Med Educat Curri Develop 10:23821205231175037

Alexander M, Hall MN, Pettice YJ (1994) Cinemeducation: an innovative approach to teaching psychosocial medical care. Fam Med 26(7):430–433

Alismail A, Meyer NC, Almutairi W et al (2018) CPR in medical TV shows: non-health care student perspective. Adv Med Educ Pract Volume 9:85–91

Allen KA, Charpentier V, Hendrickson MA et al (2023) Jargon be gone–patient preference in doctor communication. J Patient Experienc 10:23743735231158942

Aryanty N, Raudhoh S (2023) Cinemedicine: the use of movies to teach bioethical principles for undergraduate medical students. In: 4th green development international conference (GDIC 2022). Atlantis Press, pp 58–64

Baños JE (2007) How literature and popular movies can help in medical education: applications for teaching the doctor-patient relationship. Med Educ-Oxf 41(9):918

Baños JE, Lucena MI, Farré M (2019) The usefulness of TV medical dramas for teaching clinical pharmacology: a content analysis of House, MD. Educ Méd 20(5):295–303

Baram-Tsabari A, Wolfson O, Yosef R et al (2020) Jargon use in public understanding of science papers over three decades. Public Underst Sci 29(6):644–654

Blasco PG, Moreto G (2012) Teaching empathy through movies: reaching learners' affective domain in medical education. J Educ Learn 1(1):22

Blasco PG, Moreto G, Roncoletta AF et al (2006) Using movie clips to foster learners' reflection: improving education in the affective domain. Family Med-Kansas City 38(2):94

Blasco PG, Blasco MG, Levites MR et al (2011) Educating through movies: how Hollywood fosters reflection. Creat Educ 2(03):174

Braun V, Clarke V (2006) Using thematic analysis in psychology. Qual Res Psychol 3(2):77–101

Braun V, Clarke V, Hayfield N (2022) 'A starting point for your journey, not a map': Nikki Hayfield in conversation with Virginia Braun and Victoria Clarke about thematic analysis. Qual Res Psychol 19(2):424–445

Cape G (2009) Movies as a vehicle to teach addiction medicine. Int Rev Psychiatry 21(3):213–217

Casarett D, Pickard A, Fishman JM et al (2010) Can metaphors and analogies improve communication with seriously ill patients? J Palliat Med 13(3):255–260

City Hospital (1951) Directed by Cort Steen [Television Show]. American Broadcasting Company, Burbank, CA
Clarke V, Braun V (2013) Successful qualitative research: a practical guide for beginners
Czarny MJ, Faden RR, Nolan MT et al (2008) Medical and nursing students' television viewing habits: potential implications for bioethics. Am J Bioeth 8(12):1–8
Dahms K, Sharkova Y, Heitland P et al (2014) Cobalt intoxication diagnosed with the help of Dr House. Lancet 383(9916):574
Darbyshire D, Baker P (2012) A systematic review and thematic analysis of cinema in medical education. Med Humanit 38(1):28–33
Datta V (2009) Madness and the movies: an undergraduate module for medical students. Int Rev Psychiatry 21(3):261–266
De Souza D, Adams GC (2023) Practical pedagogical tips for using movies in medical education. Acad Psychiatry 47(1):90–94
ER (1994) Directed by Christopher Chulack et al. [Television Show]. National Broadcasting Company, New York, NY
Forrest Gump (1994) Directed by Robert Zemeckis [Feature Film]. Paramount Pictures, Los Angeles, CA
Fritz GK, Poe RO (1979) The role of a cinema seminar in psychiatric education. Am J Psychiatry 136(2):207–210
Galesic M, Garcia-Retamero R (2013) Using analogies to communicate information about health risks. Appl Cogn Psychol 27(1):33–42
Gallagher P, Wilson N, Edwards R et al (2011) A pilot study of medical student attitudes to, and use of, commercial movies that address public health issues. BMC Res Notes 4:1–7
Grey's Anatomy (2005) Directed by Rob Corn et al. [Television Show]. American Broadcasting Company, Los Angeles, CA
Ha JF, Longnecker N (2010) Doctor-patient communication: a review. Ochsner J 10(1):38–43
Haboubi HN, Morgan H, Aldalati O (2015) Hospital doctors' opinions regarding educational utility, public sentiment and career effects of medical television dramas: the HOUSE MD study. Med J Aust 203(11):462–466
Hildenbrand GM, Perrault EK (2022) The influence of physician use of analogies on patient understanding. Commun Q 70(5):495–518
Hirt C, Wong K, Erichsen S et al (2013) Medical dramas on television: a brief guide for educators. Med Teach 35(3):237–242
Hoffman BL, Shensa A, Wessel C et al (2017) Exposure to fictional medical television and health: a systematic review. Health Educ Res 32(2):107–123
Hoffman BL, Sidani JE, Jonassaint CR et al (2023) Utilizing television clips for graduate medical education anti-racist curricula: an acceptability study. Cureus 15(7)
Hospital Playlist (2020) Directed by Shin Won-Ho [Television Show]. Netflix, Los Angeles, CA
House MD (2004) Directed by Greg Yaitanes et al. [Television Show]. Fox Network, Los Angeles, CA
IMDB (2024) House (TV Series 2004–2012). Available at https://www.imdb.com/title/tt0412142/. Accessed 1 Apr 2024
It's a Wonderful Life (1946) Directed by Frank Capra. [Feature Film]. RKO Pictures, Los Angeles, CA
Jain P, Slater MD (2013) Provider portrayals and patient–provider communication in drama and reality medical entertainment television shows. J Health Commun 18(6):703–722
Jerrentrup A, Mueller T, Glowalla U et al (2018) Teaching medicine with the help of "Dr. House". PLoS One 13(3):e0193972
Jubas K (2013) Grey ('s) identity: complications of learning and becoming in a popular television show. Rev Educ Pedagogy Cult Stud 35(2):127–143
Kadivar M, Mafinejad MK, Bazzaz JT et al (2018) Cinemedicine: using movies to improve students' understanding of psychosocial aspects of medicine. Ann Med Surg 28:23–27
Kelly PJ (2012) Using popular movies and reflections to increase the cultural sensitivity of students. J Physician Assist Educ 23(4):39–42
Klemenc-Ketis Z, Kersnik J (2011) Using movies to teach professionalism to medical students. BMC Med Educ 11:1–5
Lapostolle F, Montois S, Alhéritière A et al (2013) Dr House, TV, and reality…. Am J Med 126(2):171–173
Law M, Kwong W, Friesen F et al (2015) The current landscape of television and movies in medical education. Perspect Med Educ 4:218–224
Lee SH, Pandya RK, Hussain JS et al (2022) Perceptions of using infographics for scientific communication on social media for COVID-19 topics: a survey study. J Vis Commun Med 45(2):105–113
Lumlertgul N, Kijpaisalratana N, Pityaratstian N et al (2009) Cinemeducation: a pilot student project using movies to help students learn medical professionalism. Med Teach 31(7):e327–e332
Merriam-Webster (2024a) Analogy. Available at https://www.merriam-webster.com/dictionary/jargon. Accessed 1 Apr 2024
Merriam-Webster (2024b) Jargon. Available at https://www.merriam-webster.com/dictionary/jargon. Accessed 1 Apr 2024
New Amsterdam (2018) Directed by Michael Slovis et al. [Television Show]. National Broadcasting Company, New York, NY
One Flew Over the Cuckoo's Nest (1975) Directed by Milos Forman [Feature Film]. United Artists, Beverly Hills, CA
Painter DL, Kubala A, Parsloe S (2020) Playing doctor on TV: physician portrayals and interactions on medical drama, comedy, and reality shows. Atlantic J Commun 28(5):322–336
Parrish-Sprowl J, Ziberi L, Parrish-Sprowl S (2023) Complexity-based approaches in health communication. Front Commun 8:1115983

Parsi K, Elster N (2020) "You took an oath!": engaging medical students about the importance of oaths and codes through film and television. In HEC Forum. Springer, Cham, vol 32, No. 2, pp 175–189
Petkari E (2017) Building beautiful minds: teaching through movies to tackle stigma in psychology students in the UAE. Acad Psychiatry 41:724–732
Philadelphia (1993) Directed by Jonathan Demme [Feature Film]. Tristar Pictures, Culver City, CA
Qualtrics (2024) Qualtrics (Version 4.2024) [Software]. Qualtrics, Provo, Utah
Ranjan P, Kumari A, Chakrawarty A (2015) How can doctors improve their communication skills? J Clin Diagn Res 9(3):JE01
Rattani A, Kaakour D, Syed RH et al (2020) Rethinking TV and movies in medical ethics and professionalism education. Med Teach 42(4):477–478
Riggs EE, Shulman HC, Lopez R (2022) Using infographics to reduce the negative effects of jargon on intentions to vaccinate against COVID-19. Public Underst Sci 31(6):751–765
Rodríguez Estrada FC, Davis LS (2015) Improving visual communication of science through the incorporation of graphic design theories and practices into science communication. Sci Commun 37(1):140–148
Scrubs (2001) Directed by Michael Spiller et al. [Television Show]. American Broadcasting Company, Los Angeles, CA
Seinfeld (1989) Directed by Andy Ackerman et al. [Television Show]. National Broadcasting Company, Ney York, NY
Shankar PR (2019) Cinemeducation: facilitating educational sessions for medical students using the power of movies. Arch Med Health Sci 7(1):96–103
Shankar PR, Rose C, Balasubramanium R et al (2016) Using movies to strengthen learning of the humanistic aspects of medicine. J Clin Diagn Res 10(1):JC05
Shapiro J, Rucker L (2004) The Don Quixote effect: why going to the movies can help develop empathy and altruism in medical students and residents. Fam Syst Health 22(4):445
Shrivastava SR, Shrivastava PS (2022) Incorporating movies and cinema in the medical education delivery: a curricular innovation. Med J Dr DY Patil University 15(5):662–665
Shulman HC, Bullock OM (2020) Don't dumb it down: the effects of jargon in COVID-19 crisis communication. PLoS One 15(10):e0239524
SocioCultural Research Consultants, LLC (2024) Dedoose (Version 9.2.007) [Software]. SocioCultural Research Consultants, LLC, Los Angeles, CA
Spartacus (1960) Directed by Stanley Kubrick [Feature Film]. Universal Pictures, Universal City, CA
Stanek A, Clarkin C, Bould MD et al (2015) Life imitating art: depictions of the hidden curriculum in medical television programs. BMC Med Educ 15:1–8
Sumioka H, Nakae A, Kanai R et al (2013) Huggable communication medium decreases cortisol levels. Sci Rep 3(1):3034
Taylor C, Dewsbury BM (2018) On the problem and promise of metaphor use in science and science communication. J Microbiol Biol Educ 19(1):10–1128
The Last Samurai (2003) Directed Edward Zwick [Feature Film]. Warner Bros, Burbank, CA
The Lion King (1994) Directed by Roger Allers and Rob Minkoff [Feature Film]. Buena Vista Pictures Distribution, Burbank, CA
The Resident (2018) Directed by Rob Corn et al. [Television Show]. Fox Network, Los Angeles, CA
This is Going to Hurt (2022) Directed by Lucy Forbes and Tom Kingsley [Television Show]. AMC Networks, New York, NY
Thon FM, Jucks R (2017) Believing in expertise: how authors' credentials and language use influence the credibility of online health information. Health Commun 32(7):828–836
Tobia A, Bisen V, Zimmerman A et al (2015) Psy-feld: an innovative didactic using the TV show Seinfeld to teach delusional disorder subtypes. Acad Psychiatry 39:482–486
Weaver R, Wilson I (2011) Australian medical students' perceptions of professionalism and ethics in medical television programs. BMC Med Educ 11:1–6
Weaver R, Wilson I, Langendyk V (2014) Medical professionalism on television: student perceptions and pedagogical implications. Health 18(6):597–612
White GB (2008) Capturing the ethics education value of television medical dramas. Am J Bioeth 8(12):13–14
Williams DJ, Re D, Ozakinci G (2014) Television viewing habits of preclinical UK medical undergraduates: further potential implications for bioethics. AJOB Empirical Bioethics 5(2):55–67
Wong RY, Saber SS, Ma I et al (2009) Using television shows to teach communication skills in internal medicine residency. BMC Med Educ 9:1–8
Zeppegno P, Gramaglia C, Feggi A et al (2015) The effectiveness of a new approach using movies in the training of medical students. Perspect Med Educ 4:261–263

GPSR Compliance

The European Union's (EU) General Product Safety Regulation (GPSR) is a set of rules that requires consumer products to be safe and our obligations to ensure this.

If you have any concerns about our products, you can contact us on ProductSafety@springernature.com

In case Publisher is established outside the EU, the EU authorized representative is:

Springer Nature Customer Service Center GmbH
Europaplatz 3
69115 Heidelberg, Germany

Batch number: 10755744

Printed by Printforce, the Netherlands